MODELLING AND PREDICTION OF THE UPPER LAYERS OF THE OCEAN

PERGAMON MARINE SERIES
Volume 1

Editor: Mrs. J. C. SWALLOW

Other titles of interest

Journals:
Computers & Geosciences

Books:
McLellan: Elements of Physical Oceanography
Parsons & Takahashi: Biological Oceanographic Processes
Pickard: Descriptive Physical Oceanography, 2nd edition

MODELLING AND PREDICTION OF THE UPPER LAYERS OF THE OCEAN

Edited by

E. B. KRAUS

Rosenstiel School of Marine and Atmospheric Science
University of Miami, Coral Gables, Florida, U.S.A.

Proceedings of a NATO Advanced Study Institute

PERGAMON PRESS

OXFORD · NEW YORK · TORONTO · SYDNEY · PARIS · FRANKFURT

U.K.	Pergamon Press Ltd., Headington Hill Hall, Oxford OX3 0BW, England
U.S.A.	Pergamon Press Inc., Maxwell House, Fairview Park, Elmsford, New York 10523, U.S.A.
CANADA	Pergamon of Canada Ltd., 75 The East Mall, Toronto, Ontario, Canada
AUSTRALIA	Pergamon Press (Aust.) Pty. Ltd., 19a Boundary Street, Rushcutters Bay, N.S.W. 2011, Australia
FRANCE	Pergamon Press SARL, 24 rue des Ecoles, 75240 Paris, Cedex 05, France
WEST GERMANY	Pergamon Press GmbH, 6242 Kronberg-Taunus, Pferdstrasse 1, Frankfurt-am-Main, West Germany

First edition 1977

Library of Congress Cataloging in Publication Data

Advanced Study Institute on Modelling and Prediction
of the Upper Layers of the Ocean, Urbino, Italy, 1975
Modelling and prediction of the upper layers of the
ocean.

(Pergamon marine series)
Institute sponsored by the NATO Scientific Affairs Division, Air-Sea Interaction
Panel.
Includes bibliographical references and index.
1. Oceanography--Mathematical models--Congresses.
2. Ocean-atmosphere interaction--Mathematical models--
Congresses. I. Kraus, Eric Bradshaw, 1913-
II. North Atlantic Treaty Organization. Division of
Scientific Affairs. Air-Sea Interaction Panel.

GC10.4.M36A35 1977 551.4'6'00184 76-40958
ISBN 0-08-020611-5
ISBN 0-08-020610-7 pbk.

In order to make this volume available as economically and rapidly as possible the author's typescript has been reproduced in its original form. This method unfortunately has its typographical limitations but it is hoped that they in no way distract the reader.

Printed in Great Britain by A. Wheaton & Co., Exeter

CONTENTS

III. PHYSICAL MODELS OF THE UPPER OCEAN

IV. BIOLOGICAL MODELS OF THE UPPER OCEAN

V. EXPERIMENTAL CONSIDERATIONS

CONTRIBUTORS

Niels E. Busch
Research Establishment RISØ
Roskilde, Denmark

R. Michael Clancy
Rosenstiel School of Marine and Atmospheric Science
University of Miami
Coral Gables, Florida, U.S.A.

Allan J. Clarke
Department of Applied Mathematics and Theoretical Physics
University of Cambridge
Cambridge, U.K.

Michel Crepon
Laboratoire D'Oceanographie Physique
Muséum National d'Historie Naturelle
Paris, France

Kenneth L. Denman
Marine Ecology Laboratory
Bedford Institute of Oceanography
Dartmouth, Nova Scotia, Canada

Russell Elsberry
Department of Meteorology
Naval Postgraduate School
Monterey, California, U.S.A.

Tor Gammelsrød
Geofysisk Institutt
University i Bergen
Bergen, Norway

William R. Holland
National Center for Atmospheric Research
Boulder, Colorado, U.S.A.

Alexandre Ivanoff
Laboratoire d'Oceanographie Physique
Université de Paris VI
Paris, France

L. H. Kantha
Department of Earth and Planetary Sciences
The Johns Hopkins University
Baltimore, Maryland, U.S.A.

E. B. Kraus
Rosenstiel School of Marine and Atmospheric Science
University of Miami
Coral Gables, Florida, U.S.A.

Malcolm MacVean
Department of Oceanography
University of Southampton
Southampton, U.K.

P. P. Niiler
School of Oceanography
Oregon State University
Corvallis, Oregon, U.S.A.

J. J. O'Brien
Ocean Science and Technology Division
Office of Naval Research
Arlington, Virginia, U.S.A.

O. M. Phillips
Department of Earth and Planetary Sciences
The Johns Hopkins University
Baltimore, Maryland, U.S.A.

Trevor Platt
Marine Ecology Laboratory
Bedford Institute of Oceanography
Dartmouth, Nova Scotia, Canada

R. T. Pollard
Department of Oceanography
University of Southampton
Southampton, U.K.

Lars Petter Röed
Matematisk Institutt
University i Oslo
Oslo, Norway

François C. Ronday
Institut de Mathematique
Université de Liège
Liège, Belgium

Richard C. J. Somerville
National Center for Atmospheric Research
Boulder, Colorado, U.S.A.

John Steele
Marine Laboratory
Department of Agriculture and Fisheries for Scotland
Aberdeen, U.K.

J. Dana Thompson
JAYCOR, Inc.
Arlington, Virginia, U.S.A.

Conley R. Ward
Fleet Numerical Weather Central
Monterey, California, U.S.A.

J. D. Woods
Department of Oceanography
University of Southampton
Southampton, U.K.

PARTICIPANTS

Miss Isabel Ambar, Portugal
Dr. Laurence Armi, U.S.A.
Dr. A. Artegiani, Italy
Cmdr. M. M. Barnett, U.K.
Dr. Dominique Begis, France
Miss Kay Ruth Burnett, U.S.A.
Mr. Christoph Brockmann, Germany
Prof. Niels E. Busch, Denmark
Mr. Norman T. Camp, U.S.A.
Dr. Hassan Hage Chehade, France
Mr. R. Michael Clancy, U.S.A.
Dr. Allan Clarke, U.K.
S. Colacini, Italy
Dr. Michel R. Crepon, France
Dr. Kriton Curi, Turkey
Dr. Maxence R. D'Allonnes, France
Prof. Kenneth L. Denman, Canada
Mr. Gabriel Chabert D'Hieres, France
Mr. Ernst Dreisigacker, Germany
Prof. Russel Elsberry, U.S.A.
Mr. Armando Fiuza, Portugal
Dr. Roberto Frassetto, Italy
Dr. Hans J. Friedrich, Germany
Mr. Tor Gammelsrød, Norway
Mr. Roland W. Gardwood, U.S.A.
Dr. Peter R. Gent, U.K.
Dr. Louis Goodman, U.S.A.
Dr. Paul Hamblin, Canada
Mr. Howard P. Hanson, U.S.A.
Prof. M. Henderschott, U.S.A.
Dr. R. Falconer Henry, Canada
Dr. J. Brackett Hersey, U.S.A.
Dr. Heinrich Hoeber, Germany
Dr. William R. Holland, U.S.A.
Mr. Edward P. W. Horne, Canada
Prof. A. Ivanoff, France
Dr. Lakshmi Kantha, U.S.A.
Dr. Rolf Kase, Germany
Dr. Jack Kaiser, U.S.A.
Dr. Jeong-Woo Kim, U.S.A.
Mr. P. Klein, France
Prof. E. B. Kraus, U.S.A.
Dr. Søren E. Larsen, Denmark
Dr. William R. Lindberg, U.S.A.
Mr. M. K. MacVean, U.K.
Cmdr. Thomas Marshall, Belgium
Mr. Paul J. Martin, U.S.A.
Dr. Miles A. McPhee, U.S.A.
Dr. John Mitchell, U.K.
Mr. Paul Moersdorf, U.S.A.
Mr. R. Molcard, U.S.A.
Prof. J. C. J. Nihoul, Belgium
Prof. Pearn P. Niiler, U.S.A.
Prof. James J. O'Brien, U.S.A.
Prof. Clayton A. Paulson, U.S.A.
Mr. Hartmut Peters, Germany
Mr. Thomas P. Peterson, U.S.A.
Prof. Owen M. Phillips, U.S.A.
Dr. Trevor R. Platt, Canada
Dr. Raymond T. Pollard, U.K.
Dr. Geert Prangsma, Netherlands
Mr. James R. Price, U.S.A.
Mr. Lars Petter Røed, Norway
Dr. Paola Rizzoli, Italy
Mr. Francois C. Ronday, Belgium
Prof. Bernard Saint-Guily, France
Mr. James Salmon, U.K.
Dr. Sallustio Salusti, Italy
Mr. John Schedvin, U.S.A.
Mr. Paul S. Schopf, U.S.A.
Dr. David Shonting, U.S.A.
Mr. James Simpson, U.S.A.

Dr. Richard C. J. Somerville, U.S.A.
Dr. John H. Steele, U.K.
Prof. Harald Svendsen, Norway
Mr. Jean-Claude Therriault, Canada
Dr. J. Dana Thompson, U.S.A.
Capt. C. R. Ward, U.S.A.
Prof. John D. Woods, U.K.
Ms. Lucy Wyatt, U.K.

PREFACE

The initiative for holding an Advanced Study Institute on "Modelling and Prediction of the Upper Layers of the Ocean" came from the NATO Scientific Affairs Division, Air-Sea Interaction Panel. I would like to thank the Panel for having selected me as the Director.

To a very large extent, the success of the Institute was due to Dr. Roberto Frassetto who negotiated its location at Sogesta, Urbino. Afterwards he organized a remarkable bus trip which brought many of the participants from the I.U.G.G. meeting at Grenoble to Urbino and which provided the scientific and gastronomic prelude to the ASI. Together with Dr. Enrico Catani he saw to it that chalk and pencils and recorders and conference rooms and projectors were unfailingly available when needed, and that checks were cashed when anyone's money ran out. He knew about all the delightful little inns in neighboring farms. He organized the grand finale of a party at some romantic old castle in the Appenines. In between, he took an active interest in the scientific proceedings - by any criterion,Frassetto was indispensable.

The Institute worked through plenary lectures and through working group sessions which were led by Pearn Niiler, Jim O'Brien, Owen Phillips, John Steele and John Woods. As an Organizing Committee the same group together with Frasseto and myself were responsible also for the arrangement of the program.

Sogesta University made their facilities available to us at a price which cannot possibly have been profitable to them. In particular, we all owe a debt of gratitude to their manager, Signor Fossa Margutti, and his staff for their warm hospitality and courtesy.

It would have been difficult for me to prepare this book without Howard Hanson who was an Institute participant in Urbino and who has since acted selflessly and conscientiously as my associate editor in spite of other calls on his time. Lynn Zakevich-Gheer produced the manuscript, and Peggy Nemeth did an enormous amount of work behind the scene, corresponding with applicants for the Institute, keeping the books, and finally typing all the scripts - with good humor and patience in the face of illegible manuscripts and other odds. Last but not least, we had most valuable editorial assistance from Mary Swallow of the Institute of Oceanographic Sciences and from Pergamon Press.

The money granted by the NATO Scientific Division, though generous, would not have covered all expenses. Jim O'Brien, the Institute's godfather in Washington, helped very much by allowing some Office of Naval Research grant funds to be used in support of several students and lecturers. We are indebted also to the United States National Science Foundation for similar support during the Institute and the subsequent preparation of this book. Many of the participants were supported individually by institutions in their own countries; some came at their own expense.

Initially I had invited about fifty participants, expecting that perhaps forty might turn up. In fact over eighty people from almost every NATO country came to Urbino. I was concerned that it might be difficult to organize such a large group in a way which would allow individual participation by all those who wanted to be heard. However, things fell into place easily with the help of evening sessions and little individual meetings in conference rooms or around the bar. Everybody contributed not only some scientific experience, but also good will and consideration - in other words, they could not have been a nicer bunch.

Coconut Grove, Florida
April 1976

E. B. Kraus

Chapter 1

INTRODUCTION

E. B. Kraus

The long bibliography at the end of this book is a measure of scientific effort which has gone recently into the modelling of the upper ocean. With its apparent vast regularity - so much more repetitive than the land - the upper ocean seems to invite analysis in mechanical terms. It is left to the poet to warn us that . . .

> There are, there are, in a place of foam and green water
> as in the clearings aflame of mathematics,
> truth more restive of our approach
> than the necks of fabled beasts,
> and suddenly there we lose footing . . .
>
> St. John Perse (1958).

The state of the sea surface and the waters immediately below concern us for many reasons. Ship traffic, naval operations, fisheries, the well-being and the very existence of coastal settlements have all been affected by it. Beyond its immediate influence on human affairs, the upper ocean modulates almost all the processes which shape the terrestrial environment. More than eighty per cent of all the solar radiation absorbed below the bottom of the atmosphere is stored temporarily in the top twenty meters of the sea. The first stirrings of life occurred in these sunlit waters. Photosynthesis there was not only the original source of free oceanic and atmospheric oxygen, but continues to play an essential role in the cycling of oxygen and carbon dioxide through the planetary system. By the same token, it constitutes the crucial link in the food chain of the sea. The dynamics of the whole terrestrial climate cannot be separated from heat storage effects in the upper ocean. They determine not only the temperature of the winds which blow across the surface, but also the moisture content of the air, the radiation balance, and all the river discharge. Last but not least, surface processes supply almost all the energy for circulations and turbulence within the ocean interior.

It is this all-pervasive influence which has provided much of the motivation for the modelling of the upper ocean. In trying to cover the ground we were concerned, in the first instance, with the question whether simple upper ocean models could be used in a sub-routine mode for the specification of the sea surface boundary conditions in numerical models of the general oceanic and atmospheric circulations. The papers by Holland, Somerville, and Ward, which deal with the present state of general circulation modelling, indicate that little has been done so far along these lines, though it is accepted as a potentially useful concept.

The following part of this book deals with some of the physical processes which provide the dynamic forcing. The most fundamental of these processes is the absorption of solar radiation which is the subject of Sacha Ivanoff's carefully documented contribution. In the following paper, Niels Busch considers the specification and the parameterization of the turbulent transports of momentum, sensible heat, and water vapor between the atmosphere and the sea. The transport of momentum across a fluid interface is related directly to a corresponding flux of kinetic energy and the water vapor flux also involves a flux of latent heat. The papers by Ivanoff and by Busch cover therefore the flux of energy in all forms across the sea surface except for infrared radiation. This does not imply that infrared radiation is not important for the heat budget of the ocean. Its magnitude depends, however, to such an extent on conditions deep in the atmosphere that its discussion here did not seem appropriate. The modeller who is concerned with this subject may find some relevant information in Kraus (1972) or in textbooks on atmospheric physics.

Apart from air-sea interactions, the temperature and the motion in the usually well mixed ocean surface layers is determined also by interaction with the ocean interior. This interaction proceeds mainly through the mixing or entrainment of cold water from below into the turbulent surface layer. Owen Phillips describes the theoretical and experimental investigations of this process.

The two papers from the University of Southampton by Raymond Pollard and by John Woods both deal with the remaining aspects of vertical mixing in the upper ocean. Pollard is concerned mainly with the localization of shearing motion and the corresponding generation of eddy kinetic energy. This process involves the Coriolis force through Ekman currents and inertial oscillations in the surface layers. It also is affected by Langmuir circulations and other phenomena which can redistribute horizontal momentum along the vertical in the upper ocean. Woods complements this with a discussion of the physics of phenomena such as internal waves and fronts which are involved in vertical mixing. He concludes that mixing is essentially an intermittent process, which cannot be simulated adequately by most of the existing parameterization schemes, but requires new approaches that allow for the actual mechanics of vertical motions in the upper ocean.

Upper ocean modelling - the central theme of the Advanced Study Institute is the subject of Part III, which is appropriately the longest part of the book. It is divided into reviews of one-dimensional models, horizontally variable upwelling models, and models of sea surface height. The differing style of the three reviews reflects differences in the manner in which the working groups concerned with these topics conducted their business in Urbino.

Since they were first developed ten years ago, integral one-dimensional mixed layer models have been rather successful in providing general explanations of the seasonal and diurnal evolution of the surface layer or of the local effects of storms. I suspect that too much further elaboration and refinement will yield diminishing returns, because models of this type cannot be expected to yield detailed and accurate results without sacrifice of their essential simplicity. The work group associated with one-dimensional models attracted the majority of Institute members. This very fact, together with the internal coherence of the subject, did not allow us to associate everyone who had something to say about this topic with the writing of the review, which was finally prepared by Pearn Niiler and myself. After it had been written, L. H. Kantha penned a short note which suggested a parameterization scheme for the loss of energy to internal waves in mixed layer models.

Upwelling has important climatic and biological consequences. It also is an interesting subject for modellers. Jim O'Brien orchestrated and stimulated a group of rather lively co-workers who dealt with different aspects of the subject including two and three-dimensional upwelling models; interactive, sea breeze produced upwelling models; the effect of coastally trapped waves; storm induced upwelling in the open ocean; upwelling along ice edges and also along fronts.

Modelling of the sea state and of the sea surface is a very large subject in its own right. For the oceanic modeller, the sea surface is primarily an upper boundary through which the atmospheric forcing is communicated to the water. This communication is complicated by the presence of wind driven waves. The resulting questions are considered by Owen Phillips in his lucid review. Phillips' paper is complemented by Francois Ronday's short note on the surface deflections which can occur as the result of large-scale circulations in the North Sea.

Biological and ecological models by John Steele and by Kenneth Denman and Trevor Platt are brought together in Part IV. Steele's paper, which was meant originally to be included in Part I, deals with the general problem of modelling the variable abundance of different species who interact with each other and with the environment. The paper by Denman and Platt, on the other hand, is concerned more specifically with the production and abundance of phytoplankton.

The value of any evolving model depends on the specification of initial state and again upon the verification of the model prediction. These have to be supplied by appropriate observational data. The problems and some of the pitfalls involved in the assembly of these data sets are addressed in John Woods' paper on the theory of experiments.

It is emphasized once more that much else besides the papers assembled below was said and presented in Urbino. Particularly memorable were Jacques Nihoul's witty introduction on the first day of the Institute, the demonstrations by Chabert d'Hieres of the facilities which are available in Grenoble for laboratory simulations, and Roberto Frassetto's account of the studies concerned with the environmental problems of the city of Venice.

Howard Hanson, as mentioned in the Preface, spent much effort in replacing the different symbols used by various authors by a uniform notation. We have used vector notation throughout to apply to oceanic variables; vectors may either be two-dimensional (x,y) or three-dimensional (x,y,z). Since the horizontal direction generally plays a different role from the vertical near the sea surface, the use of horizontal vectors with separate symbols for vertical quantities predominates. Tensor notation is used in Chapter 12, since wave-wave interaction coefficients require it, and in Chapter 6, in order to serve as a reminder that variables there refer to the atmosphere. A list of the commonly used symbols as well as a common bibliography can be found below.

It was our joint purpose in Urbino to give a reasonably complete account of the present state of the art. The present book is meant to reflect this aim. We hope that it will be found useful for the next few years by students who are concerned with the modelling of the upper layers of the ocean.

Part I

MOTIVATION FOR UPPER OCEAN MODELLING

Chapter 2

THE ROLE OF THE UPPER OCEAN AS A BOUNDARY LAYER IN MODELS OF THE OCEANIC GENERAL CIRCULATION

William R. Holland

2.1 Introduction

In this paper I would like to describe numerical models of the large-scale features of the oceanic circulation and attempt to show that the role of the upper layers of the ocean is a crucial one for determining the rest of the general circulation. Motions in the upper ocean layers provide the means for the exchange of momentum, mass, heat, and energy between the atmosphere and the large bulk of the underlying ocean. Since the lower ocean is driven only from above, a correct representation of these processes is absolutely essential for determining the large scale circulation. On the global scale, the upper ocean is also one of the most important factors in determining climate. This is due to the transports of moisture and energy upwards from the sea surface which exert a governing influence on the general atmospheric circulation.

Although much has been learned during the past few decades, a complete description of the general circulation of the ocean is not yet available. This is due, largely, to a lack of knowledge about the effect of small-scale processes upon the large-scale flow. In other words, although the physical laws governing the system (the Navier-Stokes equations) are familiar, valid approximations necessary for model studies are still unknown. The parameterizations of processes which are characterized by short time and small space scales are too crude at present to allow for their proper representation. This lack of knowledge is the primary reason why such a large part of the present oceanographic effort is concerned with the surface layers of the ocean (a vertical mixing problem) and with mesoscale eddies (a horizontal mixing problem). Until valid parameterizations of these processes are included in large-scale models, we can have little confidence that these models will be useful for prediction purposes, for climate studies, or for studies of the fate of man-made and natural pollutants in the ocean.

The major task of this paper is to set the stage for the following discussions of mixed-layer modelling. The treatment of vertical mixing of heat and momentum in the surface layers of current oceanic general circulation models will be discussed with the aid of a few examples, which illustrate the simple parameterizations of vertical mixing processes which are now being used. In a sense this can be thought of as the starting point for mixed layer modelling, and, in fact, the vertical diffusion-convective adjustment model to be described may be thought of as the archetypal upper ocean model.

It is clear that some degree of success in matching observations is possible with very simple models. However, it can also be shown that the derived large-scale fluxes of heat and energy across the sea surface are quite sensitive to the particular upper ocean model used. As indicated below, current models need significant improvement in this regard. Although it is easy to convince oneself that a proper parameterization of vertical mixing in the surface layers is necessary in a large-scale ocean model, it is more difficult to test various alternative formulations. This is partly due to the large computational requirements of numerical general circulation models. It is best, therefore, first to test the current one-dimensional models against each other and to compare model results with observations. Then, when realistic one-dimensional models have been achieved, they can be extended to three dimensions and incorporated into the framework of current general circulation models. Finally, by comparing various models, we can decide just how important details of the upper ocean are in determining the general circulation.

The final part of this paper will discuss briefly a problem that is analogous to the problem of parameterizing vertical mixing processes, namely, the problem of horizontal mixing of heat, salt, and momentum by mesoscale eddies in the ocean. Recent and future field programs (MODE and POLYMODE) are contributing and will continue to contribute substantially to solving this problem. Recent theoretical efforts to understand the role of mesoscale eddies in the general circulation of the ocean will be described. It is clear that these two problems, the proper parameterizations of vertical and horizontal mixing, constitute the most important next step to further progress in understanding the large-scale ocean circulation.

2.2 Models of the oceanic general circulation

The last decade has seen a rapid development of prognostic, wind and thermohaline driven numerical ocean circulation models by a number of investigators. These models seek to reproduce the large-scale, three-dimensional density distribution (i.e., the temperature and salinity distributions) as well as the three-dimensional flow field within an ocean basin, given the wind and thermohaline conditions at the sea surface. The model equations are based upon the Navier-Stokes equations and conservation equations for heat and salt suitably averaged over the grid scale in space and a corresponding time interval. These averaged equations then contain Reynolds stress terms which account for a diffusion of momentum, heat, and salt by small-scale processes not directly included in the model. In the calculations to be discussed, a simple turbulent viscosity hypothesis has been used. For horizontal momentum mixing, it is identical to that of Munk (1950). A similar assumption is made for horizontal eddy mixing of heat and salt and for vertical mixing of all three quantities.

A correct representation (or closure hypothesis) for these eddy fluxes remains one of the major roadblocks to further progress in ocean modelling. Horizontal and vertical mixing of heat and momentum is accomplished by complex turbulent motions whose nature is still unknown in detail. Whether a simple prescription of mixing in terms of large-scale variables is possible remains to be demonstrated, but, as this symposium shows, the problem is under vigorous attack. The work on mesoscale eddy modelling described briefly in later chapters represents a beginning at explicitly including the small-scale (mesoscale) effects that the horizontal eddy parameterization is supposed to represent. The results of that work suggest that fluxes of momentum, heat, and salt which are assumed proportional to the gradient (and down gradient) are an inadequate parameterization, and in fact it is not clear that such a parameterization can even be found. If this turns out to be the case, general circulation calculations will have to include mesoscale phenomena explicitly in order to achieve valid results.

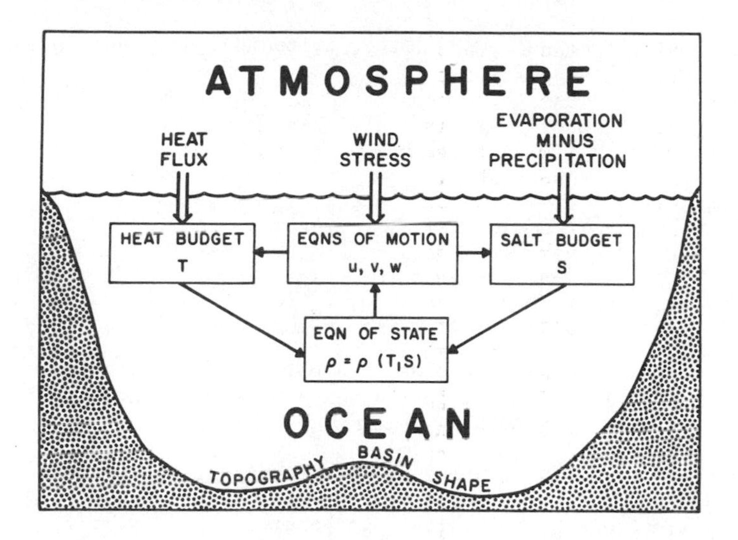

Fig. 2.1 - A schematic diagram showing the various elements in large-scale ocean models.

Our discussion is based upon the ocean model developed by Bryan and co-workers (see Bryan, 1969). Figure 2.1 shows schematically the important elements of this model. Most prognostic, three-dimensional models contain similar physics, and in fact many of the numerical experiments to be mentioned have been carried out with variants of that basic model. The equations of motion are based upon two assumptions in addition to the eddy viscosity hypothesis already mentioned: (a) the Boussinesq approximation is made, in which density variations are ignored except in buoyancy terms; (b) a hydrostatic balance is assumed. These approximations are valid for large-scale oceanic motions, but, as we shall see, certain aspects of the circulation (e.g., convective overturning) have to be treated implicitly because they are inherently small-scale.

Let $\vec{v}$ be the horizontal velocity vector (with eastward and northward components u and v) and let w be the vertical velocity. Then the equations of motion and continuity are

$$\frac{\partial \vec{v}}{\partial t} + \vec{v}\cdot\nabla_H\vec{v} + w\frac{\partial \vec{v}}{\partial z} + f\,\vec{k}\times\vec{v} = -\frac{1}{\rho_r}\nabla_H p + \vec{F} + K_{MV}\frac{\partial^2 \vec{v}}{\partial z^2}; \qquad (2.1)$$

$$\rho g = -\frac{\partial p}{\partial z}; \qquad (2.2)$$

$$\nabla_H\cdot\vec{v} + \frac{\partial w}{\partial z} = 0. \qquad (2.3)$$

Here f is the vertical component of the Coriolis parameter, equal to $2\Omega \sin\phi$; $\vec{k}$ is a unit vector in the vertical direction z; p is the pressure; and ρ is the density. K_{MV} is the vertical coefficient of eddy viscosity, which is assumed to be depth independent, and $\vec{F}$ is a horizontal body force caused by lateral friction. In Bryan's model, the latitudinal and longitudinal components of $\underset{\rightarrow}{F}$ are given by

$$F^{\lambda} = K_{MH}\left[\nabla_H^2 u + \frac{1-\tan^2\phi}{r_e^{\,2}}\, u - \frac{2\tan\phi}{r_e^{\,2}\cos\phi}\frac{\partial v}{\partial\lambda}\right] \tag{2.4}$$

$$F^{\phi} = K_{MH}\left[\nabla_H^2 v + \frac{1-\tan^2\phi}{r_e^{\,2}}\, v + \frac{2\tan\phi}{r_e^{2}\cos\phi}\frac{\partial u}{\partial\lambda}\right] \tag{2.5}$$

where λ and ϕ are longitude and latitude, r_e is the radius of the earth, and K_{MH} is the coefficient of lateral mixing of momentum, also assumed constant. These terms represent the simple closure hypothesis which parameterizes the effects of Reynolds stresses due to scales of motion (subgrid scale) not explicitly included in the model.

The equation of state relating density to potential temperature θ, salinity S, and pressure p is complicated, but can be expressed by various analytic approximations or in tables. In general

$$\rho = \rho(\theta,S,p). \tag{2.6}$$

Neglecting internal sources and sinks, the equations for the conservation of heat and salt are

$$\frac{\partial}{\partial t}(\theta,S) + \underset{\rightarrow}{v}\cdot\nabla_H(\theta,S) + w\frac{\partial}{\partial z}(\theta,S) =$$
$$K_H\nabla_H^2(\theta,S) + K_V\frac{\partial^2}{\partial z^2}(\theta,S) + CA_{\theta,S}, \tag{2.7}$$

where it has been assumed that the eddy mixing coefficients K_H and K_V are constant. $CA_{\theta,S}$ symbolically represents a contribution to the local rate of change of θ and S due to a parameterization of vertical convection. $CA_{\theta,S}$ is identically zero wherever the water column is stable (there is no forced convection) but is non-zero wherever the water column becomes unstable (free convection is permitted). This latter process will be called convective adjustment.

In many studies of the dynamics of ocean circulation, a simple equation of state (2.6) has been used, to relate the density to the temperature T, i.e.,

$$\rho = \rho_r[1-\alpha(T-T_r)], \tag{2.8}$$

where T_r and ρ_r are reference values and α is the coefficient of thermal expansion. T (or alternatively the density) is then calculated from an equation of the form (2.7), i.e.,

$$\frac{\partial T}{\partial t} + \underset{\rightarrow}{v}\cdot\nabla_H T + w\frac{\partial T}{\partial z} = K_H\nabla_H^2 T + K_V\frac{\partial^2 T}{\partial z^2} + CA_T. \tag{2.9}$$

We shall henceforth make use of this single prediction equation for density rather than the more complex pair (2.6) and (2.7) in order to simplify the discussion. The appropriate surface boundary conditions for T are then either that the total buoyancy flux is given ($\partial T/\partial z$ given), or that values of T based upon distributions of θ and S are given (T given), or some mix of these two conditions. We shall come back to this question of boundary conditions later.

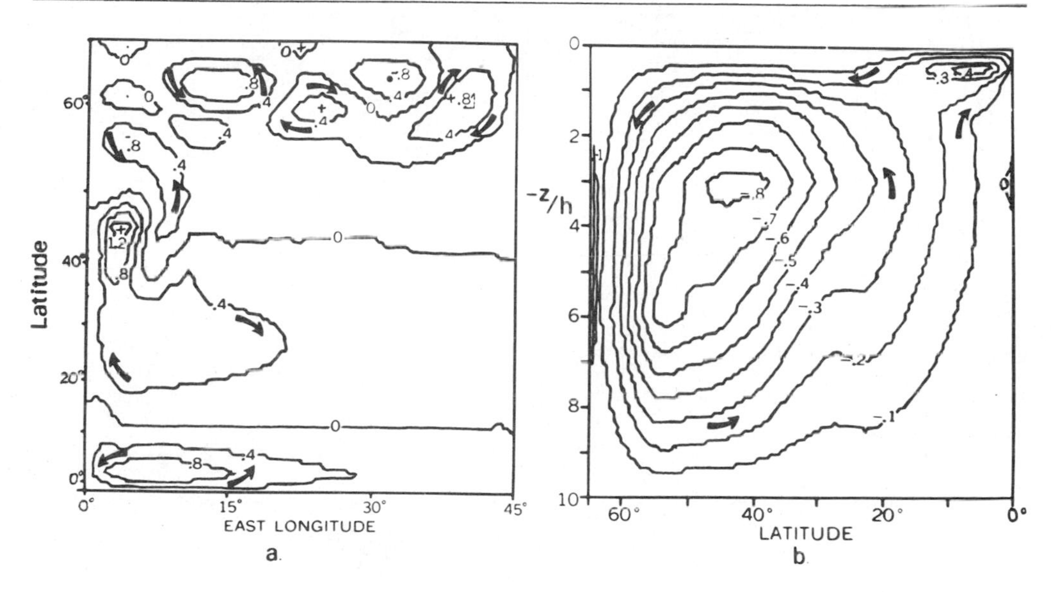

Fig. 2.2 - The pattern of total mass transport in non-dimensional units (a) in the horizontal plane, and (b) in the meridional plane (from Bryan and Cox, 1968a).

An example of the kinds of studies made with these large-scale numerical models is shown in Figs. 2.2-2.4 (Bryan and Cox, 1968a, b). The model ocean is driven by prescribed density and wind stress distribution at the sea surface. The basin is bounded laterally by two meridians 45° apart, extends from the equator to a latitude of 67°N, and is of constant depth. Starting from an initial state of uniform stratification and no motion, the finite-difference equations of motion are integrated in time until a near equilibrium state is reached (about 200 years). Figures 2.2 show the final patterns of mass transport in the horizontal and meridional planes, respectively, while Fig. 2.3 shows the horizontal temperature patterns at various depths and Fig. 2.4 shows the computed and observed western boundary currents at a latitude of 28°. Many features of the real ocean are apparent in these results. For example, the model produces a swift, northward flowing surface Gulf Stream and a weaker, southward flowing undercurrent, a Gulf Stream transport value enhanced by nonlinear effects above the Sverdrup transport value, a temperature distribution not unlike the observed thermocline structure, and a meridional overturning suggestive of deep water formation at high latitudes. It is clear, however, that a large number of such calculations are necessary to examine the relevant parameter space. In particular, the dependence of results on the mixing parameters K_{MH}, K_{MV}, K_H,

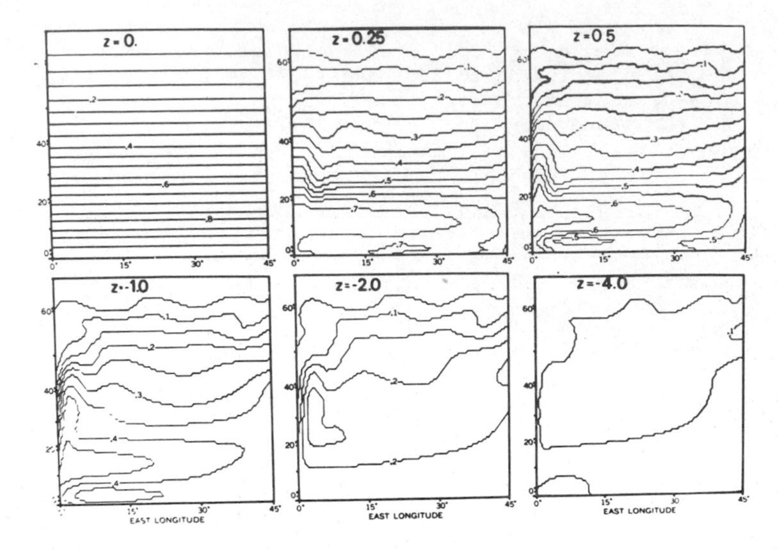

Fig. 2.3 - The horizontal temperature patterns normalized by the maximum sea surface temperature at various nondimensional depths in the final solution. The pattern at z=0.0 is specified by the surface boundary condition (from Bryan and Cox, 1968a).

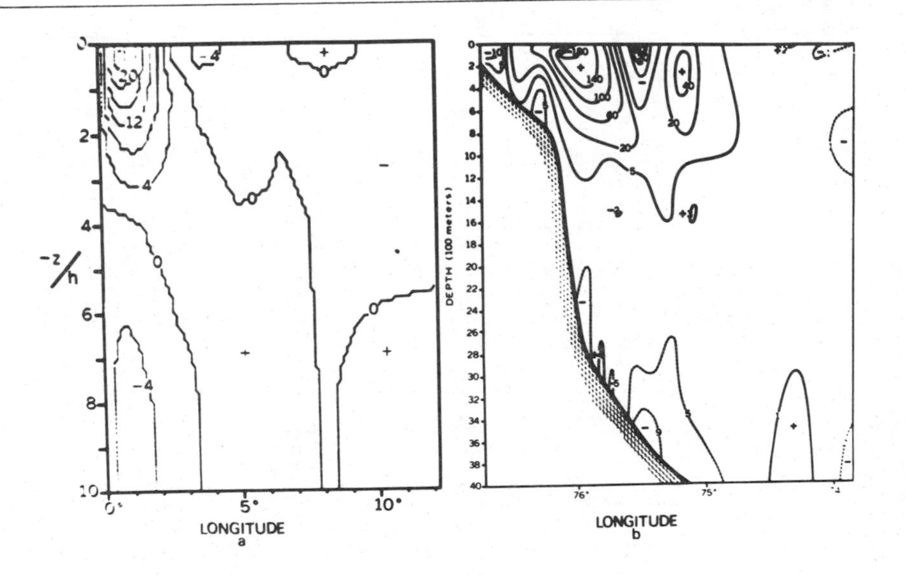

Fig. 2.4 - (a) The computed western boundary current at a latitude of 28°, given in non-dimensional units (from Bryan and Cox, 1968a). (b) The observed profile of the Gulf Stream off Cape Hatteras in cm s^{-1} (Swallow and Worthington, 1961).

K_V needs to be understood and, in fact, alternative formulations of mixing need to be tried.

2.3 The parameterization of free convection

Let us now describe the convective adjustment process (Bryan and Cox, 1967) which is the major parameterization, along with simple diffusion, in the upper layers of oceanic general circulation models. Equation (2.9) is used in a straightforward way to predict density in regions of static stability ($CA_T=0$). The vertical diffusive heat flux is just $-K_V\ \partial T/\partial z$, where usually K_V is treated as a constant (as has been assumed in Eq. 2.9). However, whenever a portion of the water column is predicted to be statically unstable, that is, $\partial\rho/\partial z>0$, a convective adjustment is made to bring that portion of the water column back to neutral stability. This is done in such a way that heat is conserved and vertical gradients of T vanish. Convective adjustment leads locally to large buoyancy fluxes upward and hence to a net decrease in the potential energy of the water column. Net energy is not conserved in this process. Note that while heat and salt are vigorously mixed by the convective adjustment process, momentum is usually assumed to mix vertically only by means of the diffusive momentum flux, $-K_{MV}\ \partial\underset{\rightarrow}{v}/\partial z$. In the large-scale, numerical general circulation models, then, convective adjustment leads to a rapid upward heat flux (ignoring salinity effects) at high latitudes, and this is balanced in the equilibrium state by a diffusion of heat downward into the ocean at low latitudes.

To illustrate how this vertical diffusion-convective adjustment model works, let us examine a simple, one-dimensional model with advection ignored. The governing equation is then

$$\frac{\partial T}{\partial t} = K_V \frac{\partial^2 T}{\partial z^2} + CA_T \qquad (2.10)$$

Figures 2.5-2.7 show the results of a time-dependent numerical calculation for which the bottom boundary condition (at z = -500m) is T_b=16°C and the surface boundary condition (at z=0) is $T_0 = 22+4\cos(2\pi t/\tau)$. The surface heat flux relates the surface temperature T_0 to the temperature in the first layer T_1 such that $K_V\ \partial T/\partial z = K_V(T_0-T_1)/(0.5\Delta z)$. The initial condition is $T = 22+6z/500$, τ is one year, the grid spacing Δz is 10m, and K_V=1 cm^2/s^{-1}. During the heating season downward diffusion is the only process; the water column is stable. During the cooling season, however, the surface becomes unstable with respect to the underlying water and convective mixing (free convection) reaches down to a certain depth. Figure 2.5 shows the temperature as a function of depth and time over the final 4 years of integration (years 116-120); Fig. 2.6 shows the heat content as a function of surface temperature and Fig. 2.7 shows the surface heat flux as a function of time.

Friedrich and Holland have performed a seasonal calculation, as yet unpublished, for the North Atlantic ocean with a three-dimensional model including the above convective adjustment process. In the central part of the basin, where horizontal advective processes for heat are small, the

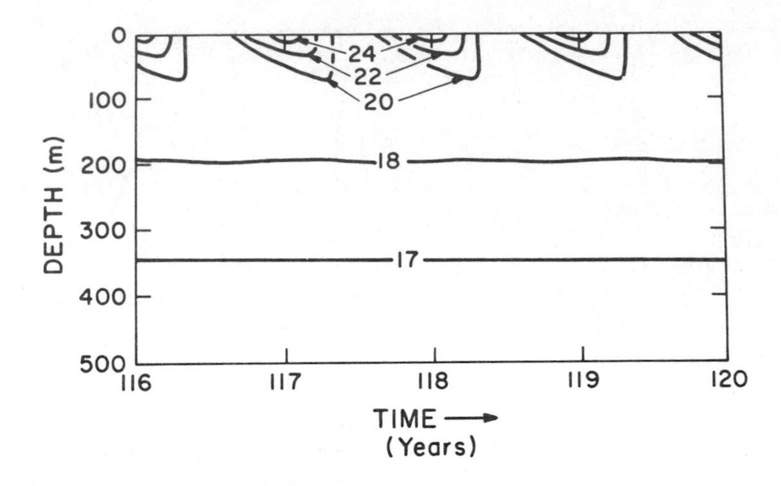

Fig. 2.5 - The seasonal cycle of temperature in °C as a function of depth and time over the last four years of the integration with a fine resolution, one-dimensional model.

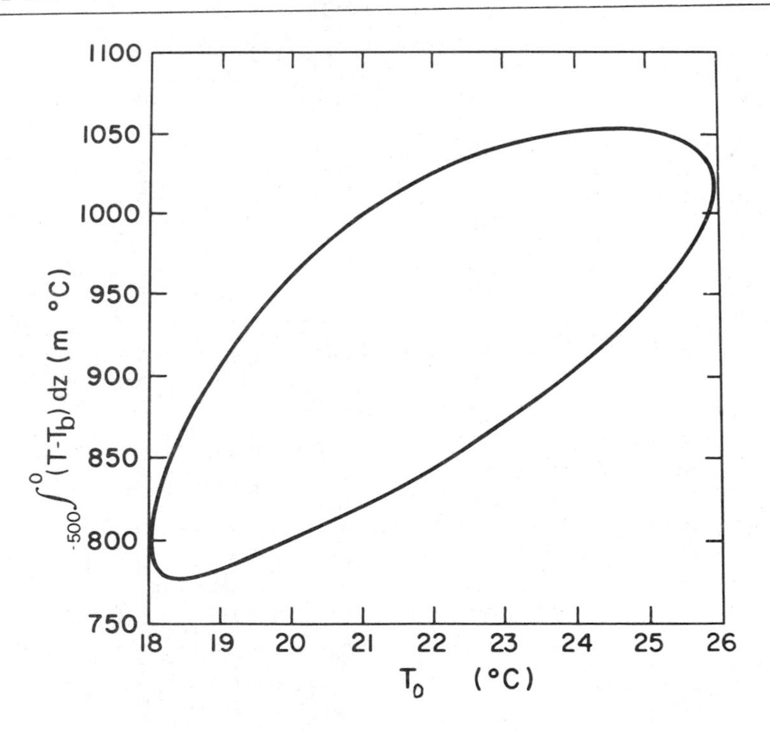

Fig. 2.6 - The heat content as a function of sea surface temperature T_0 during the last four years of the integration of a fine resolution, one-dimensional model. Note that since the four cycles overlay each other, a seasonal equilibrium has been reached.

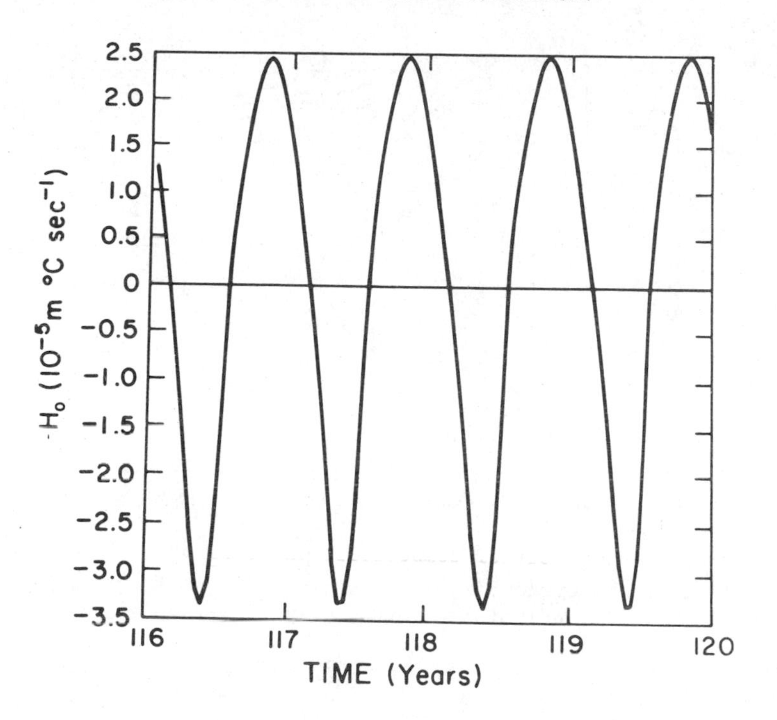

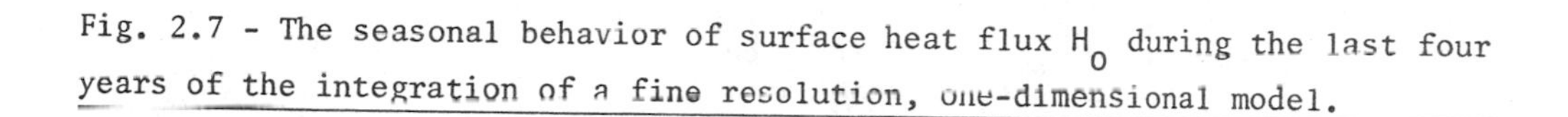
Fig. 2.7 - The seasonal behavior of surface heat flux H_0 during the last four years of the integration of a fine resolution, one-dimensional model.

seasonal cycle has been examined and temperature as a function of depth and time is shown in Fig. 2.8. The results are presented in another way in Fig. 2.9 where the temperature in various layers is shown as a function of time, together with some observations presented in the same fashion.

From the point of view of the large-scale circulation, it is not a local balance in heating that is important but rather the net heating or cooling across the sea surface averaged over many years, for that, together with direct wind forcing, is what drives the mean meridional circulation. It may be then that observations of the seasonal signal, while useful in ascertaining the local properties of vertical mixing and heat storage on the seasonal time scale, will not be able to determine the rather small, net heat flux necessary to calculate the large-scale thermohaline forcing.

Wetherald and Manabe (1972) have analyzed the response of a joint ocean-atmosphere model to the seasonal variation of the solar radiation, using the ocean model described above. They show that there is a significant warming of the lower troposphere in high latitudes, compared to the case with annual mean solar radiation, and find that this warming is in part due to a net rise in the temperature of the ocean surface in high latitudes. This latter effect results from the seasonal variation of convective activity (convective adjustment) in the surface layers of the ocean. The main consequences of this high latitude warming include a reduction in the mean

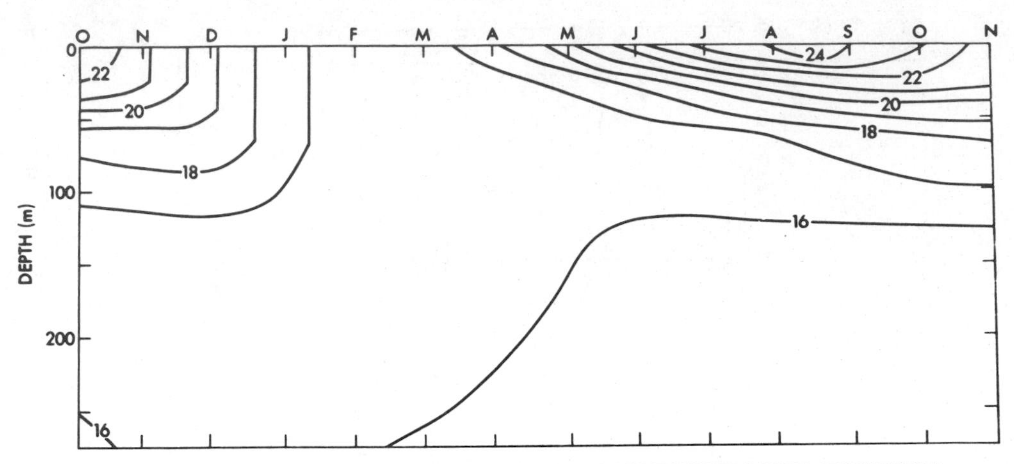

Fig. 2.8 - The seasonal cycle of temperature (°C) at 35°N, 36°W as a function of depth in a model of the North Atlantic (from Friedrich and Holland, unpublished).

atmospheric north-south temperature gradient, a reduction in the mean oceanic meridional circulation, and a reduction in the total poleward heat transports by both ocean and atmosphere.

A simple test was made by Wetherald and Manabe (1971) to show that convective activities in the surface layer of the ocean in high latitudes was responsible for the warming there. A one-dimensional model (Eq. 2.10) was integrated for a very long time period with a zero mean, sinusoidal heat flux imposed at the sea surface. Several experiments were performed (with and without convective adjustment, with differing bottom boundary conditions), and the results show a net warming of the sea surface by convective activity when compared to the non-convective case. Figure 2.10 shows the time variation of the surface temperature along with the average running mean surface temperature over each yearly cycle. Figure 2.11 shows the final annual mean temperature profiles for the upper 500 m, contrasting the two cases with and without convective adjustment. The temperature difference at the surface is about 4°C and the warming is confined to the upper 150 m.

This study indicates the importance of knowing how to parameterize the complex behavior of the upper ocean. The simple model treated here shows quite important results for the large scale circulation and its interaction with the atmosphere, but it is not clear how valid the convective adjustment approximation is and whether other effects such as forced mixing by wind generated turbulence will modify these results.

We must mention the most ambitious model calculations to date. These are carried out with the coupled ocean-atmosphere model described in Manabe, et al., (1975) and Bryan, et al., (1975). Here, in the ocean component of the model, convective adjustment is used to simulate free convection. In addition, a further mixing in the surface layers is carried out to simulate forced convection, that is, mixing due to the wind. The model is suggested

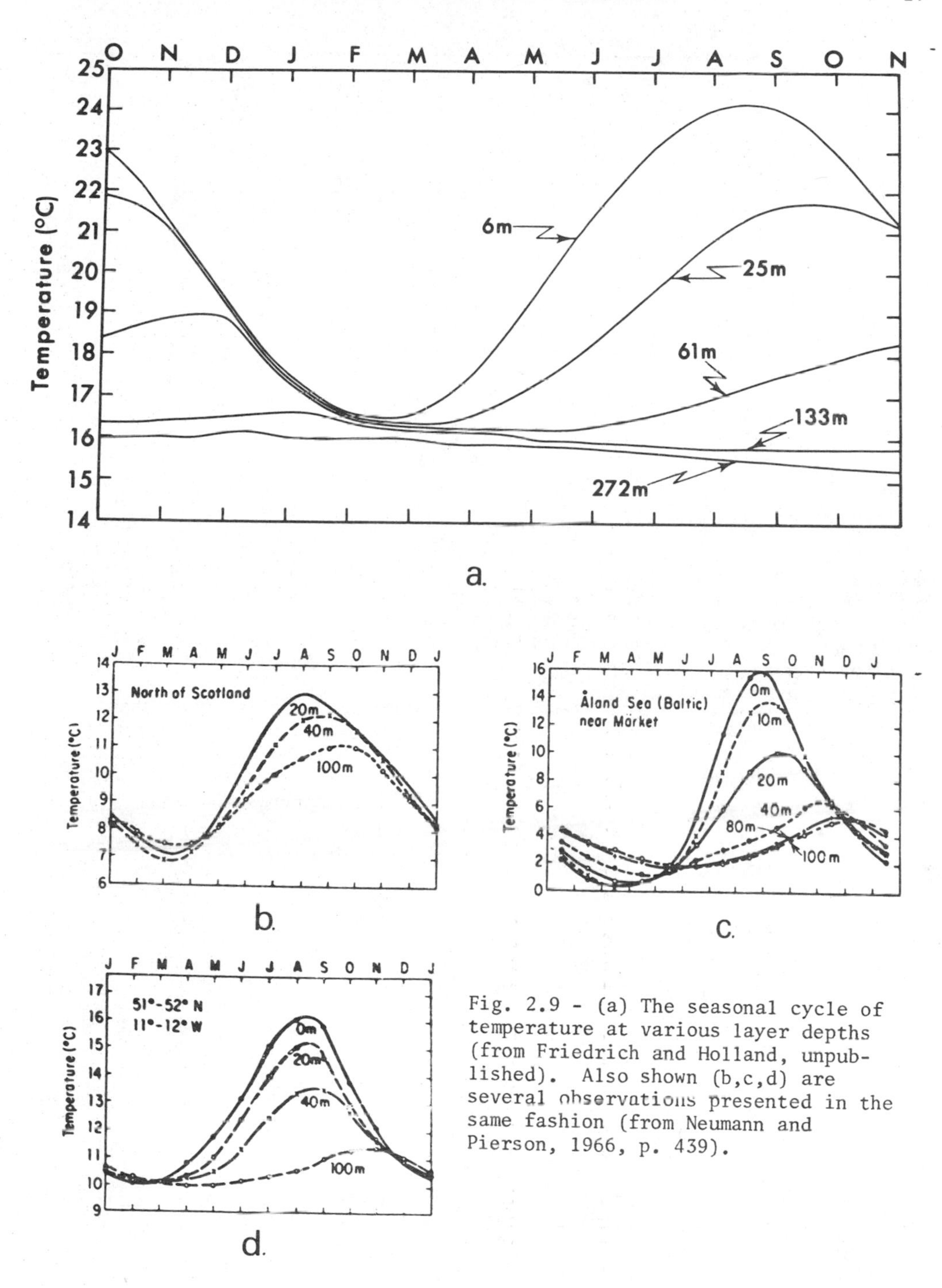

Fig. 2.9 - (a) The seasonal cycle of temperature at various layer depths (from Friedrich and Holland, unpublished). Also shown (b,c,d) are several observations presented in the same fashion (from Neumann and Pierson, 1966, p. 439).

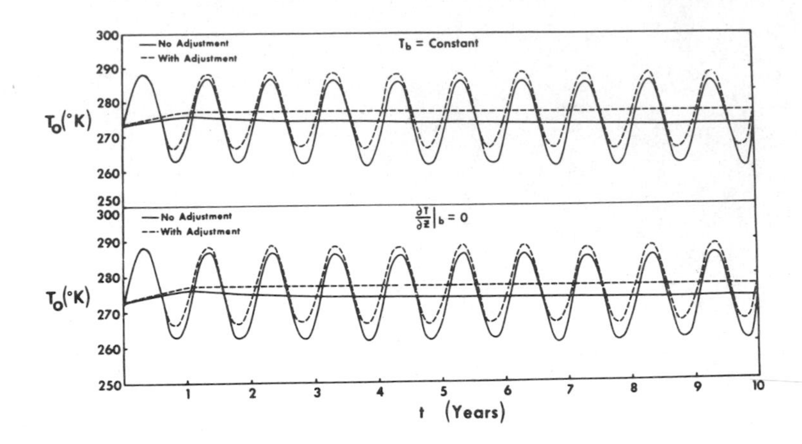

Fig. 2.10 - Time variation of the surface temperature (°K) for the first 10 years of integration of the one-dimensional diffusion model for (top) fixed bottom temperature and (bottom) zero bottom heat flux. Solid lines indicate runs made with a convective adjustment; dashed lines indicate runs made without a convective adjustment. The subscript "b" denotes bottom. From Wetherald and Manabe (1972).

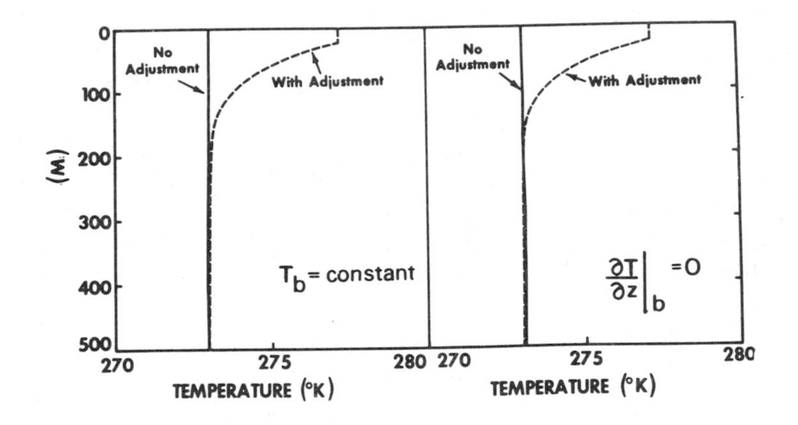

Fig. 2.11 - Final temperature profiles (°K) for the upper 500m ocean depth obtained from the one-dimensional diffusion model: (left) fixed bottom temperature and (right) zero bottom heat flux. The small vertical portion at the top of the curves corresponds to the surface mixed layer. The same graphical conventions are used as in Fig. 2.10. From Wetherald and Manabe (1972).

by the energetic approach of Kraus and Turner (1967). The energy fed into the ocean by the wind over two time steps ($2\Delta t$) is estimated as $\Delta KE = 2\Delta t \rho_a (u_*)^3$ where u_* is the friction velocity, equal to $(\tau_o/\rho_a)^{1/2}$, τ_o being the surface stress and ρ_a the air density. The potential energy of the water column is $PE = \int_{-h}^{0} \rho g z dz$. In this simple parameterization of forced convection (see Bryan, et al. (1975) for details) the depth of the mixed layer is increased, conserving heat and salt, such that $\Delta PE = \Delta KE \times$ constant; that is, some fraction of the wind energy goes to increase the potential energy, deepening the mixed layer.

Bryan, et al. (1975) do not discuss the effect which this forced convection has upon the ocean-atmosphere coupling. Presumably a number of experiments (but probably not with the coupled model) are necessary to determine the influence of such a mixed layer model on the large scale. This task remains to be done. It is probably advantageous to await the critical comparison with observations of the one-dimensional mixed layer models now under development in order that a valid (tested) parameterization is included into the large-scale models.

2.4 Surface boundary conditions and questions of resolution

Large-scale numerical ocean models have made use of a variety of surface thermal boundary conditions for the heat equation (2.9). There are several possible choices including (i) prescribed surface temperature, (ii) prescribed surface heat flux, or (iii) some relationship specified between the surface temperature and the heat flux. Bryan and Cox (1968) chose the latter condition, that is,

$$H_o = \rho_r c K_V \frac{T_1 - T_o}{0.5\Delta z_1}, \tag{2.11}$$

where H_o is the heat flux (defined positive upwards), T_o is the known surface temperature, T_1 is the temperature of the upper layer (to be predicted), c is the specific heat of sea water, and Δz_1 is the thickness of the first layer (see Fig. 2.12). Haney (1971) showed that a more realistic representation of the heat flux into the ocean, based upon an examination of what is known about net radiative heating, and latent and sensible heat fluxes at the sea surface, could be put in the form

$$H_o = m^* (T_o - T_a^*), \tag{2.12}$$

where T_a^* is an apparent atmospheric equilibrium temperature and m^* is a coefficient determined from averaged atmospheric conditions. Haney (1971) computed the zonally averaged, annual mean values of T_a^* and m^* shown in Fig. 2.13. It would be of interest to compute these quantities also for seasonal time scales and to include longitudinal variations but this has not yet been done.

There still is the problem of how to relate T_o to the first layer predicted temperature T_1 in oceanic global circulation models. Haney (1974)

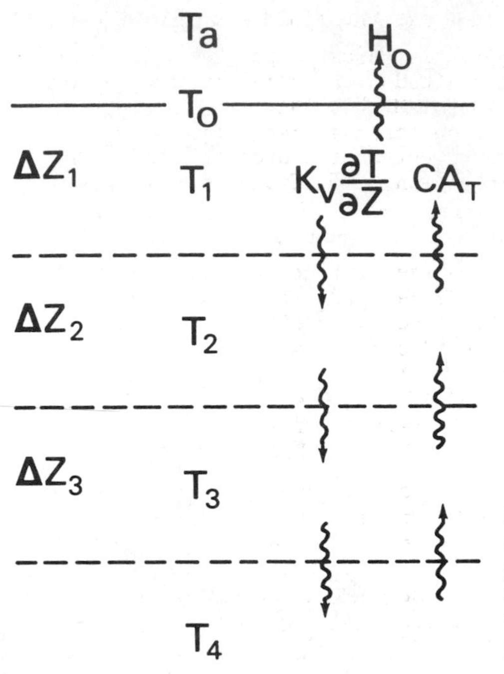

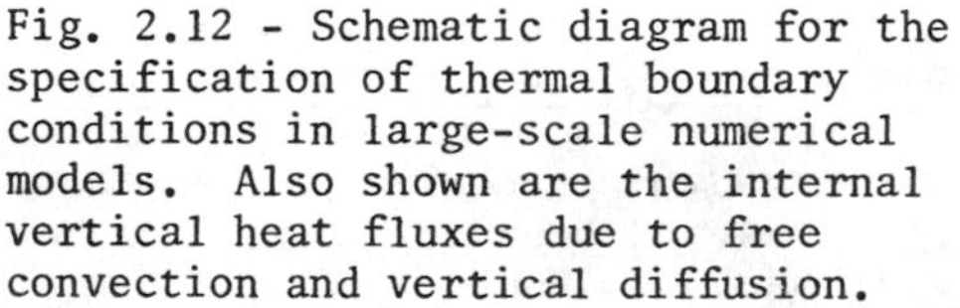

Fig. 2.12 - Schematic diagram for the specification of thermal boundary conditions in large-scale numerical models. Also shown are the internal vertical heat fluxes due to free convection and vertical diffusion.

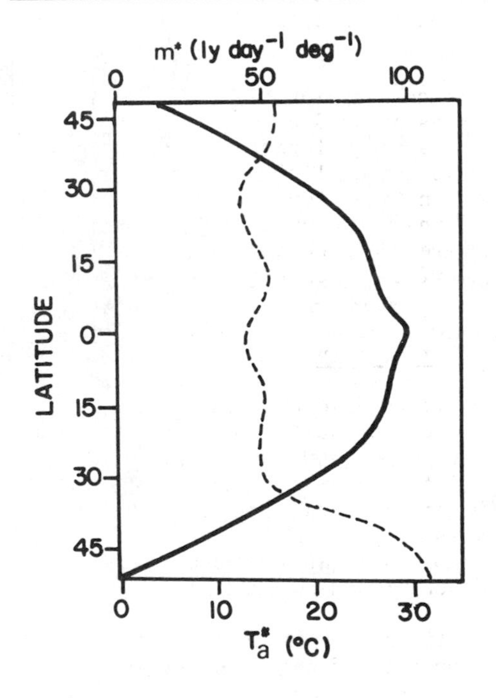

Fig. 2.13 - The apparent atmospheric equilibrium temperature T_a^* (solid line) and the coupling coefficient m* (dashed line) for zonally averaged, annual mean atmospheric conditions, from Haney (1971).

assumed that T_0 and T_1 were identical, that is, the fluid was well mixed over the depth of the first layer. When concern is focused upon the upper ocean, however, this relationship should be determined by a valid upper ocean model. Since atmospheric general circulation models (Mintz, 1964; Manabe, et al., 1965; Kasahara and Washington, 1967) need an ocean surface temperature as their bottom boundary condition, it is apparent that a proper coupling of air-sea models requires a correct determination of T_0 from the internally calculated ocean layer temperatures T_i.

In addition to the problem of relating the surface heat flux to the internally predicted temperature of the surface layer, there are also problems of vertical resolution in numerical ocean models. It is easily shown with the one-dimensional, convective adjustment model previously used that current vertical resolutions in large-scale models are not adequate for many problems. Let us repeat the calculations, the results of which were shown in Figs. 2.5-2.7, but this time with a vertical resolution of 50 m (Δz_i=50m) rather than the 10 m used before. The same boundary and initial conditions are used and the physical parameters and model are the same. The results of the coarse resolution case are shown in Figs. 2.14-2.16. As before the temperature is shown as a function of depth and time for the final four years of the calculation (years 236-240), as well as the heat content as a function of surface temperature T_0 and the heat flux as a function of time. Note that the surface boundary condition (2.11) will have much larger finite difference errors in the coarse resolution case than in the fine resolution one, causing significant discrepancies in the heat flux into the ocean. As a result the seasonal cycle of heat storage and the heat flux and the annual mean temperature profiles (Fig. 2.17) are significantly different in the two experiments.

A second aspect of this problem of vertical resolution has been discussed by Bryan and Cox (1968). That is a problem associated with the lateral advection of heat by the drift currents in the upper layers. In most large-scale calculations the surface Ekman layer is not explicitly resolved by the vertical resolution, and the effect of the finite differencing may be interpreted physically as a smearing out of the drift current throughout the entire depth of the upper layer. If there are rapid changes with depth in horizontal temperature gradients near the surface, then the advective heat flux will be seriously underestimated.

2.5 Energetics of the large-scale oceanic circulation

It is useful to examine the energetics related to the model Eqs. (2.1-2.9) in order to isolate the importance of the surface layers for the large scale. In various numerical studies of atmospheric and oceanic motions, energy budgets have contributed a great deal to the present understanding of large-scale motions. Although observations in the ocean do not yet exist which allow a comparison with reality in this respect, a comparison of budgets from alternative experiments is useful especially to understand the vital role played by motions just below the ocean-atmosphere interface.

Let us develop the relevant energy equations. If the kinetic energy per unit volume is $KE=\rho_r(\vec{v}\cdot\vec{v})/2$ then the time rate of change of kinetic energy is found by forming the scalar product of $\vec{v}$ with the terms in the

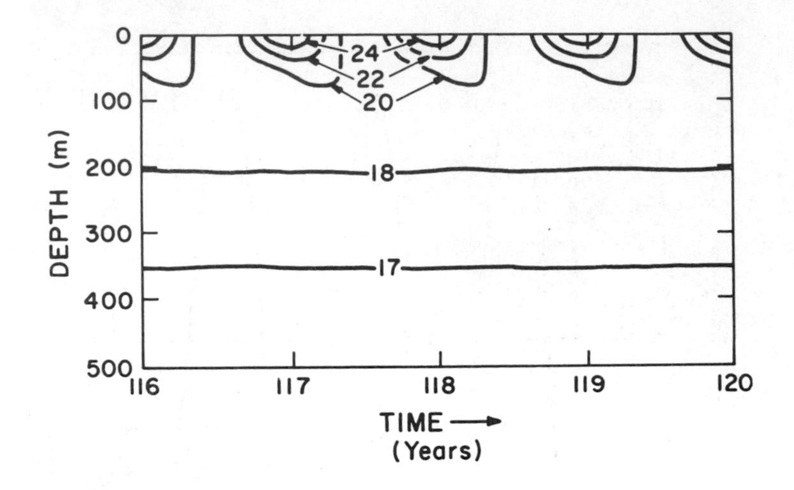

Fig. 2.14 - The seasonal cycle of temperature in °C as a function of depth and time over the last four years of the integration with a coarse resolution, one-dimensional model.

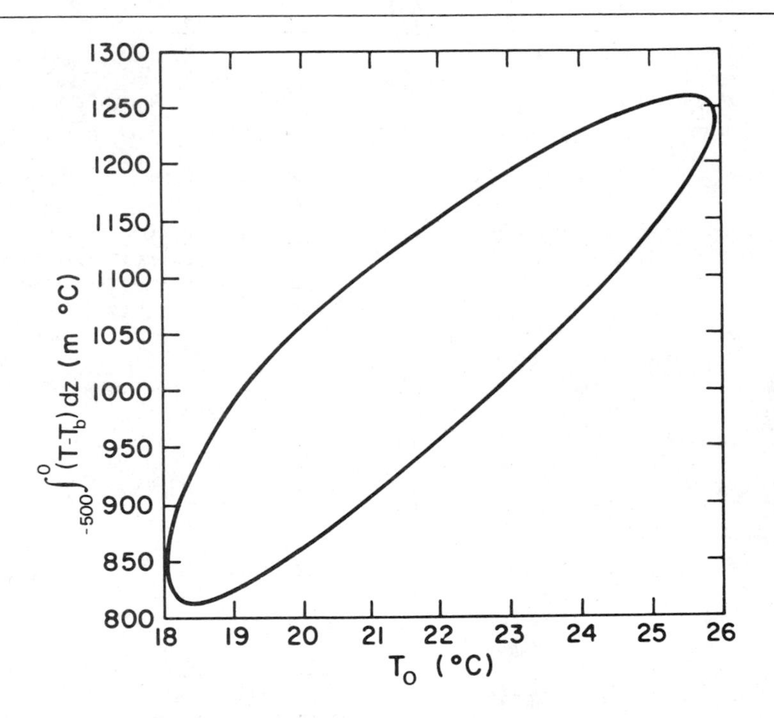

Fig. 2.15 - The heat content as a function of sea surface temperature T_0 during the last four years of the integration of a coarse resolution, one-dimensional model.

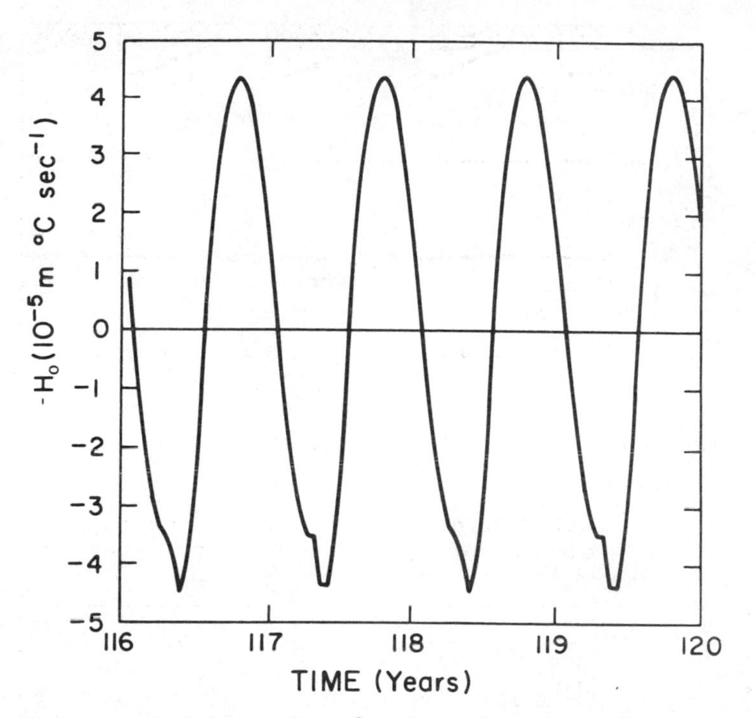

Fig. 2.16 - The season behavior of surface heat flux H_0 during the last four years of the integration of a coarse resolution, one-dimensional model.

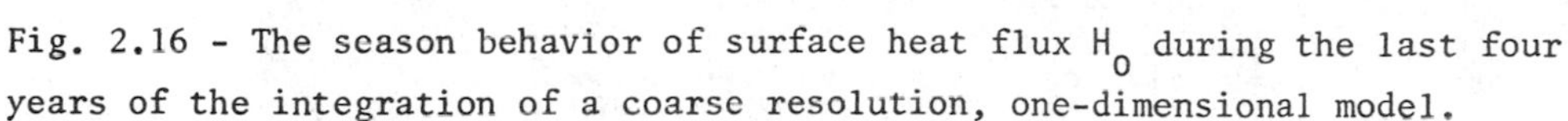

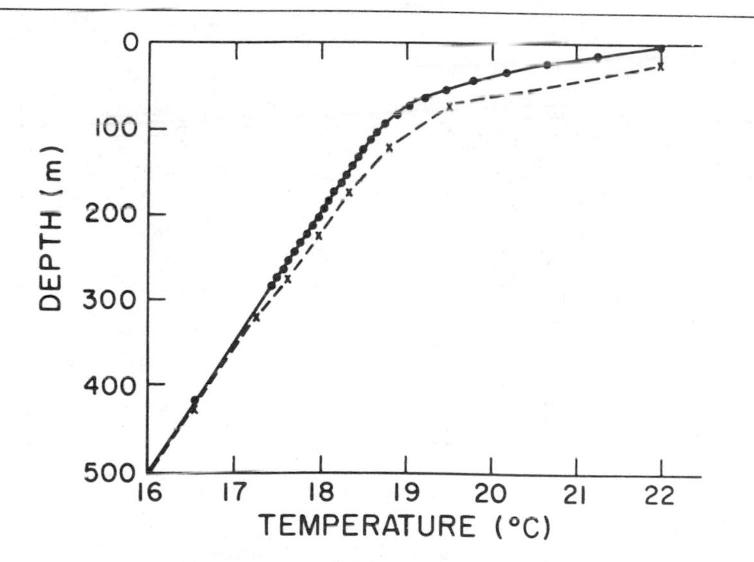

Fig. 2.17 - A comparison of the annual mean temperature profiles for fine resolution (solid line) and coarse resolution (dashed line) one-dimensional convective adjustment models.

horizontal momentum equations (2.1). The result can be put in the form

$$\frac{\partial KE}{\partial t} = -\nabla_H \cdot [\underset{\rightarrow}{v}(KE+p)] - \frac{\partial}{\partial z}[w(KE+p)] + \rho g w + \underset{\rightarrow}{v} \cdot \underset{\rightarrow}{F}$$

$$+ \rho_r K_{MV} \frac{\partial}{\partial z}(\underset{\rightarrow}{v} \cdot \frac{\partial \underset{\rightarrow}{v}}{\partial z}) - \rho_r K_{MV} \frac{\partial \underset{\rightarrow}{v}}{\partial z} \cdot \frac{\partial \underset{\rightarrow}{v}}{\partial z}. \tag{2.13}$$

For simplicity we will look only at the global energetics in a closed basin of constant depth D. Taking a volume integral of the terms in Eq. (2.13); dividing by the surface area A_s; and defining < > as $A_s^{-1}\iiint()dxdydz$, we have

$$\frac{\partial}{\partial t}\langle KE\rangle = g\rho_r\alpha\langle w(T-T_r)\rangle + \langle \underset{\rightarrow}{v} \cdot \underset{\rightarrow}{F}\rangle - \rho_r K_{MV}\langle \frac{\partial \underset{\rightarrow}{v}}{\partial z} \cdot \frac{\partial \underset{\rightarrow}{v}}{\partial z}\rangle$$

$$- A_s^{-1}\iint_{\text{bottom}} \underset{\rightarrow}{v} \cdot \underset{\rightarrow}{\tau}_b dxdy + A_s^{-1}\iint_{\text{surface}} \underset{\rightarrow}{v} \cdot \underset{\rightarrow}{\tau}_o dxdy. \tag{2.14}$$

The term on the left is the rate of change of kinetic energy per unit area per unit time (ergs/$cm^{-2}s^{-1}$) and the terms on the right are the work done per unit area. We have already assumed the basin is mechanically closed, that is, there is no flow in or out of the basin, either laterally or vertically.

The first term on the right is the work done by buoyancy forces. As we shall see, this term is responsible for an energy transfer between kinetic and potential energy but leads to no net production of energy. The other terms are (i) the dissipation of kinetic energy by lateral viscosity, (ii) the internal dissipation due to vertical shear stresses, (iii) the work done on the fluid by bottom stresses, and (iv) the work done on the fluid by the wind stress at the sea surface.

The equation governing the rate of change of the potential energy of the system is derived by multiplying Eq. (2.9) by $g\rho_r\alpha(z+D)$, where the depth z varies from -D at the bottom to zero at the sea surface. Here again, for a simplicity of interpretation, we make use of the temperature equation (2.9) rather than the more complex and accurate set of equations governing the potential temperature, Eqs.(2.6)-(2.7). Then, if $PE=-g\rho_r\alpha(z+D)(T-T_r)$ is the potential energy per unit volume (with a zero reference value at z=-D):

$$\frac{\partial PE}{\partial t} = -\underset{\rightarrow}{v} \cdot \nabla_H PE - w\frac{\partial PE}{\partial z} - g\rho_r\alpha w(T-T_r) - K_V g\rho_r\alpha(z+D)\frac{\partial^2 T}{\partial z^2}$$

$$+ K_H \nabla_H^2 PE - g\rho_r\alpha(z+D)CA_T. \tag{2.15}$$

Integrating Eq. (2.15) over the volume, dividing by the surface area as above and rearranging,

$$\frac{\partial}{\partial t}\langle PE\rangle = -g\rho_r\alpha\langle w(T-T_r)\rangle + A_s^{-1}\iint\limits_{\text{surface}} \frac{g\alpha H_0 D}{c}\,dxdy$$

$$+ A_s^{-1}\iint\limits_{\text{bottom}} K_V g\rho_r\alpha(T_0-T_b)dxdy - g\rho_r\alpha\langle(z+D)CA_T\rangle \qquad (2.16)$$

Here H_0 is the upward heat flux at the sea surface, and T_0 and T_b the temperatures at the surface and bottom of the ocean respectively. The term on the left is the rate of change of potential energy per unit area per unit time (erg cm^{-2} s^{-1}), and the terms on the right are the work done per unit area in changing the potential energy. Note again that we have made use of the assumption of a closed basin.

The first term on the right is the work done per unit area by buoyancy forces and is the term which transfers energy from potential to kinetic form. The other terms on the right are the rates of change in potential energy due (i) to surface heating, (ii) to net downward diffusion, and (iii) to upward heat convection due to convective adjustment, respectively.

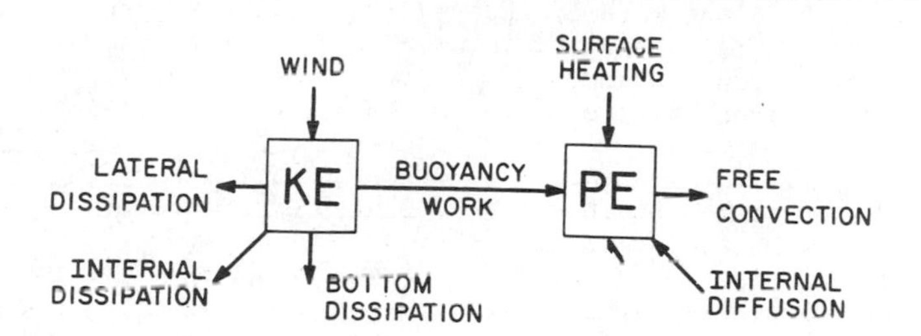

Fig. 2.18 - Schematic diagram of the energetics of a large-scale, numerical ocean model identifying the various inputs and losses to the system.

Figure 2.18 shows the energy flow diagram associated with Eqs. (2.14) and (2.16) that governs the energetics of the large-scale circulation. Note now the important role played by the upper ocean and surface boundary conditions. Basic energy inputs are accomplished at the sea surface by wind forcing and surface heating. The wind forcing depends upon knowing the surface currents in the direction of the wind stress, and the surface heating depends upon the heat flux distribution H_0. If the heat flux can be approximated by Eq. (2.12), then the potential energy input depends upon a knowledge of the surface temperature T_0. In addition to these surface energy inputs, both the kinetic energy loss due to dissipation by vertical shear stresses and the potential energy loss due to free convection occur mainly in the upper ocean. Thus the large-scale circulation is intimately tied to what goes on in the upper ocean, and new parameterizations of vertical mixing for both momentum and heat will modify the modelled general circulation of the ocean in a basic way.

2.6 Mesoscale eddies and horizontal mixing

Although there exists neither an observational nor a theoretical basis for looking into the interaction of mesoscale eddy motions and important upper layer motions at the present time, an analogy exists between the need to understand the behavior of the upper ocean (to be parameterized as a vertical mixing process in large-scale ocean models) and the need to understand the behavior of mesoscale eddies (to be parameterized as a horizontal mixing process in large-scale ocean models). At some future time, the coupled problem, that is, how the mixed layer affects mesoscale eddy activity and how mesoscale eddies affect vertical mixing of heat and momentum, will have to be solved. For the present they must be considered as independent processes affecting the large-scale circulation.

Let us just mention here the modelling attempts to understand the role of mesoscale eddies in the general circulation. In contrast to the mixed layer problem, no parameterization of the mixing properties of mesoscale eddies has yet been proposed; the properties of the eddy field and the source of the eddy energy are just beginning to be explained. Progress has been made on this problem largely by developing eddy resolving, general circulation models, that is, ones with fine horizontal resolution. In such models the eddies can be explicitly dealt with and their influence on the large-scale circulation explicitly ascertained.

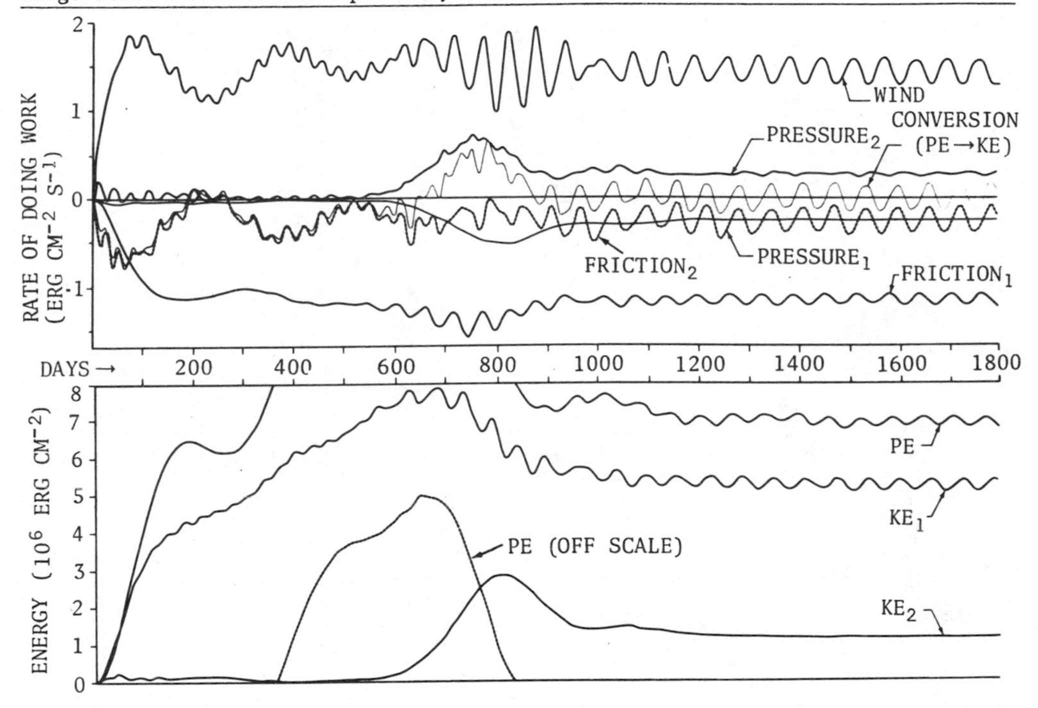

Fig. 2.19 - Results from an eddy resolving general circulation model (Holland and Lin, 1975a, b). The time history of the volume averaged energies and energy rates. The upper panel shows various rates of doing work including the rate of energy conversion between potential and kinetic. The lower panel shows the kinetic energy in each layer and the potential energy of the system.

Recently Holland and Lin (1975a, b) have begun a series of investigations of this kind. They examined a rectangular, wind driven ocean with two layers and fine enough horizontal resolution to resolve horizontal scales on the order of the internal radius of deformation, which is the preferred scale for the production of baroclinic eddies. The model is driven by a simple sinusoidal wind stress pattern and the stratification and layer thicknesses are chosen to give realistic values of the internal radius of deformation. The results of several such experiments show that the circulation pattern and the development of eddies depend strongly on the coefficient of lateral friction as well as the boundary conditions imposed at the side boundary.

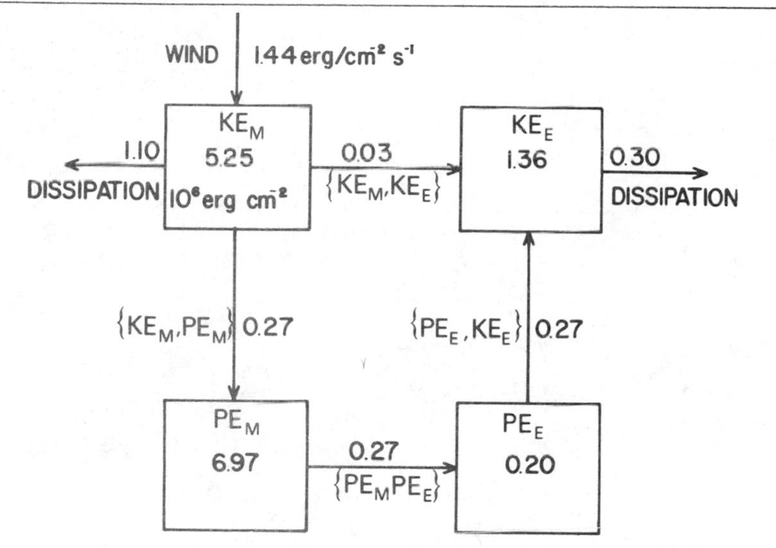

Fig. 2.20 - The energy budget diagram for the final, statistically steady state. The energy is distributed between eddy (E) and mean (M) kinetic (KE) and potential (PE) energies by the energy fluxes shown.

An example of one case is shown in Figs. 2.19-2.23. The model ocean is spun up from rest by a single gyre wind stress pattern. The density contrast between the upper and lower layers, $\Delta\rho/\rho_r$, is .002 and the upper and lower layer thicknesses are 1000 m and 4000 m, respectively. This gives a radius of deformation of 50 km. The wind stress amplitude is 1 dyne/cm^2, and the lateral eddy viscosity is 3.3×10^6 cm^2 s^{-1}, and free slip (stress free) boundary conditions are used in this calculation. During the initial phase the energy put in by the wind increases the kinetic energy of the upper layer and the available potential energy of the system (Fig. 2.19). At about 500 days the flow becomes baroclinically unstable and the available potential energy falls rapidly to some lower equilibrium value. During that time the kinetic energy of the lower layer rapidly increases as mesoscale eddies, with horizontal wavelengths of 400 km and periods of 64 days, appear in the region of instability. After some adjustment time the ocean reaches a statistically steady state in which the eddies are superimposed upon a mean circulation.

The energetics of the mean and eddy flows for the final statistically steady state are shown in Fig. 2.20. The energy equations governing the mean kinetic energy (KE_M), eddy kinetic energy (KE_E), mean potential energy (PE_M), and eddy potential energy (PE_E) for the case of a wind driven ocean can be written symbolically

$$\frac{\partial KE_M}{\partial t} = -\{KE_M, KE_E\} - \{KE_M, PE_M\} + F_M + W_M;$$

$$\frac{\partial KE_E}{\partial t} = +\{KE_M, KE_E\} + \{PE_E, KE_E\} + F_E; \qquad (2.17)$$

$$\frac{\partial PE_M}{\partial t} = +\{KE_M, PE_M\} - \{PE_M, PE_E\};$$

$$\frac{\partial PE_E}{\partial t} = -\{PE_E, KE_E\} + \{PE_M, PE_E\};$$

where {A,B} means an energy flux from A to B, F_M and F_E are the dissipations of mean and eddy energy, and W_M is the energy input by the (steady) wind. In this case the process of baroclinic instability releases mean potential energy to supply potential energy to the eddies. A smaller amount goes directly from mean to eddy kinetic energy. Note the large drain of energy from the total mean energy to the eddies.

Fig. 2.21 - The instantaneous fields of interfacial depth, mass transport streamfunction, and pressure fields in the upper (P_1) and lower (P_2) layers. The contour interval (CI) is shown on each diagram. Stippled areas represent negative values.

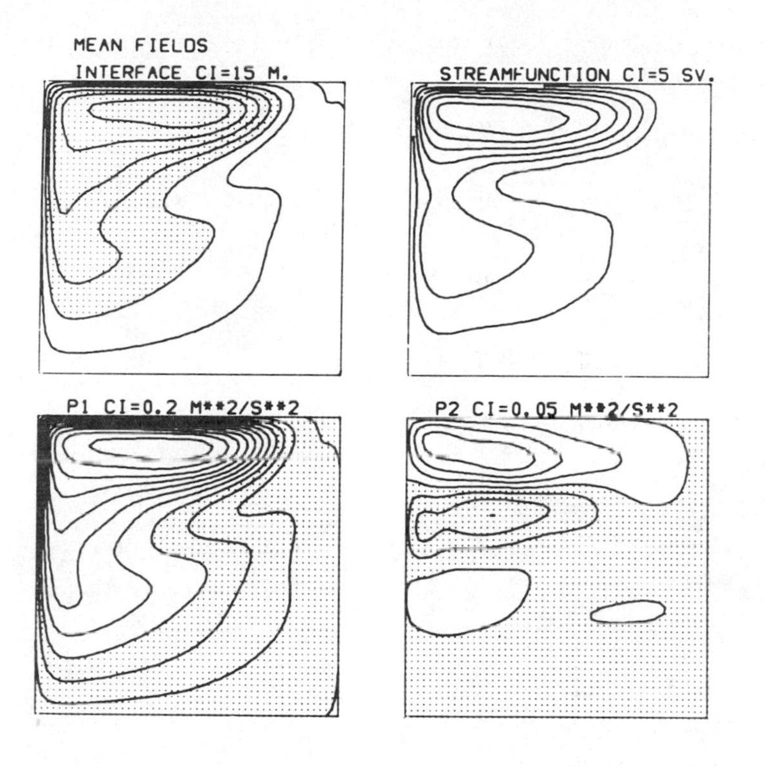

Fig. 2.22 - As Fig. 2.21, for mean (time averaged) fields.

Figures 2.21-2.23 show the instantaneous, mean, and eddy fields, respectively. Note the important result that the baroclinic eddies set up a secondary circulation in the deep water, entirely driven by the Reynolds stresses. Such a result would not be possible in steady models which parameterize the Reynolds stresses as in earlier studies.

These very preliminary experiments do show that mesoscale processes alter the general circulation in the ocean just as transient disturbances do in the atmosphere. The difficulty for the ocean modeller is that these oceanic eddies have very small horizontal scales and thus are difficult to include in models of entire ocean basins. It is not yet known whether some parameterization of these eddy processes in terms of large-scale flow properties is possible, but it is clear that the simple closure laws previously used are not adequate.

2.7 Concluding remarks

In this paper we have tried to make the case that a physically correct parameterization of mixing in the upper ocean is a crucial ingredient in models of the large-scale ocean circulation. Although there is little experience as yet with alternative formulations of this parameterization within the context of three-dimensional model studies, one-dimensional analogs show that the surface temperature, the heat content, and the heat and energy inputs at the sea surface all are strongly influenced by the

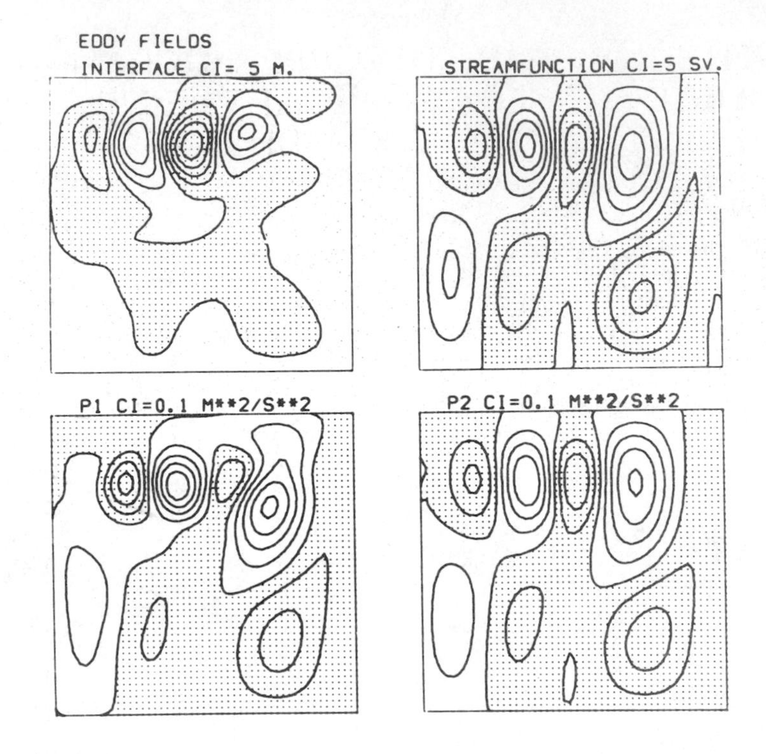

Fig. 2.23 - As Fig. 2.21 for eddy fields (instantaneous minus mean).

details of the mixing model. We can have little confidence therefore in large-scale model results that contain as yet untested parameterizations of these mixing processes.

The future of large scale ocean modelling depends upon the success of mixed layer model development in the next few years. Similarly the mesoscale eddy problem must be solved if we are to have confidence in global ocean circulation studies and climate studies with coupled air-sea models. Thus far, progress has been rapid in both of these areas. Hopefully, new observations will be made that can critically test various hypotheses about the nature of these mixing processes, and we will be able soon to incorporate new and better mixing parameterizations into models of the general circulation of the ocean.

Acknowledgement: NCAR is sponsored by the U. S. National Science Foundation.

Chapter 3

THE ROLE OF THE UPPER OCEAN IN LARGE-SCALE NUMERICAL PREDICTION OF THE ATMOSPHERE

Richard C. J. Somerville

3.1 Introduction

One of the most important aspects of progress in dynamic meteorology in the last thirty years has been the development of comprehensive numerical models of the large-scale circulation of the atmosphere. These models are based on conservation laws for momentum, mass, thermodynamic energy, and water substance, which govern the time evolution of atmospheric velocity, pressure, temperature, and humidity. The computational domain of such models may be as large as the entire global atmosphere, spanned by a three-dimensional finite-difference mesh of tens of thousands of grid points. The more physically complete models include representations of a myriad of interacting physical processes, such as solar and terrestrial radiative transfer, small-scale turbulent transports, surface interactions, and hydrological cycle, as depicted in Fig. 3.1. These processes, parameterized in terms of the prognostic variables, represent sources and sinks of momentum, heat, and water for the conservation equations.

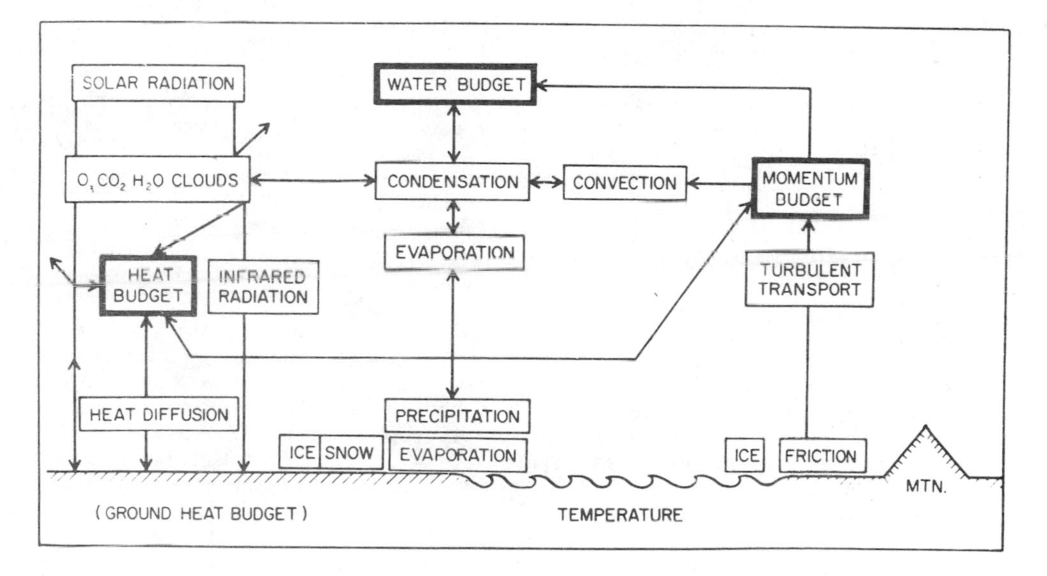

Fig. 3.1 - Schematic diagram of interactions of physical processes in a typical atmospheric model (Somerville, 1975).

Applications of such models fall into two broad categories. The first is weather prediction, the calculation of the detailed transient evolution of the large-scale circulation from an observed initial state. Routine operational forecasts by the meteorological services of many countries are currently based on such model integrations. Over the time ranges involved, typically a few days, model forecasts of the large-scale atmospheric features are consistently more skillful than subjective predictions by experienced meteorologists (Jensen, 1975).

The second principal type of model application is the simulation of the statistical equilibrium state of the atmosphere by integrations of at least several weeks in length. The initial state, which is all-important in weather prediction, is usually regarded as irrelevant to this latter application, knows as general circulation modelling. Instead, it is the properties of the long term statistical behavior of the atmosphere, and their dependence on boundary conditions and external forcing, that are of interest. In short, it is understanding climate, rather than predicting weather, which largely motivates general circulation modelling.

Constructing a model for weather prediction or general circulation simulation involves making critical choices in defining the boundary of the system and the degree to which the model atmosphere interacts with its surroundings. It is only in recent years, for example, that a comprehensive, coupled, global ocean-atmosphere model has been attempted (Bryan, et al., 1975; Manabe, et al., 1975). Instead, nearly all large-scale atmospheric circulation models have employed the artifice of a non-interactive oceanic boundary condition. The sea is typically represented as exerting a frictional drag on the air, providing an infinite source of water for evaporation, and possessing a prescribed geographical and temporal distribution of surface temperature.

Within this constraint, it is of course still possible to vary the specification of sea surface temperature (SST) and to study the resulting effects on the model atmosphere. Such numerical experiments have now been carried out with several major atmospheric models. The purpose of this paper is to survey these experiments and to summarize what has been learned about the sensitivity of atmospheric models to SST variations. In particular, let us ask the following question: if an adequate coupled upper ocean model were available to provide accurate predictions of the SST field, what improvement would result in our ability to predict the large-scale behavior of the atmosphere?

3.2 Predictability

In answering this question, it is essential to be aware of a fundamental limitation on the skill of weather forecasts. It is clear that present-day forecasts suffer from both imperfect models and inadequate observational data. Yet even a perfect model, which exactly represented the dynamics of the real atmosphere, and which was initialized with data from the best conceivable observing system, could not produce an accurate forecast of indefinitely long range. This limit arises because unavoidable small errors in the initial conditions will inevitably amplify as the forecast evolves. Put another way, any two atmospheric states, initially closely resembling one another, will eventually diverge and ultimately differ from each other as much as two randomly chosen states.

This error growth is a fundamental nonlinear property of turbulent flows (Lorenz, 1969). Its magnitude may be estimated by performing two integrations (from slightly differing initial conditions) of the same atmospheric model. One integration represents the evolution of the atmosphere and the other represents the forecast. The initial difference represents the observational error. Such experiments have been performed with several models, and best current estimates are that small errors double in about two or three days and that the maximum possible forecast range is a few weeks. These results are consistent with empirical data and with turbulence theories, as summarized by Leith (1975).

This limitation on the predictability of the detailed evolution of the atmosphere ("weather") does not preclude the possibility of seasonal and longer range forecasts of means and other statistical properties ("climate"). At this point, however, a striking difference between weather forecasts and climate forecasts should be stressed. It is simply that skillful weather forecasts are presently made routinely, while climate forecasting is in a much earlier stage of development.

The currently attainable skill of extended range numerical weather forecasts is represented by results of Miyakoda, et al. (1972) for the GFDL model and of Druyan, et al. (1975) for the GISS model. The models themselves are described, respectively, by Manabe, et al. (1965) and Somerville, et al. (1974). Both models have been tested on an ensemble of experimental predictions begun from and verified against actual data. The skill of both models decreases rapidly in the first few days of the forecasts but has not entirely disappeared at one week or even ten days. Several other major models (Global Atmospheric Research Program, 1974) are probably capable of similar performance, and the outlook for continued progress, through improvements in models and data, is promising.

Recent surveys of climate modelling include Schneider and Dickinson (1974), Smagorinsky (1974), Global Atmospheric Research Program (1975), and National Academy of Sciences (1975). There currently is extremely vigorous activity in this field. Nevertheless, it is not yet known what components of climate beyond the seasonal cycle may be predictable, and there are presently no reliable means of making consistently skillful climatic forecasts, at least by modelling methods, although some encouraging results have been achieved by statistical and synoptic techniques (e.g., Namias, 1968).

We are thus faced with a potentially serious difficulty in determining the value of SST information for improving large-scale atmospheric predictions. Our ability to assess the atmospheric forecasting impact of SST data will depend critically on the time scale required for the ocean to influence the model atmosphere. If this time scale is shorter than the predictability limit, we may be able to detect an impact on the forecast weather. If, on the other hand, the time scale is longer than the predictability limit, then we must seek an impact on the model climate. Finding such an impact may yield valuable insight into the mechanisms involved in atmospheric sensitivity to SST, but will not necessarily provide any basis for prediction.

3.3 Experiments

With this background, we now examine briefly a number of numerical experiments designed to explore the sensitivity of atmospheric models to SST anomalies. These experiments typically involve the intercomparison of model integrations which differ only in the specification of SST. Such studies have now been carried out with several different major models, using very different SST patterns, and they have been evaluated in rather different manners.

These experiments have in most instances been motivated to a considerable extent by a series of intriguing ideas and observations concerning possible atmospheric effects of sizable SST anomalies, including effects which may be far removed from the anomaly in time or space or both. Bjerknes (1966, 1969) and Namias (1962, 1963, 1965, 1968, 1969, 1970, 1971) are notable and representative examples of this provocative and fascinating research which has captured the interest of the numerical modellers.

The first numerical experiment which we shall consider is that of Rowntree (1972), who used a version of the hemispheric form of the GFDL model (see, e.g., Manabe, et al., 1965). His experiment was designed to test Bjerknes' (1966) idea that the position and intensity of the Aleutian low in the surface pressure field are influenced by variations in SST in the eastern tropical Pacific. Rowntree's integrations were for thirty simulated days, he used both warm and cool SST anomalies with maximum amplitudes of 3.5C, and he found both tropical and extratropical effects.

In the tropics, Rowntree found that the warm anomaly induced a surface low near the location of the SST maximum, accompanied by low level convergence, upward vertical motion, and enhanced rainfall. Extratropically, the Aleutian low moved eastward and deepened, in rough accordance with Bjerknes' hypothesis. The tropical effects occurred sufficiently quickly that Rowntree concluded that it might be important to use correct local SST values in tropical short range prediction, but mid-latitude changes did not become significantly large until after one week.

Subsequently, several investigators have examined the effects on model atmospheres of anomalies located in different parts of the world ocean. For example, Houghton, et al. (1974) used the global NCAR model (Kasahara and Washington, 1971) and prescribed anomalies of approximately 2C amplitude in the Atlantic off the coast of Newfoundland. They used both warm and cold anomalies and compared model results with observed data. The interpretation of the results was hampered by model deficiencies and by questions of statistical significance of the results, and Houghton, et al. found "physically realistic" results "only in the region of the anomaly itself (the North Atlantic sector) and only for limited time intervals."

Other examples of experiments with different locations of the prescribed SST anomalies are the studies of Shukla (1975) and Simpson and Downey (1975), both using versions of the GFDL model. Shukla employed a global domain to examine the effects on the Indian summer monsoon of a cold SST anomaly over the Somali coast and the western Arabian Sea. This anomaly (maximum amplitude 3C) was integrated for 48 days and was apparently associated with a dramatic reduction in precipitation over India.

Simpson and Downey (1975) used a southern hemisphere model version with a warm (maximum amplitude 4C) mid-latitude Pacific anomaly. Two anomaly runs were made, with the anomalies introduced at different times. They found, in 50-day integrations, systematic effects in some regions of the Antarctic trough and the mid-Atlantic and Indian Oceans, but no systematic effect in the immediate area of the anomaly.

Recently, Chervin, et al. (1976) have examined the effect in the NCAR model of wintertime SST anomalies in middle latitudes of the north Pacific. They made an especially diligent effort to test the statistical significance of their results. They concluded, within the constraints of a somewhat limited experiment, and verifying only the temperature field in the lowest model layer, that the model's response was statistically significant, "but primarily over the region of the anomaly pattern and primarily in the case of an anomaly of unrealistically large magnitude."

We may also quote a caveat from Chervin, et al.:

> "*However, the fact that the existence of downstream teleconnections is not well demonstrated in our limited experiment does not necessarily imply their non-existence in the real atmosphere.* Obviously, more analyses and intercomparisons . . . are required to determine more conclusively the extent to which teleconnections between north Pacific Ocean surface temperature anomalies and downstream atmospheric anomalies may improve long-range forecasting of monthly to seasonal means."

Before describing one last SST impact test with a large atmospheric circulation model, we may note that although these comprehensive models do represent the state of the art in weather forecasting and climate simulation, there are potentially serious defects in the methodology of these numerical experiments. The likelihood that the models may inadequately resemble the actual atmosphere in some important respects is one disturbing possibility. The economic difficulties of performing more than a very few integrations, and the consequent vexing questions as to the statistical significance of results, are another troublesome aspect of this approach. Finally, the simplifying expedient of specifying SST may itself obscure vital facets of the interaction between atmosphere and ocean. Thus large-scale numerical modelling should be carried out together with alternative techniques, which may be less subject to the above defects.

One such alternative is exemplified by recent work of Davis (1976), who examined the variability and statistical predictability of 28 years of sea level pressure (SLP) and SST data in the central North Pacific. In examining the association between SLP and SST anomalies on seasonal time scales, Davis is led to the conclusion that the predominant process is the atmosphere forcing the ocean rather than vice versa. He finds support for the premise that SST anomalies are the result of SLP anomalies, and not the cause, in a study by Salmon and Hendershott (1976). They employed a very simple atmospheric circulation model coupled to an even simpler ocean: a motionless global heat reservoir. Despite the severe idealizations, this technique permits some degree of two-way interaction between air and sea. Salmon and Hendershott find that their numerical experiments "suggest that the atmosphere may simply be too noisy to be much affected by mid-latitude SST anomalies on time scales over which the anomalies are themselves predictable." Both they and Davis give detailed discussions of the limitations and uncertainties in their work. These two studies may be regarded as complementary to the impact tests with detailed and comprehensive models, which, as we have noted, have their own serious limitations.

The emphasis in the studies reviewed above seems generally to have been on local effects and on effects at times longer than the detailed predictability limit. In view of this, a recent numerical experiment by Spar and Atlas (1975) is especially relevant to the question of the predictive impact of SST information. Earlier, Spar (1973a, b, c) had conducted experiments in the spirit of those of Rowntree (1972) and Houghton, et al. (1974). Spar and Atlas, however, ask

> "how sensitive are current prediction and general circulation models to real variations, as well as errors, in the SST fields? Will the use of observed SST values rather than climatological (or other time-averaged) values lead to forecast improvement in the short range (1-3 days), or in the extended range (4-14 days)? It is, in fact, not known with any certainty how important it is to specify accurately, or to predict, the SST field for the purposes of short-range and extended weather prediction. The benefits, if any, gained from more accurate specification of the SST field could, for example, be completely overwhelmed by the decay of predictability by the time the diabatic effects of the ocean surface fluxes significantly affect the atmosphere."

Spar and Atlas used the GISS global atmospheric model (Somerville, et al., 1974) in a two week prediction experiment to test the impact on forecasting skill of using an observed distribution of SST in place of a climatological one. They found a marked local effect on oceanic precipitation, associated with increased convective activity over warm SST anomalies. Inadequate observational data, however, prevented the detailed verification of the oceanic forecasts. Over continents, where verification was possible, no definite effect of the observed SST field was discernible. Spar and Atlas also found that the sea level pressure and 500 mb height fields were relatively insensitive to SST variations.

In general, this study showed that the model prediction error grows much more quickly than the effect of the SST anomalies. Thus the experiment indicates that, for time ranges shorter than the predictability limit, the use of observed SST data will contribute negligibly to the forecast skill of the model.

3.4 Conclusions

Let us recall the question posed in Section 3.1: if an adequate coupled upper ocean model were available to provide accurate predictions of the SST field, what improvement would result in our ability to predict the large-scale behavior of the atmosphere? Of course, we are still in no position to answer this question fully or confidently. But, in attempting to generalize from the results of the experiments we have cited, we might, for example, anticipate an improvement in the local forecast over the region of an SST anomaly, particularly in the precipitation forecast over a warm anomaly in the tropics.

We are also able to place some tentative requirements on an upper ocean model or observing system, if it is to provide information useful to a large-scale atmospheric numerical model. First, the model should produce a global distribution of sea surface temperature with a horizontal resolution of about 200 to 500 km (the typical finite-difference grid spacing of large atmospheric models). The experiments we have cited, and the known sensitivity of forecasting models to errors in atmospheric temperature, might lead us to estimate that the SST should be specified to within 1 degree C and perhaps better, and the time resolution should be sufficient to maintain this accuracy.

Finally, we should temper any tendency toward pessimism in estimating the value of SST information, keeping in mind not only the questions of statistical significance and natural variability which plague this subject, but also the limitations imposed by model deficiencies. It may very well be that the physical processes chiefly involved in transmitting the influence of SST - boundary layer transports and moist convection - are modelled with grossly inadequate realism. Thus as our models improve, their sensitivity to the oceanic boundary may dramatically increase.

Acknowledgements: For helpful comments on an earlier version of this paper, and for access to unpublished work, I am grateful to R. M. Chervin, R. E. Davis, M. C. Hendershott, J. McWilliams, J. Miller, J. Namias, R. Salmon, S. H. Schneider, J. Shukla, J. Spar, and W. M. Washington. NCAR is sponsored by the U. S. National Science Foundation.

Chapter 4

OCEANOGRAPHIC PREDICTION REQUIREMENTS FOR NAVAL PURPOSES

Conley R. Ward

4.1 Introduction

Synoptic prediction of the ocean environment is a national requirement. It is important to a wide range of users, from the surfer at the beach to the Navy submarine commander. Fleet Numerical Weather Central (FNWC), at Monterey, California has the responsibility for provision of basic oceanographic parameters for the entire U. S. Department of Defense through the U. S. Navy Fleet Weather Centrals located at Guam, Mariana Islands; Pearl Harbor, Hawaii; Norfolk, Virginia; and Rota, Spain.

Active in numerical, synoptic oceanography since 1962, we have applied an engineering approach toward numerical solutions to the problem of operational interest. Techniques learned in numerical weather analysis and prediction have been adapted to the oceanic environment. Availability of a large computer and an extensive digital communications system have facilitated steady progress. A variety of environmental products is distributed to civil and military organizations, including NATO countries throughout the world.

The remainder of this paper will be concerned only with the national security application; however, it is certainly true that much of the material is applicable to other categories of users. The many scientific and technical questions such as resolution, time step, coordinate system, physical model, boundary conditions, range of predictability and air-ocean model interaction are considered to be beyond the scope of this paper and are therefore only implicitly treated.

4.2 Objective

It is axiomatic that the performance of naval platforms, sensors and weapons is affected by the environment. It is true that some systems are purported to be "all-weather", but this really means that part of the system is impervious to the weather part of the time. The word "environment" in the axiom was deliberately chosen because within the context of this paper we must also look at the interaction between ocean and atmosphere which is surely part of the naval operating environment.

The objectives of environmental study, research, development, and operations supported by the Navy are to:

1. Understand oceanic properties and processes.
2. Develop the capability to forecast them.

These two objectives include not only theoretical study but also field measurements and experiments.

3. Determine the effect of the ocean on naval operations. This objective is often underemphasized. Yet we must realize that the typical mariner or naval officer, though he is attuned to sensible parameters such as wind

and sea, is usually not able to correlate directly the variability of the ocean with his weapons systems. This objective must be met through the environmentalists, engineers, and naval tacticians working together to develop suitable algorithms.

4. Provide oceanic analyses and forecasts and tactical indices for the naval forces. The objective can be partially met by the creation of atlases for mariners, but this seriously limits the level of information provided to ships. To meet the objective satisfactorily it is necessary to bring together an extensive communications facility, computer center, and staff which is needed to run an operational forecast center.

4.3 Problem Areas

Some of the operational analysis and forecast models in use at FNWC are outlined in Table 4.1, including the validity time of forecasts (FCST) and the representative horizontal scale. We forecast atmospheric pressure, surface winds, and wave spectra, but do not forecast ocean parameters (except for a rudimentary 24 hour forecast of currents and mixed layer depth), nor do we prepare routine predictions nor diagnoses of air-sea heat exchanges.

TABLE 4.1

Models in operation at FNWC.

NAME	PARAMETER	AREA	PERIOD (HR)	FCST (HR)	SCALE (NM)
INTERFACE REGIME					
Sea Level Pressure	Wind and Pres	0-90N	6	72	100
		0-90S	6	0	200
		Mediterranean	6	72	25
Planetary Boundary Layer	Wind, Temp, Moisture	0-90N	6	72	200
Sea SFC Temp	Temperature	0-90N	6	0	100
		Gulf Stream	6	0	200
		Mediterranean	6	0	25
		Kuroshio	6	0	50
Spectral Ocean Wave	Energy 12 Dir, 15 Freq	0-90N	6	72	200
		Mediterranean	6	72	25
OCEAN REGIME					
Ocean Surface Current	Current	0-90N	6	24	200
Subsurface Temperature	Temp 100,200, 300,400,600, 800,1200 Ft.	0-90N	6	0	200
Potential Mixed Layer Depth	Depth of MLD	0-90N	6	24	200

The U. S. Naval Weather Service Command has an explicitly stated requirement for global atmospheric and oceanographic prediction models of 100 nm resolution, for window models of 30-50 nm resolution and for hydro-dynamical and advanced acoustic models. There is also a demand for improvements of basic dynamic models to show skill out to 5 days when compared to climatology and persistence. At present, these requirements are largely unrealized. In order to achieve the stated capability, improvements are needed in data processing, climatology, analysis, validation, prediction, verification, and interpretation.

a. Data processing

There are three new emerging sources of observational data from the ocean areas: satellites, buoys, and automatic ship stations (data relay by satellites). These platforms will provide timely observations of clouds, sea surface temperature and pressure, surface wind, sea ice, and wave spectra. Full coverage should permit global analysis of the parameters to the resolution required. This will not be a trivial task. It will involve conversion between engineering units and environmental measurements, as well as massive communications and computer processing facilities.

b. Climatology

Climatology data in digital form are required for model support as well as other uses. Present climatologies available to FNWC are inadequate to support global analysis and forecast models. The climatological information is used to help design numerical models and to provide model control in sparse data areas. Additional climatological information is needed to handle the problem of applications programs such as the sound propagation loss models and the surf models.

c. Analysis

Numerical environmental prediction is to a large extent an "initial value" problem, so it is essential that the initial state be specified as accurately and completely as possible. This requires amalgamation of observations from different sources, with different error characteristics taken at different times, and which are bunched in the sea lanes and sparse or missing elsewhere. Conventional observations of sea state are poor and, in particular, do not supply the required quantitative initial information; and devices that sense directional wave spectra are virtually nonexistent. Synoptic observations of ocean currents are extremely scarse and of doubtful quality.

d. Validation

The problem of model validation is far from trivial. It is a large task to run various models competitively, and to select the best for operational use. The model chosen is necessarily a compromise of running time, computational stability, and verification.

e. Prediction

This is the most difficult of the program areas and requires the greatest expenditure of human and computer resources. There are three major causes of errors in numerical forecasts when performed on an operational basis: improper specification of initial conditions; incomplete model physics, and computational limitations. The research community must recognize these errors, or problems, in order to eliminate or, at least, mitigate them.

f. Verification

It is difficult to verify model output under many circumstances. The output is not necessarily the same as the parameters being observed and many areas of the ocean are without observations for extended periods. Where observations are available, they are often biased or of poor quality. However, it is only through verification and systematic error analysis that model tuning and improvement take place.

g. Interpretation

Interpretation includes such tasks as determination of sensible environmental parameters and applying platform, sensor, and weapons system response functions to environmental model output. It should be remembered that in the Navy, the customer, or user of the environmental products, in almost all cases is a general line officer as distinguished from a specialist or scientist. We must provide the decision maker with information that is understandable, timely, and germane.

4.4 Models

In view of the variability of the upper layers of the ocean and the sensitivity of various naval operations to small scale variation in the environmental parameters, it would be desirable to decrease the analysis grid mesh. This would allow consideration of operationally significant mesoscale phenomena, such as oceanic eddies and fronts, or squall lines and fronts in the atmosphere. The scale of significant activity will tend to be smaller in favored areas such as coastal regions and near current boundaries. It is in such areas, where data density, environmental variability, and military interest coincide, that fine mesh models are particularly useful.

We expect greatly increased data from environmental satellites. We also have some hope of acquiring computers adequate for the atmosphere-ocean forecasting problem. Our optimism in this respect is based upon the continuing increase in computational efficiency. While inflation has caused the consumer price index to rise some 80 per cent, computing costs in the United States have decreased dramatically as indicated in Table 4.2.

TABLE 4.2

Relative computing costs.

Year	Cost of 10^6 Multiplications
1952	$12.60
1958	2.60
1964	1.20
1970	.50
1974	.10

Projecting this rate on to 1982, it would appear that by then 10^6 multiplications would cost $0.02 to $0.06.

TABLE 4.3

Hierarchy of environmental models.

MODEL	NUMBER OF LEVELS	GRID POINTS PER LEVEL	BASIC PARAMETERS PER POINT*	MESH LENGTH (NM)	AREA (NM)
Global Atmos	10	7200-28800	p, T, M, W	90-180	Globe
Regional Atmos	10	5000	p, T, M, W	60	$(4200)^2$
Global Spectral Waves	1	5000	18-36 Dir 15-20 Freq	90-180	Globe
Regional Spectral Waves	1	1500	18 Dir 15 Freq	60	$(2300)^2$
Global Hydro-Dynamical (Ocean PE)	5-10	3200-13000	T, S, h, V	90-180	Globe
Regional Hydro-Dynamical (Ocean PE)	5-10	1000	T, S, h, V	5-60	$(160)^2$ $(2000)^2$
Local Tide/ Currents/ Pollutants	2-5	1000	h, V, CC	.1-5	$(30)^2$ $(160)^2$
Shallow Wave	1	90000	V, B	.05	$(15)^2$

* p = Pressure Height W = Wind Velocity V = Current Velocity
T = Temperature S = Salinity CC = Concentration
M = Moisture h = Water Level B = Breakers

TABLE 4.4

Proposed global ocean primitive equation model.

GRID:	SPHERICAL COORDINATE 1.5° LAT/LONG, 30 MIN TIME STEP 10 LAYERS (SFC, 3 UPPER WITH COMPLETE EQUATIONS, LOWER SIMPLIFIED) (RUN TO 5 DAYS IN 4 HOURS ON CDC 7600)
INPUT:	SURFACE WIND, TIDES, AIR TEMP, CLOUD COVER CLIMATOLOGY - CURRENTS, TEMP, SALINITY
OUTPUT:	SURFACE LAYER - SST, AIR/SEA HEAT EXCHANGE, SEA LEVEL 3 UPPER LAYERS - LEVEL THICKNESS, TEMP, SALINITY, CURRENTS LOWER LAYERS - DEPTH CHANGE OF STANDARD LEVELS; CHANGE OF TEMP AND SAL

4.5 Future modelling plans.

The hierarchy of environmental models listed in Table 4.3 shows those models which are planned for operation at FNWC over the next few years. Beyond this set of models we hope to develop a global ocean primitive equation model as specified in Table 4.4.

The output from the proposed global ocean model would provide us with improved diagnostic and prognostic temperature/salinity information from which to compute sound velocity for the sophisticated acoustic models which are now being developed. Additionally, the heat budget information could be used for weather and climate forecasting. For this purpose, the global ocean model is to be combined as a final objective with the global atmospheric model into a coupled dynamic atmosphere-ocean model.

Part II

PHYSICAL PROCESSES IN THE UPPER OCEAN

Chapter 5

OCEANIC ABSORPTION OF SOLAR ENERGY

Alexandre Ivanoff

5.1 Introduction

During the last few years geophysicists have increasingly appreciated the need for a more detailed knowledge of solar energy absorption as a function of depth. Thus, for example, it has become apparent that a study of the seasonal thermocline cannot depend only upon the eddy conductivity of the water, but that it must also take account of the absorption of solar energy at depth. Since the latter varies with the optical properties of the water, geophysicists are henceforth led to interest themselves in Optical Oceanography.

In this paper we shall present a brief account of the essentials of present day knowledge concerning the absorption of solar energy by the oceans. We shall examine in turn the solar irradiance at the sea surface, the magnitude of the albedo, and then the attenuation of daylight with depth, i.e., its absorption as a function of depth.

5.2 The solar irradiance at the sea surface

Neglecting the small fluctuations in solar constant, the irradiance on a horizontal surface outside the earth's atmosphere can be calculated purely from astronomical data. Figure 5.1 provides this flux as a function of latitude and month of the year.

The phenomena become much more complicated after passing through the earth's atmosphere, which simultaneously absorbs and scatters the sunlight. Absorption is very weak in the visible band; it acts mainly in the ultraviolet and infrared bands. The ultraviolet is absorbed essentially by ozone, which in practice limits the solar radiation at ground level to wavelengths greater than .29 μm. The infrared, on the other hand, is mainly absorbed by carbon dioxide and above all by water vapor. Figure 5.2 shows the main absorption bands of the earth's atmosphere.

Molecular scattering by gases in the atmosphere closely follows a Rayleigh law, following a $\lambda^{-4.09}$ relationship as the result mainly of the anisotropy of the molecules involved. Scattering by the solid particles suspended in the atmosphere follows approximately a λ^{-1} relationship, while that by water droplets in fog and clouds is not at all wavelength dependent (on the other hand these droplets give a strong absorption in the infrared). Figure 5.3 shows the effects respectively of molecular scattering, absorption, and scattering by aerosols, for clear, dry weather (condensed water thickness of 1 cm) and for four values of the zenith angle, ζ, of the sun. Table 5.1 provides, for the same conditions, the distribution of the direct sunlight at sea level between ultraviolet, visible, and infrared bands. Although the proportion of infrared is weaker during overcast weather, it is usually greater than 50%. For wavelengths greater than .7 μm the absorption coefficient for pure water is greater than 0.5 m^{-1} and increases rapidly with wavelength (this means that the transmission factor per meter rapidly drops below about 60%), so that, as we shall see, about half the solar energy is absorbed by the first half meter of sea water.

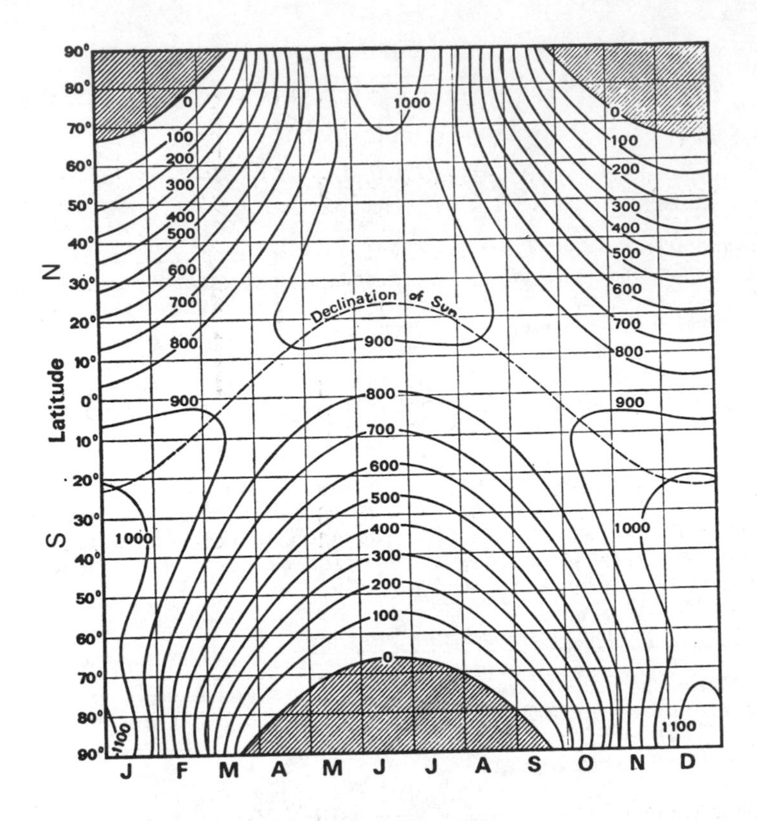

Fig. 5.1 - Solar irradiance on a horizontal surface outside the earth's atmosphere, in cal cm^{-2} day^{-1}, after Fritz (1951).

TABLE 5.1

Distribution of the direct sunlight at sea level between ultraviolet, visible and infrared bands, during clear, dry weather, for the four solar zenith angles considered in Fig. 5.3, after de Brinchambault and Lamboley (1968).

SOLAR ZENITH ANGLE	U.-V.	VISIBLE ($0.4 \mu m < \lambda < 0.7 \mu m$)	I.-R.
0°.	3 %	42 %	55 %
60°.	2 %	42 %	56 %
75°.	1 %	40 %	59 %
83°.	0 %	30 %	70 %

It remains to examine the magnitude of the energy reaching sea level, which is essentially a function of the thickness of atmosphere penetrated (i.e., of the latitude, month of the year, and time of day - in brief of the sun's elevation) and of the cloudiness. Numerous experimental results

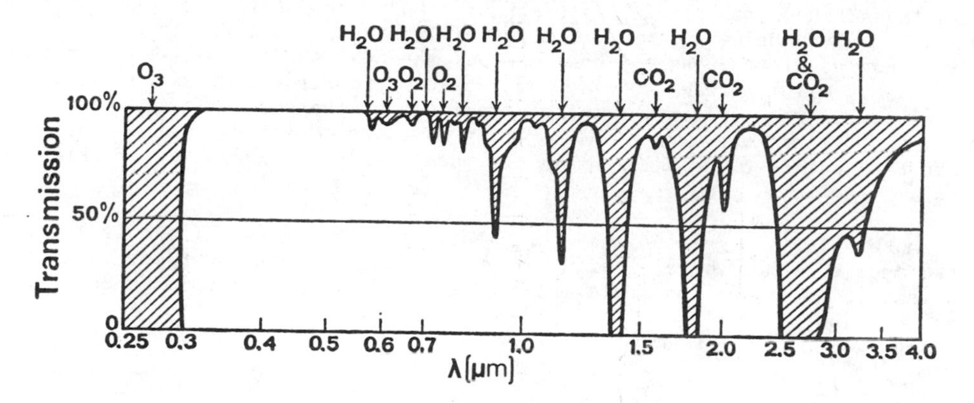

Fig. 5.2 - Spectral transmission of the atmosphere (when the sun is at zenith and the equivalent water thickness is 2 cm) in the visible and the near infrared as a function of wavelength. Principal absorption bands of water vapor, carbon dioxide, ozone, and oxygen are shown.

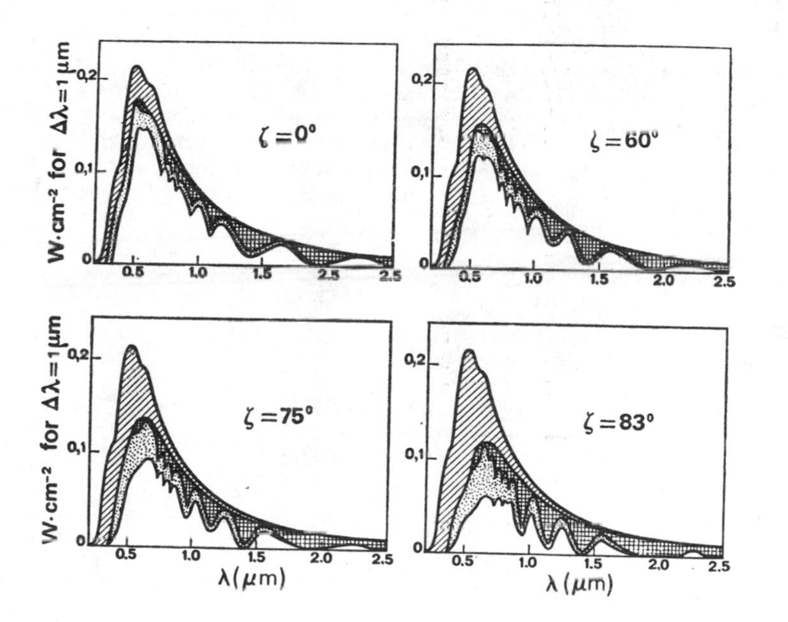

Fig. 5.3 - In each of the four diagrams (corresponding to four values of the zenith angle ζ of the sun) the upper curve shows the spectral distribution of the solar constant, while the lower curve represents the spectral distribution of the direct solar beam at the sea surface, for clear, dry weather (equivalent water thickness equal to 1 cm). The areas of oblique lines, cross hatching, and dots represent, respectively, the effects of molecular scattering, absorption, and scattering by aerosols. After de Brichambault and Lamboley (1968).

presented in form of tables or maps are available. Figure 5.4 provides an example and shows how the intervention of clouds changes the mean annual solar irradiance from being a simple function of latitude. But, for the problem that concerns us here, namely how to model the near surface layer of the ocean, it is above all important to discover, at least to first approximation, an analytical expression for the solar irradiance at the sea surface. We shall from now on limit ourselves to describing attempts to achieve this, considering first the case of clear skies, then the much more complex problem of skies partially or totally covered by clouds.

Mosby (1936) suggested that for clear skies the average (downward) solar irradiance on a horizontal surface at ground level is simply proportional to the mean value of the sun's elevation ($\xi=90°-\zeta$) during the period under consideration:

$$\overline{I}_{dc} = m_1\overline{\xi} \tag{5.1}$$

This led him to calculate, on the basis of experimental results, a constant of proportionality m_1 that varied weakly with latitude.

More recently, Matsuike, et al. (1970), after examining experimental data, reached the conclusion that during clear weather the atmospheric transmission factor τ depends only on the sun's elevation (i.e., that water vapor absorption and aerosol scattering are almost independent of location and season). Analyzing the function $\tau=f(\xi)$ into a Fourier series, they calculated from the solar constant the mean value of the irradiance at sea level as a function of latitude and month of the year: see columns M of Table 5.2. The results obtained in this way are in reasonable agreement with that published in 1960 by Berliand on the basis of observations (in clear weather) at 70 stations distributed at all latitudes (columns B of Table 5.2). This method of calculating the irradiance during periods of clear skies seems most promising.

Turning now to the question of cloudiness, Kimball proposed in 1928 the following simple relationship:

$$\overline{I}_d = \overline{I}_{dc}(1-.71\overline{n}), \tag{5.2}$$

where $\overline{n}$ is the mean value of the cloudiness on a scale 0 to 1 and $\overline{I}_{dc}$ the mean value of the irradiance under clear sky. Budyko (1956) changed the constant from 0.71 to 0.68.

Berliand (1960) proposed a quadratic relationship:

$$\overline{I}_d = \overline{I}_{dc}[1-(m_2+.38\overline{n})\overline{n}], \tag{5.3}$$

where the coefficient m_2 varies from 0.35 to 0.41 over latitudes less than 65°, but is distinctly lower at higher latitudes.

According to Matsuike, et al. (1970) one can write:

$$\overline{I}_d = \overline{I}_{dc}[1-.52\overline{n}^{1.3}]. \tag{5.4}$$

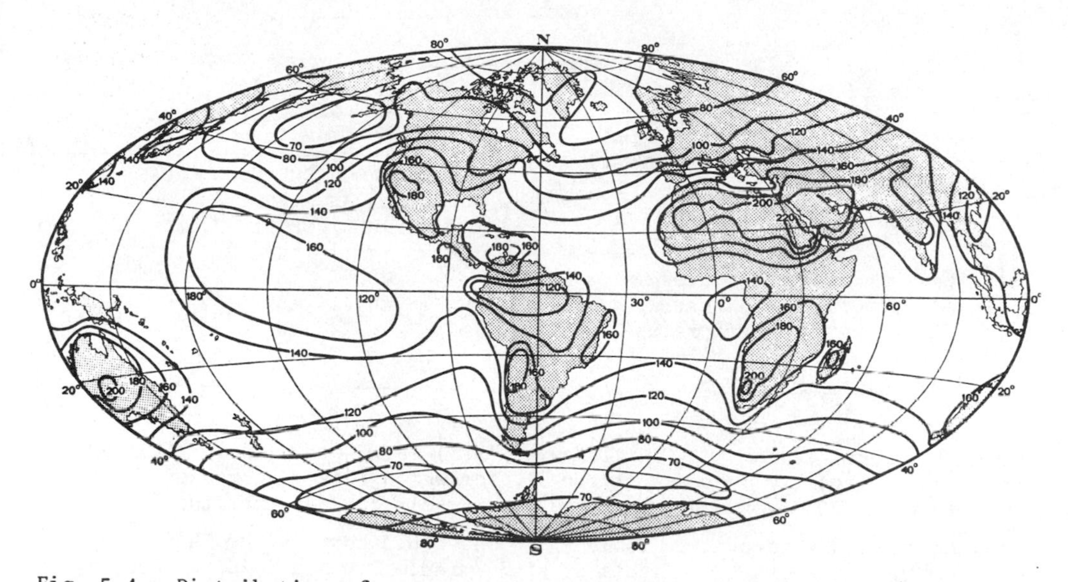

Fig. 5.4 - Distribution of mean annual solar irradiance, in kcal cm^{-2} yr^{-1}, after Landsberg (1961).

TABLE 5.2

In the columns marked B appear values of the solar irradiance at ground level during clear weather (in cal cm^{-2} day^{-1}) as a function of latitude and month of the year, obtained by Berliand (1960) from observations at 70 actinometric stations. In the columns marked M appear results calculated by Matsuike, et al. (1970) using the solar constant and assuming that for clear sky the atmospheric attenuation depends only on the mean elevation of the sun.

Latitude	Jan		Feb		Mar		Apr		May		Jun		Jul		Aug		Sep		Oct		Nov		Dec	
	B	M	B	M	B	M	B	M	B	M	B	M	B	M	B	M	B	M	B	M	B	M	B	M
90° N	0		0		4		328		720		856		780		424		78		0		0		0	
85°	0		0		24		336		716		846		771		430		100		0		0		0	
80°	0		0		69		354		706		828		754		439		140		15		0		0	
75°	0		18		132		385		690		805		727		455		191		44		0		0	
70°	0		51		198		430		675		774		700		480		248		90		16		0	
65°	24	40	92	90	264	220	478	430	672	610	751	690	692	650	513	500	311	300	142	140	45	50	8	30
60°	58		142		325		526		684		753		703		550		371		200		85		37	
55°	102	90	204	170	384	330	569	510	707	660	768	720	722	700	590	570	425	400	257	230	133	110	79	70
50°	159		270		438		608		729		780		742		628		474		318		190		131	
45°	220	180	340	280	489	430	642	590	746	700	787	740	761	720	662	630	519	500	377	340	256	210	193	160
40°	290		402		538		668		759		790		772		687		559		433		318		260	
35°	352	290	460	390	580	520	689	640	764	720	788	740	775	730	706	670	596	570	483	440	375	330	320	270
30°	410		509		613		703		763		780		771		716		628		530		430		378	
25°	463	410	552	490	640	590	710	670	754	710	768	730	760	730	719	690	653	620	572	540	482	450	431	390
20°	511		590		663		710		740		750		743		716		673		608		530		484	
15°	555	510	624	580	681	650	705	690	721	700	724	700	721	700	709	690	688	670	636	610	572	530	530	500
10°	595		650		695		698		696		692		694		698		698		661		610		575	
5°	635	610	671	650	704	670	688	670	667	650	656	640	662	640	680	660	701	670	681	660	643	620	618	590
0°	666		688		707		672		635		618		627		660		698		696		672		656	
5° S	695	680	704	680	704	680	654	640	602	590	580	550	590	560	634	610	687	660	704	680	698	680	693	680
10°	722		715		694		631		567		535		550		602		670		705		717		726	
15°	744	740	722	720	679	670	601	580	527	510	491	470	507	480	568	550	648	620	700	690	734	730	753	740
20°	762		726		660		566		485		442		464		531		622		690		746		774	
25°	776	770	726	720	639	630	527	510	439	410	395	360	416	380	491	460	519	580	679	680	755	750	793	770
30°	787		718		611		489		392		348		366		447		558		662		760		810	
35°	792	780	701	700	578	570	447	420	343	300	297	260	316	270	398	360	522	500	642	640	760	750	822	790
40°	792		680		540		401		287		241		265		350		482		616		752		830	
45°	789	760	654	650	499	490	353	310	233	190	180	150	206	170	297	250	439	410	584	580	738	740	832	790
50°	779		622		454		302		178		125		150		241		393		547		720		824	
55°	762	730	586	590	406	390	245	210	125	110	75	70	98	80	180	160	340	300	507	500	702	680	811	770
60°	743		548		353		184		79		32		52		124		280		464		690		804	
65°	735	690	505	490	298	280	127	100	36	40	3	30	15	30	75	80	220	210	416	400	685	630	807	750
70°	742		469		240		74		4		0		0		32		165		375		688		820	
75°	763		442		187		31		0		0		0		0		115		343		701		838	
80°	792		420		140		0		0		0		0		0		69		318		721		856	
85°	811		408		96		0		0		0		0		0		30		303		736		874	
90°	820		404		56		0		0		0		0		0		0		296		742		886	

TABLE 5.3

Effect of cloud type on the attenuation of irradiance during overcast weather, after Haurwitz (1948).

CLOUD TYPE	IRRADIANCE DURING OVERCAST WEATHER AS A FRACTION OF ITS VALUE FOR CLEAR SKY
Cirrostratus and Cirrocumulus	0.65 to 0.85
Altocumulus and Altostratus	0.41 to 0.52
Stratocumulus and Stratus	0.24 to 0.35
Thick Fog	0.17 to 0.19

The differences between the four formulae (5.1)-(5.4) are clearly revealed in the case of a totally overcast sky ($\overline{n}=1$), for which they give respectively $\overline{I}_d/\overline{I}_{dc}=0.29$, 0.32, 0.24 and 0.48. Now, it is self evident that the cloud type plays an important role: see Table 5.3. One could certainly do better by distinguishing between the effects of high, middle level and low clouds, as is done sometimes for the effective back radiation. Lumb (1964) has gone even further in analyzing observations made on board the Ocean Weather Ship "Juliett", situated in the North Atlantic Ocean at 52°30' North and 20° West (his results are therefore only relevant to the North Atlantic Ocean in a latitude band, say, 45° to 65°, but the method could equally well be applied to other regions). He distinguished between nine categories for the state of the sky (see Table 5.4), and derived an expression of the following type for the irradiance at sea level for each of them:

$$\overline{I}_d = m' \, \overline{\sin\xi} \, (m_3+m_4 \, \overline{\sin\xi}), \tag{5.5}$$

where ξ is the solar elevation, m' is a constant independent of the state of the sky, while m_3 and m_4 are coefficients whose values depend on the category of sky. Figure 5.5 is derived from the results that Lumb obtained by this method in the particular area of his study.

5.3 Albedo of the sea surface

We start with the relationship defining the albedo:

$$A = (I_r+I_u)/I_d, \tag{5.6}$$

where I_d is the incident solar flux (per unit surface area of the sea), I_r the solar flux reflected at the sea surface (also per unit surface area), and I_u that part which is upward back-scattered from below the sea surface. One can write

$$A = R+I_u/I_d, \tag{5.7}$$

where $R=I_r/I_d$ is the reflectance of the sea surface. The flux absorbed by

TABLE 5.4

The nine sky conditions determined by Lumb (1964) from observations made on board the Ocean Weather Ship "Juliett" (52°30'N, 20°W).

CATEGORY	DESCRIPTION
1	Virtually clear sky, less than two-eights coverage.
2	Well-broken low clouds with little or no medium or high cloud (three-to five-eights).
3	Six to eight-eights of cirrus (not cirrostratus).
4	Thin layers of altocumulus, six-to eight-eights.
5	Veil of cirrostratus over whole sky with up to four-eights.
6	Seven or eight-eights of stratocumulus without rain, with or without some cumulus and little or no medium clouds.
7	Thick altostratus six to eight-eights, with or without layers of stratocumulus beneath some rain.
8	Thick layers of stratus and stratocumulus, overcast including drizzle.
9	Thick layers of nimbostratus, overcast, also include medium clouds, and rain.

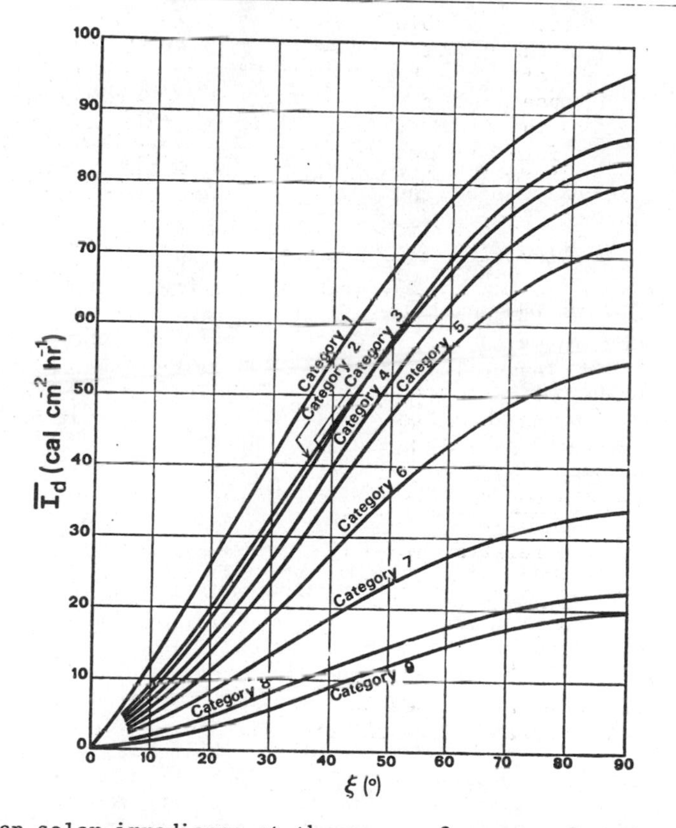

Fig. 5.5 - Mean solar irradiance at the sea surface as a function of the mean solar elevation, for the nine sky conditions of Table 5.4. These curves, derived from Lumb (1964), are valid only for the North Atlantic Ocean, between latitudes 45° and 65°.

the sea (per unit surface area) is equal to

$$I_d - (I_r + I_u) = I_d(1-A) \tag{5.8}$$

Hence our interest in the albedo.

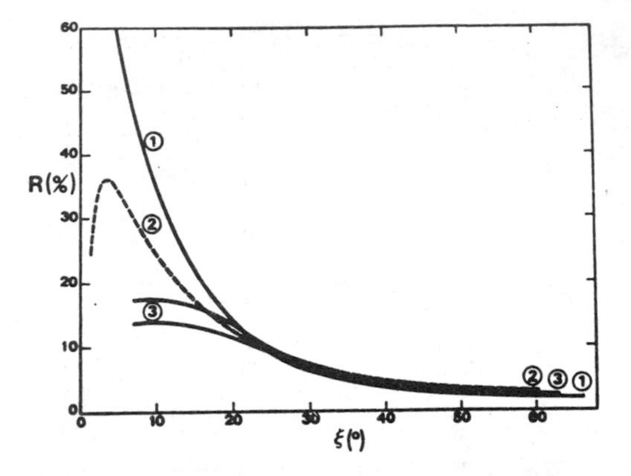

Fig. 5.6 - Curve 1 shows the effect of solar elevation ξ on the reflectance R of a perfectly plane sea surface for the direct solar beam. Curve 2 corresponds to the total solar irradiance (direct plus diffuse) for clear sky and also for a perfectly plane sea surface. Curves 3, after Cox and Munk (1956), show the effect of ξ on the reflectance for direct solar beam with a sea state 4, for two different assumptions about multiple reflections produced when the sun's elevation is relatively small.

The reflectance, R, which is independent of the absorption and scattering properties of the water, can be calculated simply from Fresnel's formula for an idealized plane sea surface. Curve 1 in Fig. 5.6 shows the results obtained from this method for the direct solar beam; the upper half of Table 5.5 provides them for the diffuse sky radiation (for a variety of assumptions concerning the distribution of radiances in the sky), and curve 2 in Fig. 5.6 shows results for the total irradiance (direct plus diffuse), by making a "reasonable" assumption about the relative importance of the diffuse radiation as a function of solar elevation (see Table 5.6). We see that for a plane sea surface and overcast sky the reflectance must lie in the range 5 to 7%, while for a blue sky it can exceed 30% when the sun is low on the horizon.

Calculating the reflectance becomes more complicated when one considers a rough sea, for which the angle of incidence of direct solar rays varies with the slope of the waves, and becomes even more difficult when the sun is low on the horizon, giving rise to shadow zones and multiple reflections. The problem has been studied by Cox and Munk (1956) and by Saunders (1967). Curve 3 in Fig. 5.6 reproduces the results obtained by Cox and Munk for the direct solar beam, for sea state 4 and for two different assumptions about multiple reflections produced at small solar elevation: while for solar elevation greater than about 25° the surface roughness leads to a slight increase in the reflectance, for lower angles it produces a strong reduction. The lower half of Table 5.5 shows the results for the diffuse sky radiation, which are slightly less than for a flat sea surface. It seems then that rough seas and overcast sky lead to a reflectance in the range 4 to 6%, but when skies are blue the latter can still reach 20% for solar elevation close to about 10°.

TABLE 5.5

Reflectance (in%) for the diffuse sky radiation, for different assumptions concerning the distribution of radiances in the sky.

	JUDD (1942)	BURT (1953)	COX AND MUNK (1956)	PREISENDORFER (1957)	PAYNE (1972)
SKY			FLAT SEA		
CLEAR			10.0		
CLOUDY					
UNIFORM	6.6	6.6	6.6		6.8
CARDIOIDAL DISTRIBUTION			5.2	5.16	5.2
SKY			SEA STATE 4		
CLEAR			7.1 to 8.8		
CLOUDY					
UNIFORM		5.7	5.0 to 5.5		5.0
CARDIOIDAL DISTRIBUTION		4.8	4.3 to 4.7		4.2

TABLE 5.6

Proportion of the diffuse sky radiation to the total irradiance as a function of the solar zenith angle.

SOLAR ZENITH ANGLE	30°	48°	50°	60°	70°	76°	80°
MEAN VALUES AFTER KIMBALL (1928)	0.16	0.16		0.20		0.31	
AT WASHINGTON, D. C. IN SUMMER			0.27	0.32	0.42		0.67
IN WINTER WILLIAMS (1955)			0.15	0.20	0.28		0.59

Now let us consider the albedo. This can be derived either by adjusting the reflectance with the factor I_u/I_d, or by direct measurement of I_d and of the sum of I_r and I_u.

The ratio of upward to downward irradiance is measured just <u>below</u> the water surface as a function of wavelength. Its magnitude and spectral distribution vary considerably with the absorption and scattering properties of the water. In extremely clear water the ratio can reach 10% at around .4 μm while for strongly scattering water it may also be about 10% at wavelengths in the region .55 to .56 μm. However, it can be shown easily that the ratio I_u/I_d to be added to the reflectance in order to calculate the magnitude of the albedo has a maximum of only about 2% (when the water is very clear or on the contrary strongly scattering) but in general is less than 0.5%.

TABLE 5.7

Some experimental values of the albedo of the sea (in %) during overcast weather.

POWELL AND CLARKE (1936)	NEIBURGER (1948)	BURT (1953)	GRICHTCHENKO (1959)	CALATHAS (1970)	PAYNE (1972)
FLAT SEA					
6.5	10.5	9.4	6 to 11	6.1	5 to 7.5
SEA STATE 4					
				4.6	

Effectively, the experimental values of the albedo are very close to those seen above for the reflectance. Table 5.7 gives a number of experimental values for the albedo during overcast weather; in this case the albedo is very probably of the order 4.5% to 7.5% (while the reflectance is of the order 4% to 7%); adopting an average value of 6% is unlikely to lead to an error exceeding 1.5%. Figure 5.7 shows some experimental values for the albedo under clear blue skies, in which case, as with the reflectance, it can reach values of the order of 20% for a solar elevation of around 10°, but only for a moderate sea state, as this maximum value decreases when the sea is rough.

For solar elevations exceeding 20° or 25° (in which case the sea state plays a relatively minor role) the albedo of the sea under blue skies can be expressed in terms of the following formula, proposed by Laevastu (1960):

$$A(\%) = 300/\xi^\circ \qquad (5.9)$$

(although $250/\xi^\circ$ would perhaps be closer to the truth for solar elevations less than 50°). For solar elevations lower that 25°, the values obtained by different authors differ greatly (perhaps as the result of the strong influence of the sea state) and this presents a serious problem when calculating the rate of solar energy absorption by the oceans in latitudes

TABLE 5.8

Mean albedos for Atlantic Ocean by month and latitude, after Payne (1972).

Latitude	Jan	Feb	Mar	Apr	May	Jun	Jul	Aug	Sep	Oct	Nov	Dec
80N			0.33	0.14	0.10	0.09	0.08	0.08	0.12			
70N		0.41	0.15	0.10	0.08	0.07	0.07	0.09	0.11	0.25		
60N	0.28	0.12	0.09	0.07	0.07	0.07	0.06	0.07	0.07	0.10	0.16	0.44
50N	0.11	0.10	0.08	0.07	0.06	0.06	0.06	0.07	0.07	0.08	0.11	0.12
40N	0.10	0.09	0.07	0.07	0.06	0.06	0.06	0.06	0.07	0.08	0.10	0.11
30N	0.09	0.07	0.06	0.06	0.06	0.06	0.06	0.06	0.06	0.07	0.08	0.09
20N	0.07	0.06	0.06	0.06	0.06	0.06	0.06	0.06	0.06	0.06	0.07	0.07
10N	0.07	0.06	0.06	0.06	0.06	0.06	0.06	0.06	0.06	0.06	0.06	0.07
0	0.06	0.06	0.06	0.06	0.06	0.06	0.06	0.06	0.06	0.06	0.06	0.06
10S	0.06	0.06	0.06	0.06	0.07	0.07	0.06	0.06	0.06	0.06	0.06	0.06
20S	0.06	0.06	0.06	0.06	0.07	0.07	0.07	0.07	0.06	0.06	0.06	0.06
30S	0.06	0.06	0.06	0.07	0.08	0.09	0.08	0.07	0.07	0.06	0.06	0.06
40S	0.06	0.06	0.07	0.08	0.09	0.11	0.10	0.08	0.07	0.07	0.06	0.06
50S	0.06	0.07	0.07	0.08	0.10	0.13	0.11	0.08	0.08	0.07	0.06	0.06
60S	0.06	0.07	0.08	0.11	0.13		0.27	0.07	0.08	0.07	0.06	0.06

TABLE 5.9

Comparison of albedo values from Table 5.8 (P) with those of Budyko (B).

Latitude	January		February		March		April		May		June	
	B	P	B	P	B	P	B	P	B	P	B	P
70N			0.23	0.41	0.16	0.15	0.11	0.10	0.09	0.08	0.09	0.07
60N	0.20	0.28	0.16	0.12	0.11	0.09	0.08	0.07	0.08	0.07	0.07	0.07
50N	0.16	0.11	0.12	0.10	0.09	0.08	0.07	0.07	0.07	0.06	0.06	0.06

Latitude	July		August		September		October		November		December	
	B	P	B	P	B	P	B	P	B	P	B	P
70N	0.09	0.07	0.10	0.09	0.13	0.11	0.15	0.25				
60N	0.08	0.06	0.09	0.07	0.10	0.07	0.14	0.10	0.19	0.16	0.21	0.44
50N	0.07	0.06	0.07	0.07	0.08	0.07	0.11	0.08	0.14	0.11	0.16	0.12

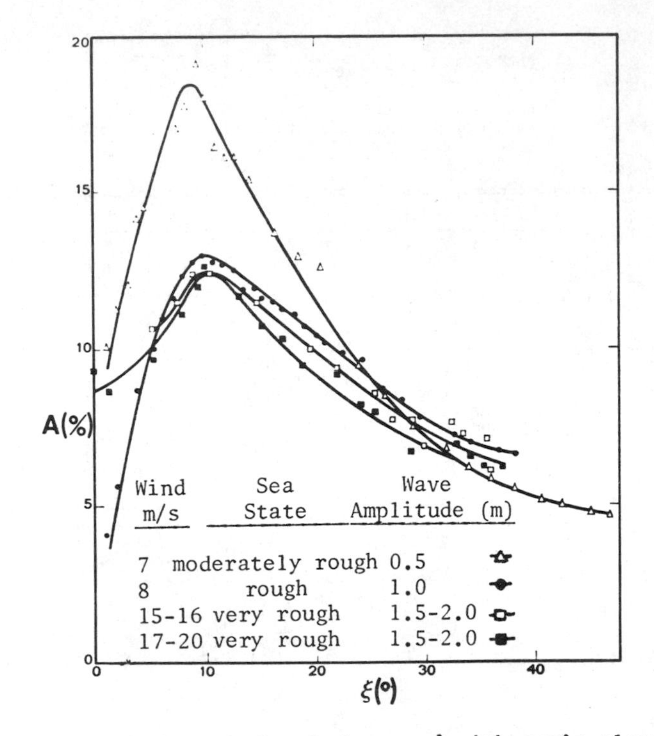

Fig. 5.7 - Variation of the albedo of the sea A with sun's elevation and roughness of the sea, for clear weather. Measurements made in the Mediterranean Sea from the "Bouée Laboratoire" (42°14'N, 5°35'E) by Calathas (1970).

higher than 50° or 60°. For these latitudes, Payne (1972) put forward mean monthly values of albedo as high as 44%, while Budyko (1963) suggested values that do not exceed 23%, which seems more reasonable. Tables 5.8 and 5.9 are reproduced from Payne (1972). They provide monthly mean values of the albedo for the Atlantic Ocean as a function of latitude, together with Budyko's mean values for latitudes higher than 50°N. In conclusion, there remains some uncertainty as to the magnitude of the albedo of the sea surface under clear skies and for solar elevations less than 25°.

5.4 Absorption of solar energy by the oceans

Just as the atmosphere, sea water simultaneously absorbs and scatters solar radiation. The absorption consists mainly of a conversion from radiant energy to heat, the remaining part of the absorbed radiant energy being involved in chemical reactions such as photosynthesis; however, in practice the fraction of solar energy absorbed by the oceans as the result of photosynthesis is, on an average, only of the order of 0.1%. Scattering consists of changes in the directions of photons without energy loss, but, as it increases the pathlength of photons between the sea surface and the depth under consideration, scattering leads to increased absorption and an additional energy loss.

Absorption is due to the water itself, to the dissolved salts, to the organic substances in solution and to the suspended matter. Figure 5.8 shows the absorption coefficient for pure water in the visible band of the spectrum, while Fig. 5.9 shows it in the near infrared. We see that the absorption coefficient for pure water increases rapidly towards the long wavelengths (its value in the ultraviolet, on the other hand, is less well known) and exceeds 2.3 m^{-1} for wavelengths greater than .8 μm, the corresponding transmittance per meter then becoming less than 10%.

Figure 5.10 shows several examples of the spectral distribution of the absorption coefficient of different sea waters in the visible part of the spectrum, while Fig. 5.11 shows the same data after subtracting the absorption coefficient of pure water; in other words, this figure represents the effects of substances in solution and suspension. One notes the two absorption bands of the chlorophyll pigments (around .44 and .675 μm) and the increasing absorption towards the short wavelengths due to organic products of decay called "yellow substances" by Kalle (1938).

Scattering is due in part to molecules of water (and those of substances in solution), but much more to the suspended matter. Since scattering by particles suspended in sea water takes place mainly at small angles (as predicted by Mie theory of electromagnetic scattering taking into account the size distribution of these particles), the sea water volume scattering function is always strongly extended in a forward direction. Examples are given in Fig. 5.12.

The molecular scattering follows approximately a $\lambda^{-4.3}$ law, while scattering by particles follows a law close to λ^{-1} (Morel, 1967). Scattering is then much more selective when the water is clear. In any case, as absorption increases strongly towards long wavelengths, scattering becomes negligible in comparison with absorption at wavelengths greater than .7 or .8 μm.

An investigation of the penetration of daylight into the oceans and its progressive attenuation with depth must clearly include simultaneous consideration of the absorbing and scattering properties of the water (as well as the angular and the spectral distributions of light incident on the sea surface). Numerous studies, as many experimental as theoretical, have been carried out with these aims. We shall limit ourselves in the following discussion to an analysis of how solar energy is absorbed as a function of depth.

At depth z the residual under water daylight is defined by the angular distribution of the open water radiance J, which is a function of wavelength. The quantity

$$I_0 = \iint_{4\pi} J \, d\omega \tag{5.10}$$

is the "scalar" irradiance, while the quantities

$$I_d = \iint_{\substack{\text{upper} \\ \text{hemisphere}}} J \cos \eta \, d\omega \tag{5.11}$$

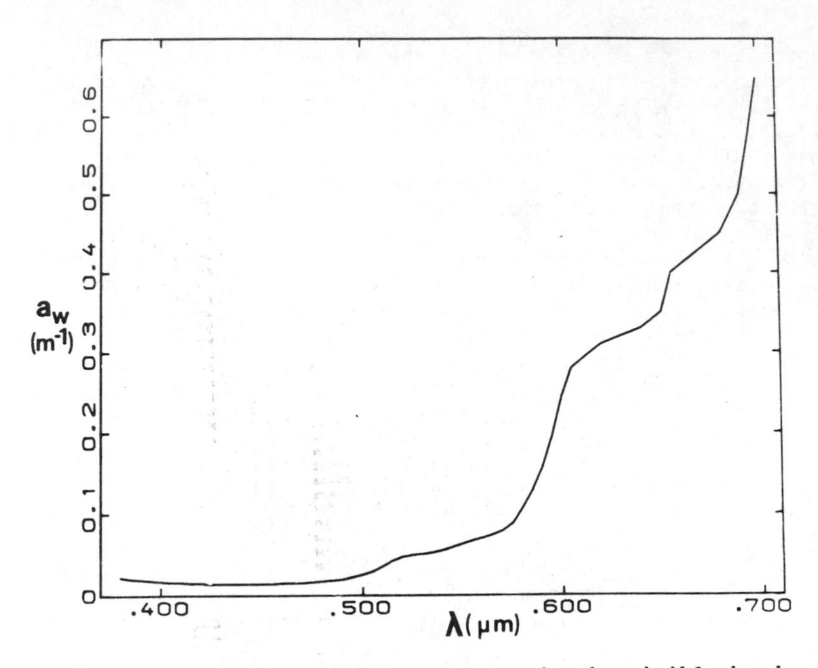

Fig. 5.8 - Absorption coefficient of pure water in the visible band, after Morel and Prieur (1975).

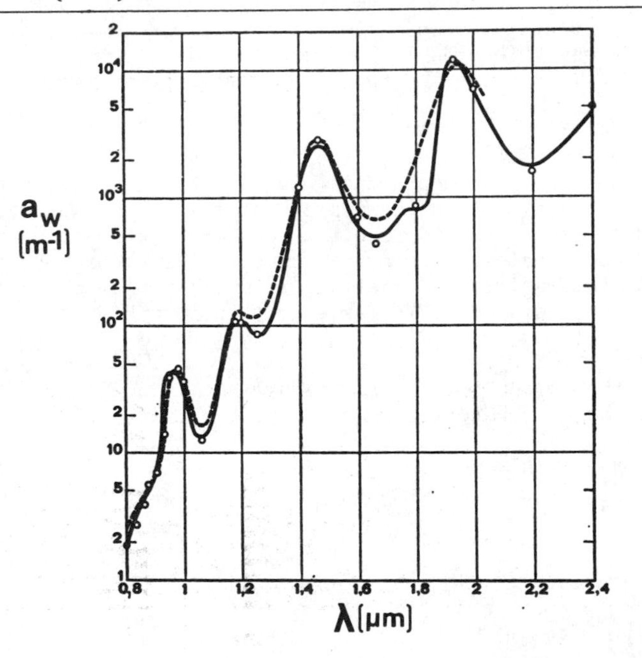

Fig. 5.9 - Absorption coefficient of pure water in the near infrared according to Collins, 1925 (dashed curve), and according to Curcio and Petty, 1951 (continuous curve). Circles show the most probable values (at 25°C) reported by Hale and Querry (1973) from a study of the published literature.

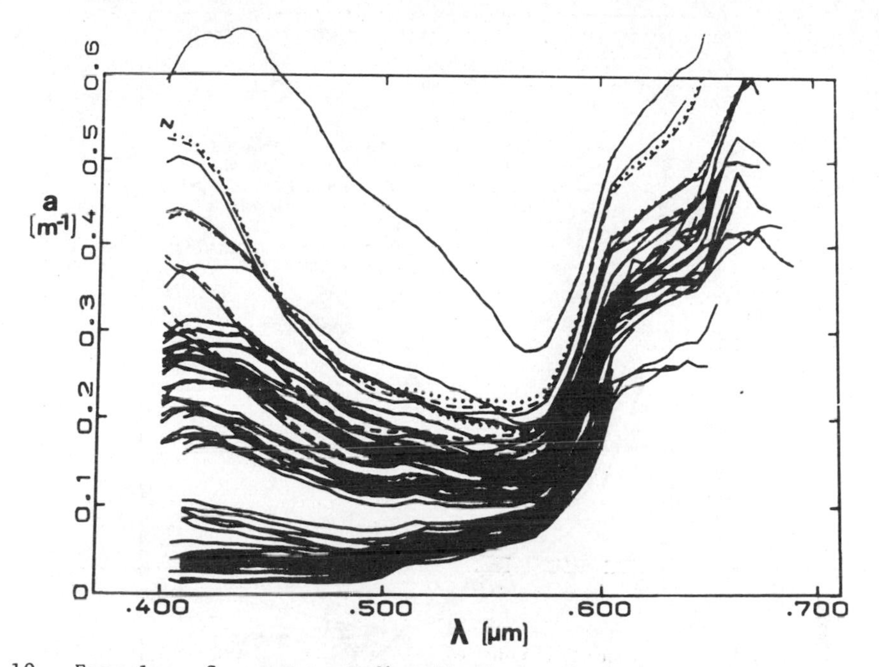

Fig. 5.10 - Examples of spectral distributions of absorption coefficient of sea water, after Morel and Prieur (1975).

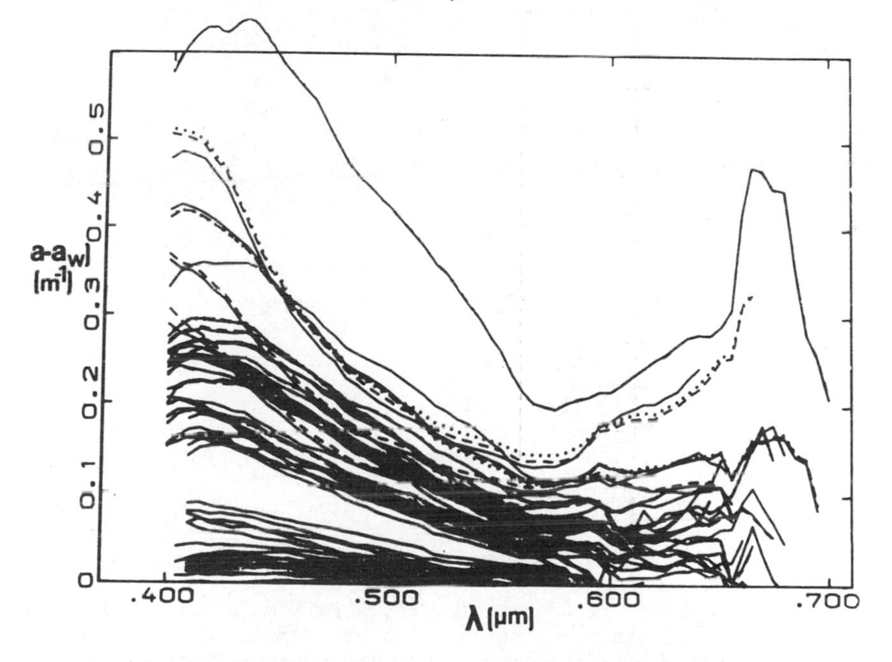

Fig. 5.11 - Examples of spectral distributions of the difference $a-a_W$ between absorption coefficients of sea water and pure water respectively, after Morel and Prieur (personal communication).

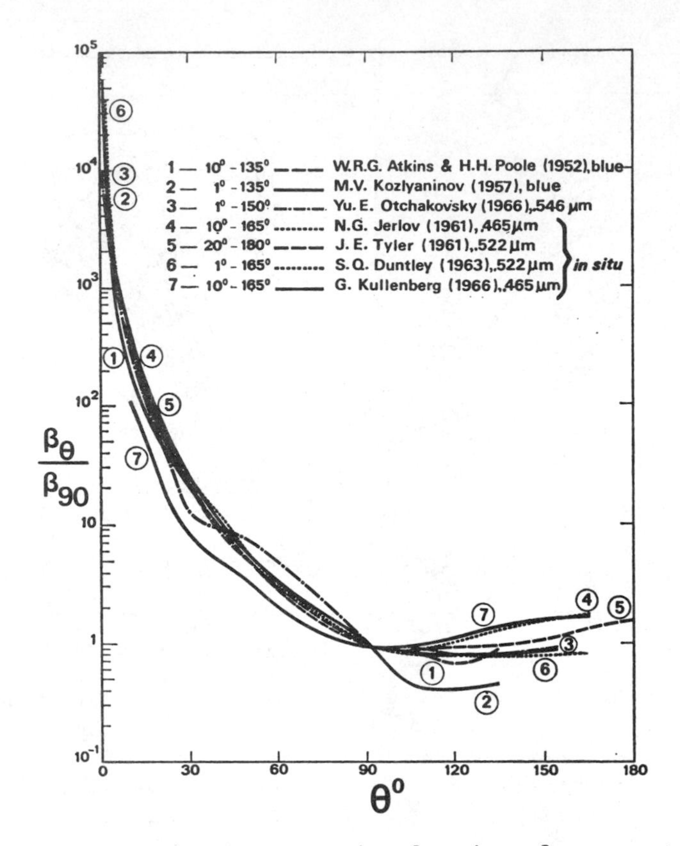

Fig. 5.12 - Examples of volume scattering functions for sea water, β_θ, measured at various wavelengths, using either samples or in situ scattering meters. Curves are normalized at $\theta = 90°$.

and

$$I_u = \iint_{\text{lower hemisphere}} J \cos \eta \, d\omega \tag{5.12}$$

(where $d\omega$ is a solid angle and η is the angle which the direction under consideration makes with the vertical) are the "downward" and the "upward" irradiances respectively, onto a horizontal plane. All these quantities diminish with depth and at the same time their spectral distributions change, due to selective absorption and scattering. Examples of the spectral distribution of downward and upward irradiances at different depths and for waters of different turbidities are shown in Figs. 5.13 and 5.14.

The attenuation of irradiance with depth follows a more or less exponential form (at least at some depth), depending of course upon the wavelength or the spectral band under consideration. Some examples of results obtained in the South Sargasso Sea concerning the downward irradiance are shown in Fig. 5.15. We see that at small depths the spectral concentration of the

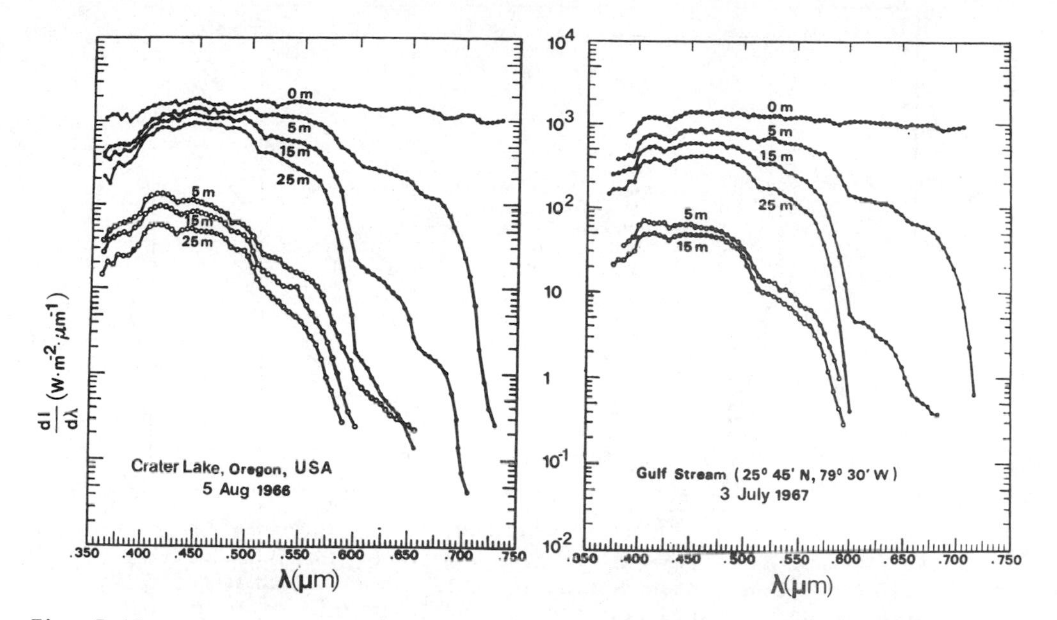

Fig. 5.13 - Spectral distributions of downwelling irradiance (dots) and upwelling irradiance (open circles) in the exceptionally clear water of Crater Lake (Oregon) and in the Gulf Stream. The downwelling irradiance at depth 0 has been measured just above the surface of the water. After Tyler and Smith (1970).

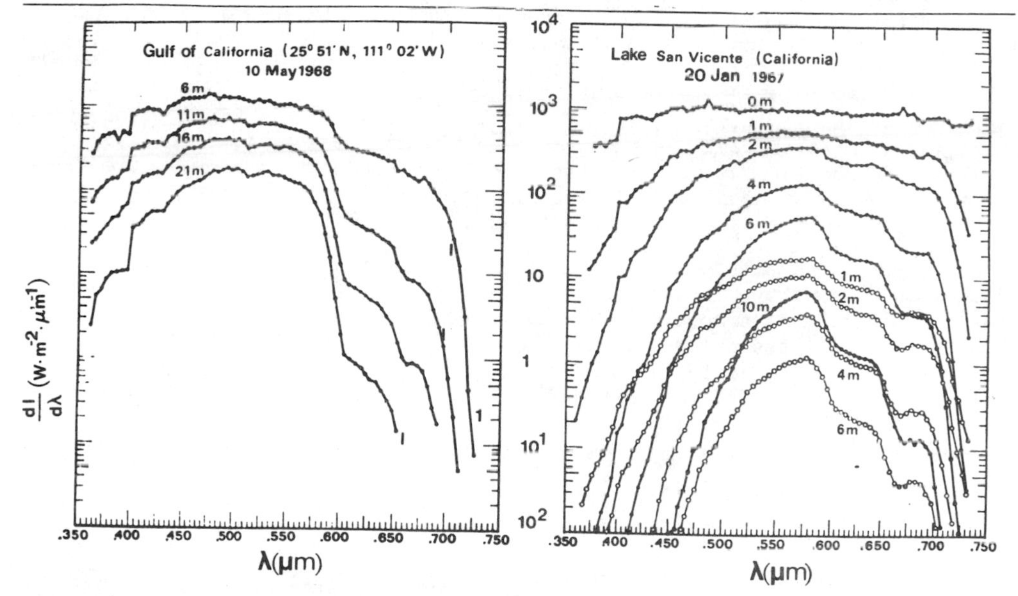

Fig. 5.14 - Spectral distribution of downwelling irradiance (dots) and upwelling irradiance (open circles) in the Gulf of California and in the turbid waters of Lake San Vicente (an artificial lake near San Diego, California). After Tyler and Smith (1970).

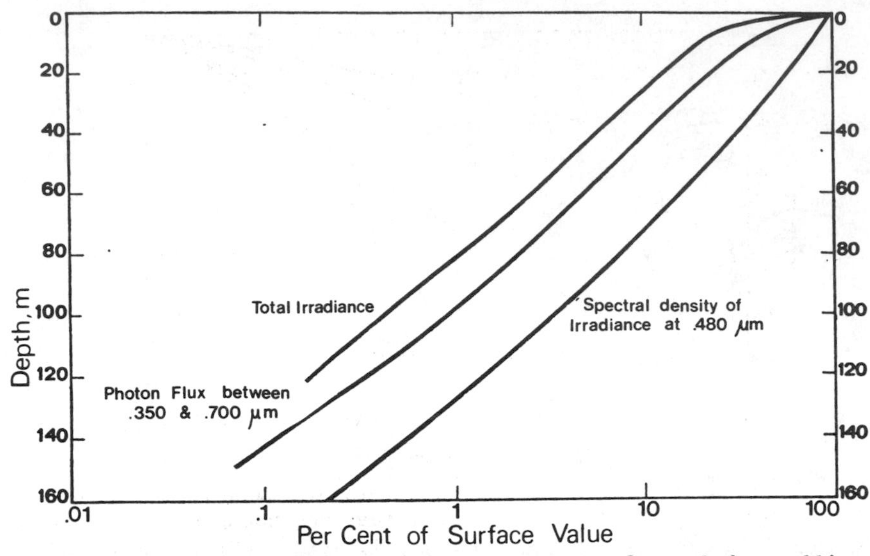

Fig. 5.15 - Comparison of the variation with depth of total downwelling irradiance, downwelling photon flux between .35 and .7 μm, and spectral concentration of downwelling irradiance at .48 μm (all these quantities measured simultaneously at each depth). Data obtained by Bethoux (personal communication) North of the Dominican Republic (20°22'N, 70°55'W) during a cruise of the "Discoverer" (May 1970).

downward irradiance at .48 μm (this wavelength penetrates particularly deep in the water under consideration) decreases very much less rapidly than the downward photon flux corresponding to the spectral band between .35 and .7 μm (which is broadly the spectral band that supplies photosynthesis) which itself decreases less rapidly than the total downward irradiance. This is easily explained by the rapid absorption of the short and above all of the long wavelengths. In contrast, at great depths only the most penetrating radiation remains and then the law of variation with depth becomes practically the same for the three cases considered (spectral concentration at .48 μm, photon flux corresponding to the spectral band between .35 and .7 μm, and total irradiance).

Values of the attenuation coefficient corresponding to the results given in Fig. 5.15 are shown in Table 5.10. The rapid decrease of the downward irradiance attenuation coefficient between 0 and 3 m is easily explained by the absorption of the infrared.

The relationship between attenuation and depth clearly depends also on the turbidity of the water. Some examples are given in Fig. 5.16; they concern the downward irradiance in the spectral band .35 to .7 μm. As illustrated in Fig. 5.17, in the top hundred meters the attenuation coefficient can vary significantly during the course of the year, notably because of changes in photosynthetic production which may modify the absorption of solar energy with depth.

On the other hand, the effect of the state of illumination at the sea surface (solar elevation, cloudiness) is not clearly understood at the present time. Let us limit ourselves to the near surface layer (the top one meter). It is then possible to the first approximation to neglect scattering and also the contribution from sky radiation during sunny weather. In these

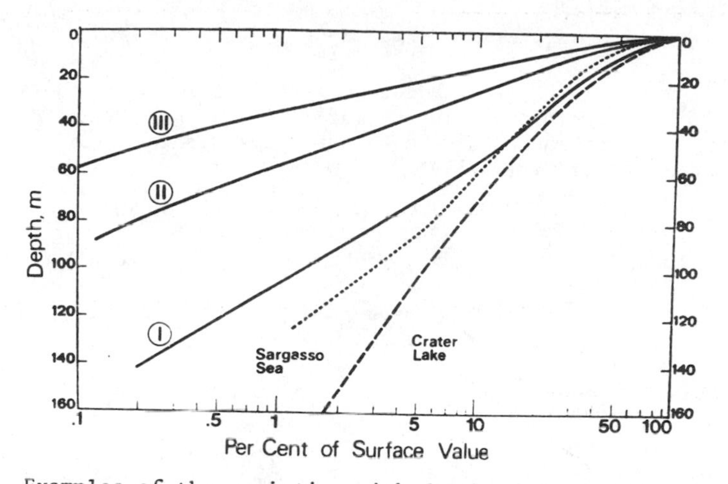

Fig. 5.16 - Examples of the variation with depth of downwelling irradiance (between wavelengths .35 and .7 μm). The three continuous curves represent types I, II and III of Jerlov's (1968) classification, while the dashed line corresponds to the spectral distribution curves for downwelling irradiance obtained by Tyler and Smith in Crater Lake (Fig. 5.13). The dotted line describes the results of Bethoux (personal communication) obtained in the Sargasso Sea (25°36'N, 69°45'W, cruise of the "Discoverer", May 1970) using a quantameter (also between wavelengths of about .35 and .7 μm).

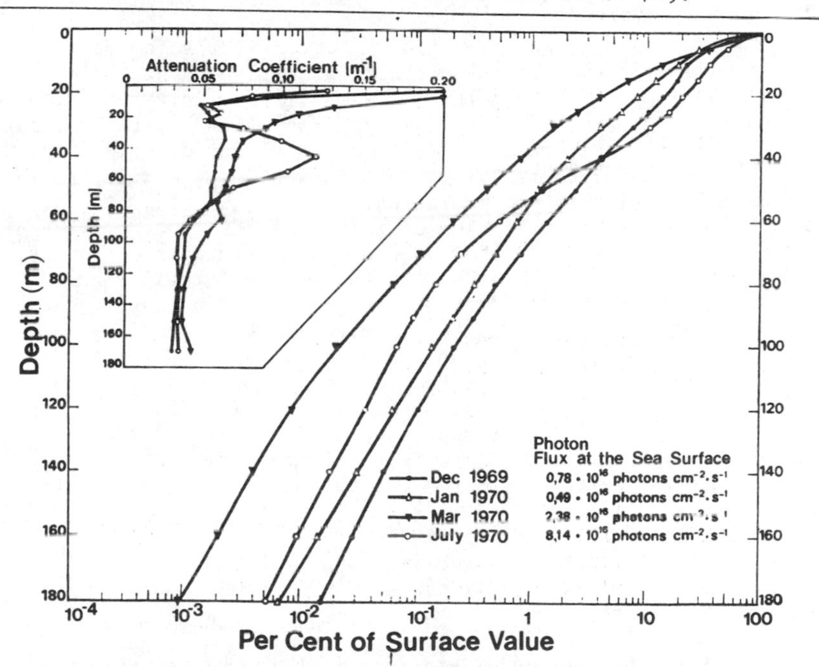

Fig. 5.17 - Variation of the downwelling photon flux (between .35 and .7 μm) from December 1969 to July 1970, about half-way from Nice to Corsica (43°12'N, 7°55'E). The inset shows the vertical distribution of the corresponding attenuation coefficient for the months of December, March, and July. Measurements made on board the "Korotneff" by Prieur (personal communication).

TABLE 5.10

Variation with depth of the attenuation coefficients for total downward irradiance, for downward photon flux between wavelengths .35 and .7 μm, and for the spectral concentration of the downward irradiance at .48 μm, taken from the three curves of Fig. 5.15.

DEPTH (m)	APPROXIMATE VALUES OF THE ATTENUATION COEFFICIENT (m^{-1}) FOR		
	TOTAL DOWNWARD IRRADIANCE	DOWNWARD PHOTON FLUX (BETWEEN .350 AND .700 μm)	SPECTRAL CONCENTRATION OF THE DOWNWARD IRRADIANCE AT .480 μm
0-1	0.57		
1-2	0.35		
2-3	0.14		
3-4	0.14		
4-5	0.12		
0-5	0.26	0.17	0.03
5-10	0.04	0.06	0.03
10-20	0.04	0.04	0.03
20-40	0.04	0.04	0.03
40-60	0.04	0.04	0.04
60-80	0.04	0.04	0.04

conditions, since the distance travelled by the direct solar rays is equal to $z/\cos\zeta'$ (where ζ' is the angle of refraction corresponding to the solar zenith angle ζ), one can express the downward irradiance attenuation coefficient as $\gamma = a/\cos\zeta'$, where a is the average absorption coefficient for the range of wavelengths being considered. Such an effect of solar zenith angle has not been shown experimentally (perhaps because of the relatively large contribution of the diffuse sky radiation when the sun sinks towards the horizon). Equally, because the flux of infrared radiation arriving at the sea surface is weaker in cloudy weather, the downward irradiance attenuation coefficient for the top one meter should be correspondingly reduced. This effect also has not been demonstrated experimentally (perhaps because the angular distribution of sky radiance is quite different during cloudy weather from that in clear weather). So, at the present state of our knowledge it is not yet possible to quantify the effect of solar zenith angle or of cloudiness. Therefore in the following discussion the downward irradiance attenuation coefficient is assumed to be not dependent on the state of the sky. But it may be that the solar zenith angle effect becomes important at high latitudes (where, as we have seen, the magnitude of the albedo in clear weather is also uncertain).

We now come to the central problem, to know the absorption of solar energy as a function of depth. It can be easily shown that the solar energy flux absorbed by a unit volume of sea water, $\frac{dF}{dA\,dz}$, where dA is horizontal surface area, can be expressed either as a function of the absorption coefficient and the magnitude of the scalar irradiance at the depth under consideration:

$$\frac{dF}{dA\,dz} = a\,I_o, \qquad (5.13)$$

or as a function of the variations of downward and upward irradiances with depth (at the depth under consideration):

$$\frac{dF}{dA\ dz} = -\frac{dI_d}{dz} + \frac{dI_u}{dz} \tag{5.14}$$

(this is simply a statement of conservation of energy). The ratio I_u/I_d being only of the order of a few per cent, it is possible to write to a good approximation

$$\frac{dF}{dA\ dz} \sim - dI_d/dz = \gamma I_d, \tag{5.15}$$

where γ is the downward irradiance attenuation coefficient at the depth under consideration. This last relationship is particularly convenient, since it permits one easily to analyze the absorption of solar energy with depth from curves of $\log I_d = f(z)$.

The problem can also be tackled from a theoretical point of view, especially if one makes simplifying hypotheses consisting, for example, of neglecting diffuse sky radiation in comparison with the direct solar beam, and scattering in comparison with the absorption. The results obtained in this way by Pruvost (1972), which we shall return to later when considering in detail the first meter of water, are in satisfactory agreement, at least for the first meters, with those derived from measurements of downward irradiance made by Bethoux in September 1969 in the Western Mediterranean (on board the "Bouée Laboratoire"), using a thermopile. The results obtained (meter by meter to 10 m depth) are shown in Fig. 5.18 for five series of measurements undertaken the same day in calm, sunny weather (solar elevation varying between 18° and 45°, giving no systematic change in the results). The curve shows the average of these series of measurements. Here, I_d is the surface downward irradiance reduced by the losses corresponding to the albedo. It appears that for the water under consideration, the top meter absorbs 55% of the solar energy (corresponding well to the proportion of infrared present), which corresponds to an attenuation coefficient equal to 0.8 m^{-1} (considerably higher than the value given in Table 5.10, which is certainly an underestimate). The downward irradiance attenuation coefficient falls rapidly with depth, reaching an approximately constant value of 0.075 m^{-1} at depths greater than about 7 m. This corresponds to fairly clear water, since it is known that the downward irradiance attenuation coefficient for the clearest sea water is of the order of 0.03 m^{-1} at sufficient depth, while that for turbid coastal water and during a period of plankton bloom can reach 0.3 m^{-1} even at depth.

The continuous curve in Fig. 5.19 shows how for the conditions of Fig. 5.18 the solar energy flux is absorbed at all depths by one meter of the considered sea water, starting with a surface irradiance of I_d (after taking the albedo into account). While the first meter absorbs 55% of the solar energy flux, the second absorbs only 9%, the third 5%, etc., the tenth less than 1%. The continuous curve in Fig. 5.20 shows these same results expressed cumulatively. For the water under consideration

the first meter absorbs 55% of the solar energy flux
the first 2 meters absorb 65% of the solar energy flux
" 3 " 70% "

the first 5 meters absorb	76%	of the solar energy flux
" 10 "	84%	"
" 20 "	92%	"

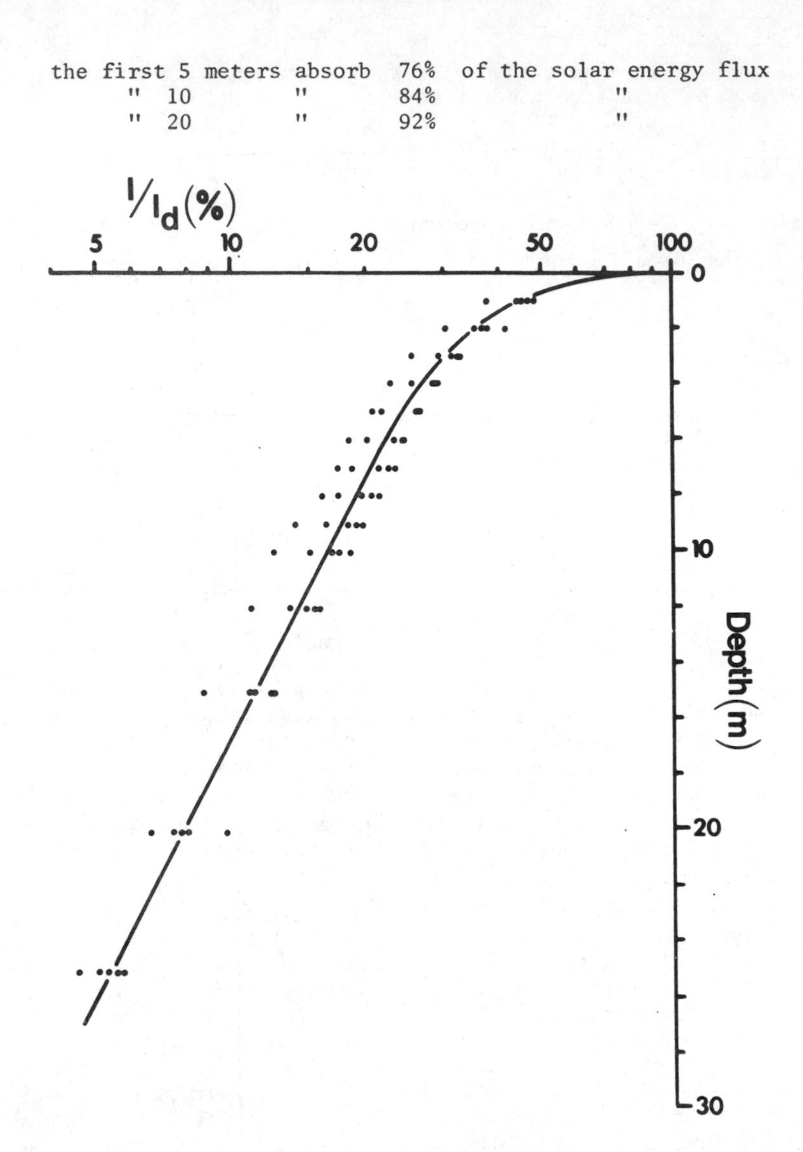

Fig. 5.18 - Variation of downwelling irradiance (as a fraction of the surface value after taking account of the albedo) as a function of depth. The curve represents the average of five series of measurements by Bethoux (personal communication) in the Mediterranean Sea on board the "Bouée Laboratoire" (42°14'N, 5°35'E), on September 23, 1969.

This only takes account of one special case, that of rather clear water, for which the downward irradiance attenuation coefficient is equal to 0.075 m^{-1} at sufficiently great depth. Regrettably, we do not have experimental data for clearer or more turbid water. One can nevertheless extrapolate "reasonably" the results presented above to clearer or more turbid water maintaining the 55% absorption of solar energy in the first meter regardless of the water type being considered (since in the infrared substances in

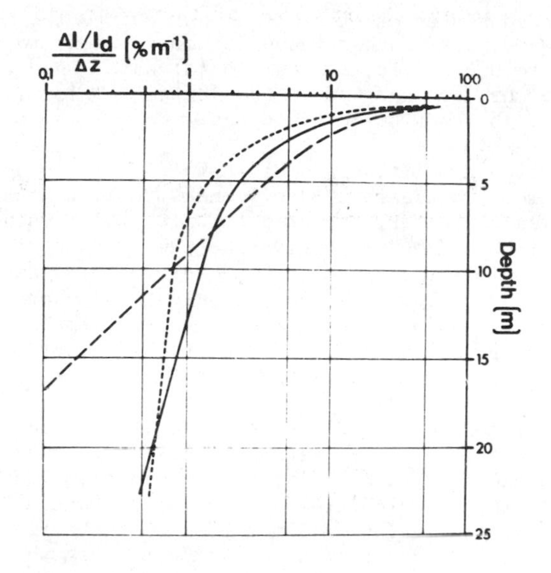

Fig. 5.19 - Fraction of incident solar energy (after taking account of the albedo) absorbed per meter of water as a function of depth. The continuous line is based on experimental data by Bethoux shown in Fig. 5.18. The other two curves were obtained by extrapolation for exceptionally clear water (short dashes) and for very turbid coastal water (long dashes).

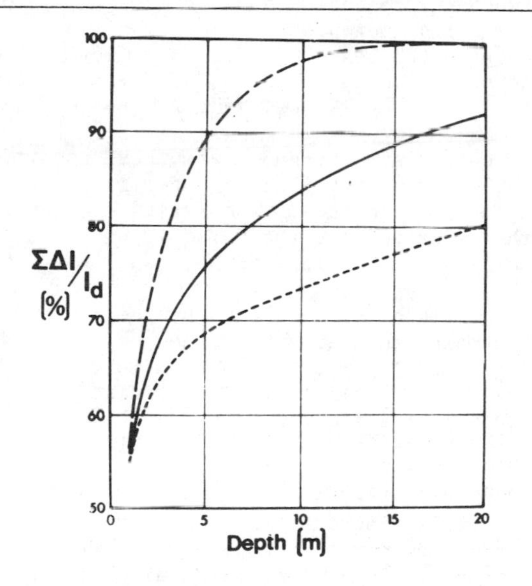

Fig. 5.20 - Curve of integrated energy absorption as a function of depth (after taking account of the albedo). The continuous line is based on experimental data by Bethoux shown in Fig. 5.18. The other two curves were obtained by extrapolation for exceptionally clear water (short dashes) and for very turbid coastal water (long dashes).

solution and suspended matter practically do not change the absorption), and adjusting the downward irradiance attenuation coefficient to 0.03 m^{-1} for the clearest water and to 0.3 m^{-1} for turbid coastal water when the depth exceeds, say, 10 m in the former case and 5 m in the latter. This will provide the extreme situations to be considered in oceans.

These two limiting conditions are drawn in Figs. 5.19 and 5.20 as lines with short dashes (for the clearest water) and with long dashes (for turbid water). For the clearest water, the proportion of solar energy absorbed is weaker to a depth of 20 m (see Fig. 5.19) and then stronger at greater depths. For turbid water, the proportion of solar energy absorbed is, in contrast, greater down to about 8 m and consequently weaker at greater depths. The first 5 m of water absorb 69% of the solar energy in the case of the clearest water (see Fig. 5.20) and 89% in the turbid water; the first 10 m absorb 73% in the clearest and 98% in the turbid.

One can next try to analyze in greater detail the absorption of solar energy by the first meter of water, although this is usually well mixed. We shall first examine the results calculated by Pruvost (1972). As we have already pointed out, these neglect the sky radiation in comparison with the direct solar beam, and scattering in comparison with absorption (this is justified for the layer nearest to the surface, at least while the solar elevation is large); in other words, Pruvost takes the downward irradiance attenuation coefficient equal to $a/\cos \zeta'$ (for the near surface layer, as already stated above, a is independent of the properties of the sea water; it is practically the absorption coefficient of pure water). The results vary then with the solar elevation. The fraction of solar energy absorbed by the first meter varies, in Pruvost's calculations, between 56.5% and 65%, whereas in those that we have presented above it is 55%. In order to avoid contradiction with the results quoted earlier, we shall restrict ourselves to those of Pruvost which agree the best with our own. It is also worth remarking that when one starts to take account of solar elevation, it becomes then necessary to consider the wave slopes, that is to say the sea state.

The continuous curve in Fig. 5.21 shows for these conditions the cumulative absorption of solar energy flux in the first meter of water (for sunny weather with the sun not too low). The first centimeter of water absorbs about 17% of the solar energy, the first 10 cm about 35%, the first 20 cm about 41% and the first 50 cm close to half.

A study of the absorption of solar energy by the first 20 cm of water has also been carried out experimentally by Bethoux (1968), using a thermopile in a plastic tank of about 300 liters. The dashed curve in Fig. 5.21 shows the results obtained in this way. They are in satisfactory agreement with those calculated by Pruvost. The first centimeter absorbs about 13% of the solar energy, the first 10 cm about 31% and the first 20 cm about 41%.

Taking for the first centimeter a mean value of the absorption of 15%, for a surface flux (after taking account of the albedo) of 0.08 W cm^{-2} (equivalent to midday during summer in the Western Mediterranean), the first centimeter of water absorbs 0.012W cm^{-3} or 0.17 cal cm^{-3} min^{-1}, equivalent to a heating of 0.17°C min^{-1} or 10.2°C hr^{-1}! This allows us to estimate the roles of exchange with the atmosphere and of mixing with the underlying waters (using the same line of argument for the first meter rather than the first centimeter, we estimate a heating rate of only 0.37°C per hour). The phenomena that we have just been analyzing constitutes only one of the basic

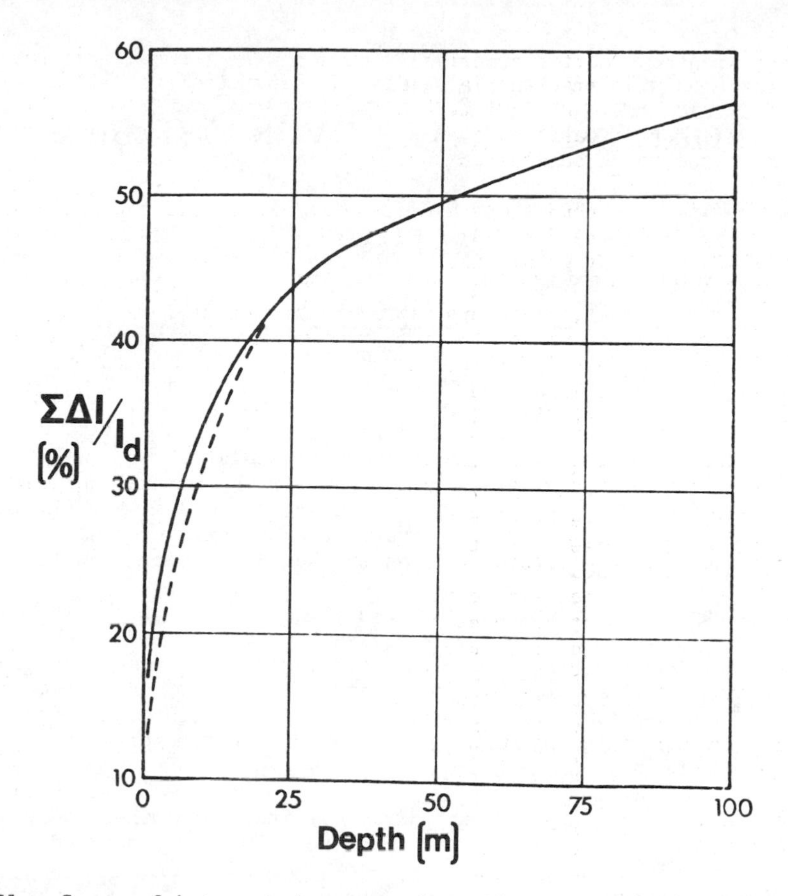

Fig. 5.21 - Curve of integrated energy absorption as a function of depth in the first meter below the surface (after taking account of the albedo). The continuous line is based on calculations by Pruvost (1972); the dashed line on measurements by Bethoux (1968).

factors that must be included in a model of the upper ocean. For that, it will be convenient to express the results presented in Figs. 5.19, 5.20 and 5.21 in analytical form. We leave that to the modellers. It remains also to clarify the influence of the sun's elevation and that of cloudiness.

Chapter 6

FLUXES IN THE SURFACE BOUNDARY LAYER OVER THE SEA

Niels E. Busch

6.1 Introduction

Since Roll (1965), Kitaigorodskii (1970), and Kraus (1972) produced their very different and important books on air-sea interaction, and since Pond (1971, 1972) reviewed the particular aspect which is the subject of this paper, there has been a proliferation of experiments and results, especially in connection with large international experiments such as BOMEX, JONSWAP, ATEX, GATE, JASIN, AMTEX, etc. However, this has not led to dramatic changes in our physical insight. Some questions have become more sharply focused and some ideas have taken firmer shape, but it seems an inappropriate time for a new and complete review, and the present paper should not be considered as such. Those interested in an up to date bibliography for the period 1971-1975 (for the western world, at least) are referred to Businger (1975). A relatively comprehensive review of the surface boundary layer has been presented by Busch (1973a) with emphasis on the boundary layer over land.

In this paper we shall give a brief outline of some of the methods available for determination of the surface fluxes of momentum, heat, and water vapor. We shall confine ourselves to consideration of the surface boundary layer (SBL) on micrometeorological space and time scales, and direct our attention mainly towards the bulk parameterization of fluxes in terms of the bulk transfer coefficients pertaining to standard anemometer height, which is normally chosen to be 10 m.

Of the eight to nine different direct or indirect methods of estimating surface fluxes which come to mind, we shall be primarily concerned with three:

1) The eddy correlation or direct method.

2) The variance budget or dissipation method.

3) The profile or gradient method.

Five other methods have been briefly reviewed by Pond (1972) and Businger (1975). They are:

4) The ageostrophic method, which depends on integration over height of the difference between the geostrophic wind and wind velocity up to the height where the stress vanishes, hence yielding the total momentum loss that is the surface stress (Hasse, 1970).

5) The total energy budget at the surface for determination of the fluxes of sensible and latent heat. This method seems mostly applicable on quite large space and time scales.

6) The line integral or divergence method, in which the flux of a quantity is determined as a residual by determining the change within a rather large volume and the fluxes through the sides and the top surface. This method, which is also called the budget method, requires several ships and aircraft. It is feasible, but

still quite uncertain (Holland and Rasmussen, 1973).

7) The "wind set-up" method is based on the possibility of observing the tilt or "set-up" of the mean water level by the wind stress. This method has recently been applied by Wieringa (1974) and Donelan, et al. (1974). Many difficulties, such as taking the geometry of the water body, the bottom stress, and the bottom topography properly into account, make the method unreliable in many cases.

8) Remote sensing techniques, e.g., radiometers (McAlister, et al., 1971; Katsaros and Businger, 1973), airborne lasers (Schule, et al., 1971), sun glitter and satellites (Levanon, 1971), radar returns from the sea surface (Pierson, et al., 1971), and infrared images of the sea surface (McGrath and Osborne, 1973). Several of these techniques look promising, but they are still dependent on comparison with ground truth data.

6.2 Basic equations for the mean flow; the "constant flux" layer

Starting from the Bousinesq approximations to the Navier-Stokes equation (e.g., Busch, 1973b), which are valid for a planetary boundary layer (PBL) without phase transformations, we may write the equations for the mean flow as

$$\frac{\partial \overline{u}_i}{\partial t} = -\frac{1}{\rho_0}\frac{\partial \overline{p}}{\partial x_i} - \frac{\partial}{\partial x_i}[\overline{u_i' u_j'} + \overline{u}_i \overline{u}_j] + g\frac{\overline{T_V'}}{T_{Vo}}\delta_{i3} - \overline{C}_i, \qquad (6.1)$$

where viscous fluxes of momentum are neglected and $\overline{C}_i$ denotes the Coriolis accelerations: $\overline{C}_i = 2\Omega\varepsilon_{ijk}n_j\overline{u}_k$. Here ε_{ijk} is the alternating tensor, n_i is the unit vector parallel to the earth's rotation axis. In Eq. (6.1) i=1,2,3 is a right-handed Cartesian coordinate system with i=3 antiparallel to gravity. i=1 is taken parallel to the surface stress. u is wind velocity, p pressure, T_V is virtual temperature and ρ_0 is a reference density. δ_{ij} is the Kronecker delta. Here, $T' = T - T_0$, with T_0 the <u>reference</u> temperature, so $\overline{T'} \neq 0$ in this case.

The Coriolis terms will be of the order fu, where f is the Coriolis parameter ($2\Omega n_3 \simeq 1.2 \cdot 10^{-4} s^{-1}$ for temperate latitudes). The horizontal pressure terms will be of the order fG, where G is the geostrophic wind speed, which, for the present purpose, we will evaluate by assuming that it is not too different from the mean wind speed measured at a height of 10 m.

If we assume that the mean flow is horizontal and horizontally homogeneous, Eq. (6.1) reduces to

$$\frac{\partial \overline{u}_i}{\partial t} \simeq -\frac{\partial \overline{u_i' u_j'}}{\partial x_j} + g\left(\frac{\overline{T_V'}}{T_{Vo}} - \frac{1}{\rho_0}\frac{\partial \overline{p}}{\partial x_i}\right)\delta_{i3} + O(fu_{10}). \qquad (6.2)$$

From Eq. (6.2) it is clear that in stationary mean flows the horizontal Reynolds stresses $-\rho_0\overline{u_1'u_3'}$ and $-\rho_0\overline{u_2'u_3'}$ change with height under the combined action of Coriolis and pressure accelerations.

Defining the local friction velocity u_* by

$$u_*^2 = [(\overline{u_1'u_3'})^2 + (\overline{u_2'u_3'})^2]^{1/2}, \tag{6.3}$$

we may write according to Eq. (6.2)

$$u_*^2 \simeq u_{*o}^2 - O(f\overline{u}_{10})x_3, \tag{6.4}$$

where u_{*o}^2 represents the surface stress. Equation (6.4) shows that the stress varies approximately linearly with height in the surface boundary layer. Of course, we presume that we are above the layer closest to the wave field in which there may be divergence of vertical momentum flux due to wave action. It is generally believed that this layer is only one to two wave heights thick. However, a recent, preliminary analysis by Kitaigorodskii (personal communication) indicates that the direct influence of the waves on the mean wind profile may extend considerably farther, up to a height of 7 to 8 m for moderate wind speeds.

In practice, the surface stress can be measured with an accuracy which is around ±20%. Equation (6.4) yields a 20% change of the stress over a height interval of approximately 0.2 $C_D\ \overline{u}_{10}/f$, where $C_D = u_{*o}^2/\overline{u}_{10}^2$ is the drag coefficient, which is generally about 1.3 x 10^{-3}. Hence, if we measure the stress at heights lower than

$$h_c \simeq 0.2C_D\overline{u}_{10}f^{-1} = 0.2u_{*o}C_D^{1/2}f^{-1}, \tag{6.5}$$

we may then consider the result to be representative of the surface stress. A typical stress on the ocean is 1 dyne cm^{-2}, corresponding to $u_{*o} \simeq 28\ cm\ s^{-1}$ and $\overline{u}_{10} \simeq 7.8\ m\ s^{-1}$, so h_c is about 17 m. However, the stress may be a factor of 10 smaller, and the drag coefficient may decrease by a factor of two due to the influence of stable stratification, in which case h_c is less than 4 m.

Considerations such as those presented above lead to the concept of a "constant flux" layer. This has caused a certain amount of confusion, since it is in this surface layer that the stress undergoes the most rapid changes with height (e.g., Wyngaard, 1973).

Neglecting molecular fluxes, the thermodynamic energy equations for horizontally homogeneous flows may be written

$$\frac{\partial \overline{T'}}{\partial t} = -\frac{\partial}{\partial x_3}[\overline{u_3'\theta'} + \frac{R}{c_p\rho_o}], \tag{6.6}$$

where θ' is the fluctuating part of the potential temperature; $c_p\rho_o\overline{u_3'\theta'}$ is therefore the vertical turbulent flux of sensible heat; and R is the vertical heat transfer due to radiation. It is common to neglect the divergence of R in Eq. (6.6). However, it has been shown by Coantic and Seguin (1971), and Gavrilov and Laykhtman (1973), that absorption of infrared radiation by water vapor may play an important role. They conclude that the radiative flux divergence acts in such a way as to increase the absolute value of turbulent heat flux over almost all of the surface layer, as compared to its value at the surface. On the basis of numerical models for stationary conditions (total heat flux constant with height), they find that the increase in sensible heat flux may be 40% at a height of 10 m in situations with low wind speeds (3 m s^{-1}) and high specific humidity (1.42×10^{-2}) over warm water (+20°C).

In the absence of phase transformations, the equation for the mean specific humidity field reads

$$\frac{\partial \bar{r}}{\partial t} = - \frac{\partial}{\partial x_3} [\overline{u_3' r'}] \qquad (6.7)$$

in horizontally homogeneous flows with negligible molecular diffusion.

In the quasi-stationary "constant stress" surface boundary layer we are led to adopt the surface fluxes of

$$\text{Momentum :} \quad \tau_0 = \rho_0 \overline{u_1' u_3'} = \rho_0 u_{*0}^2 ;$$

$$\text{Heat:} \quad R_0 + H_0 = R + c_p \rho_0 \overline{u_3' \theta'} = R - c_p \rho_0 u_{*0} T_* ; \qquad (6.8)$$

$$\text{Moisture:} \quad Q_0 = \rho_0 \overline{u_3' r'} = - \rho_0 u_{*0} r_* ;$$

as scaling parameters. Equations (6.8) define another, more convenient set of scaling parameters: the friction velocity u_{*0}, the characteristic temperature T_*, and the characteristic humidity r_*.

In the following discussion we shall neglect the variation of the radiative heat flux R with height, bearing in mind that this may be one serious source of errors.

6.3 The eddy correlation or direct method

By measuring the fluctuating components u_1', u_3', θ', and r' of the horizontal velocity, vertical velocity, potential temperature, and specific humidity in the "constant flux" layer, and correlating them to obtain the covariances, the surface fluxes of momentum, sensible heat, and moisture may be evaluated.

In order to obtain estimates of the total covariances, the instrumentation should respond fairly rapidly. Typically, it is necessary to employ instruments capable of measuring fluctuations up to frequencies of 10-50 Hz without appreciable attenuation, and to average the measurements over time periods of half an hour to an hour.

Great care must be taken in the analysis to evaluate departures from stationarity in the data records, and the removal of trends is often crucial. Severe difficulties are involved in obtaining reliable measurements from moving platforms (buoys, ships), and stable platforms (bottom-fast towers) are scarce. Since fast responding instruments are generally fragile and susceptible to contamination by salt, sea spray and slime, it is cumbersome and difficult to obtain high quality measurements over the sea, especially at high wind speeds.

These difficulties notwithstanding, a growing number of direct determinations of Reynolds fluxes have been reported in the recent literature: Friehe and Schmitt (1976); Smith and Banke (1975); S. D. Smith (1974); Dreyer (1974); Müller-Glewe and Hinzpeter (1974); Dunckel, et al. (1974); Misuta and Fujitani (1974); Wucknitz (1974); Wieringa (1974); Pond, et al. (1971); Phelps and Pond (1971); Pond (1971); and Hasse (1970). The measurements were made from stabilized buoys, ships and towers. Recent measurements made by means of instrumented aircraft have also been reported by Donelan and Miyake (1973), Grossman and Bean (1973), and Lenschow (1973).

6.4 The turbulent energy equation; temperature and humidity variance budgets

In horizontally homogeneous turbulence the turbulent energy balance appears as (Busch, 1973b):

$$\frac{\partial \bar{q}}{\partial t} = - \overline{u_1' u_3'} \frac{\partial \bar{u}_1}{\partial x_3} - \overline{u_2' u_3'} \frac{\partial \bar{u}_2}{\partial x_3} + \frac{g}{T_{Vo}} \overline{u_3' \theta_V'}$$

$$- \nu \overline{\frac{\partial u_i'}{\partial x_j} \frac{\partial u_i'}{\partial x_j}} - \frac{\partial}{\partial x_3} [\overline{u_3' q} + \frac{\overline{p' u_3'}}{\rho_o}], \quad (6.9)$$

where $q=u_i'u_i'/2$ is the turbulent energy per unit mass. Note that the Coriolis accelerations do not contribute to the change of energy.

The first two terms on the right-hand side of Eq. (6.9) represent the rate at which the turbulent flow, via the Reynolds stresses, extracts energy from the mean flow. In the SBL, and with the coordinate system arranged so that the x_1-axis points in the direction of the surface stress, these two mechanical production terms reduce to one:

$$M = - \overline{u_1' u_3'} \frac{\partial \bar{u}_1}{\partial x_3} = u_{*o}^2 \frac{\partial \bar{u}_1}{\partial x_3}. \quad (6.10)$$

In the third term, θ_V' denotes the fluctuating part of the virtual temperature. This term represents the rate at which work is done by the turbulence against buoyancy and is referred to as the thermal production term. The fourth term is the rate at which turbulent energy is dissipated into internal energy by viscosity. It is customary to denote this term by ε. The last term in Eq. (6.9) is the divergence of the vertical fluxes of turbulent energy. The first of these fluxes is called the turbulent energy flux, the second the pressure transport. Especially the latter flux is difficult to measure directly and, with a few exceptions (Elliott, 1972a and b; Snyder, 1974; Dobson, 1971), its significance has been evaluated by attributing the residual of Eq. (6.9) - with all other terms measured - to the pressure term. Clearly, direct experimental determination of the divergence term in Eq. (6.9) requires simultaneous and rather accurate measurements of fluxes at at least two heights, a proposal which makes heavy demands on experimental facilities, demands which as far as we know have never been met over the sea.

It has been shown (Wyngaard, 1973) that in the SBL over land the $\overline{u_3' q}$

flux is upward, and that the divergence of this flux represents a local loss of turbulent energy. The divergence of the pressure transport term represents a gain, which primarily flows into the horizontal velocity components from which it spills over into the vertical velocity component by local pressure action. Elliot (1972a), who directly measured the pressure-velocity correlation (over land), concluded that in the total energy budget below 5 m, under neutral conditions, the assumption of a small contribution by the pressure term is reasonable. Not only is the term small, but it is also partially balanced by the turbulent flux term. This is in good agreement with Wyngaard and Coté (1971), who found that in neutral and stable stratification, the divergence term in Eq. (6.9) may be insignificant whereas in unstable (except possibly very unstable) stratification the term represents a gain of turbulent energy.

Under the assumption of horizontal homogeneity, the temperature and specific humidity variance budgets may be written

$$\frac{\partial \tfrac{1}{2}\overline{\theta'^2}}{\partial t} = -\overline{u_3'\theta'}\frac{\partial \overline{T'}}{\partial x_3} - \mu_\theta \overline{\frac{\partial \theta'}{\partial x_j}\frac{\partial \theta'}{\partial x_j}} - \frac{\partial}{\partial x_3}[\overline{u_3'\tfrac{1}{2}\theta'^2}], \qquad (6.11)$$

and

$$\frac{\partial \tfrac{1}{2}\overline{r'^2}}{\partial t} = -\overline{u_3'r'}\frac{\partial \overline{r}}{\partial x_3} - \mu_r \overline{\frac{\partial r'}{\partial x_j}\frac{\partial r'}{\partial x_j}} - \frac{\partial}{\partial x_3}[\overline{u_3'\tfrac{1}{2}r'^2}], \qquad (6.12)$$

in which μ_θ and μ_r are the molecular diffusivities for heat and water vapor, respectively. The first term on the right-hand sides of (6.11) and (6.12) represents the production of temperature and humidity fluctuations by interaction of the turbulence with the temperature and humidity stratifications. The second term is the rate at which molecular diffusion smears ("dissipates") the fluctuations. The terms are traditionally denoted by N_θ and N_r. The last terms in the equations are the divergences of the vertical fluxes of temperature and humidity variance. Direct measurements of these divergences are rare. Panofsky's (1969) suggestion that the local dissipation of temperature variance balances the local production receives strong support from Wyngaard and Coté (1971), whose measurements were made over land.

Leavitt and Paulson (1975) present an analysis of measurements made during BOMEX at 8 and 30 m over the sea, but found unrealistically strong divergences of the vertical momentum and sensible heat fluxes. They attribute the divergences to calibration errors. However, on the basis of Monin-Obukhov similarity theory they conclude that within the limitations of the assumptions, the local production and dissipation of turbulent kinetic energy and humidity variance are approximately in balance, with turbulent diffusion being less that 15% of the production, whereas the temperature variance budget is not as simply balanced. In the practical use of Eqs. (6.11) and (6.12) in the SBL, it is common to neglect the divergence terms.

6.5 The variance budget or dissipation method

Assuming stationarity and neglecting the divergence terms, Eqs. (6.9), (6.11) and (6.12) may be rewritten

$$u_{*o}^2 = -\overline{u_1'u_3'} = (\varepsilon - \frac{g}{T_{Vo}}\overline{u_3'\theta'})(\partial\bar{u}_1/\partial x_3)^{-1}; \tag{6.13}$$

$$\overline{u_3'\theta'} = -N_\theta\,(\partial\bar{T}/\partial x_3)^{-1}; \tag{6.14}$$

$$\overline{u_3'r'} = -N_r\,(\partial\bar{r}/\partial x_3)^{-1}; \tag{6.15}$$

To this set of equations one may add an equation which relates the flux $\overline{u_3'\theta_V'}$ of virtual temperature to the sensible heat flux $\overline{u_3'\theta'}$, the mean specific humidity $\bar{r}$, the vertical flux $\overline{u_3'r'}$ of specific humidity, and the mean air temperature $\bar{T}$ (Busch, 1973b):

$$\overline{u_3'\theta_V'} \simeq \overline{u_3'\theta'}\,(1 + 0.61\,\bar{r}) + 0.61\,\bar{T}\,\overline{u_3'r'}. \tag{6.16}$$

Quite obviously Eqs. (6.13)-(6.16) can be used for determination of the fluxes, if mean profile measurements are available, and the dissipation rates ε, N_θ, and N_r can be obtained independently. Determination of the dissipation rates requires the evaluation of

$$\varepsilon = \nu\,\overline{\frac{\partial u_i'}{\partial x_j}\frac{\partial u_i'}{\partial x_j}}\;, \quad N_\theta = \mu_\theta\,\overline{\frac{\partial \theta'}{\partial x_j}\frac{\partial \theta'}{\partial x_j}}\;,$$

$$\text{and}\quad N_r = \mu_r\,\overline{\frac{\partial r'}{\partial x_j}\frac{\partial r'}{\partial x_j}}\;, \tag{6.17}$$

or, in other words, the measurement of fifteen spatial derivatives. However, if the eddies which contribute significantly to the dissipation are so small compared to the energy containing eddies that the turbulent motion on these scales may be considered isotropic, then Eq. (6.17) simplifies to (Tennekes and Lumley, 1972)

$$\varepsilon = 15\nu\,\overline{\left(\frac{\partial u_1'}{\partial x_1}\right)^2}\;, \quad N_\theta = 3\,\mu_\theta\,\overline{\left(\frac{\partial \theta'}{\partial x_1}\right)^2}\;,$$

$$\text{and}\quad N_r = 3\,\mu_r\,\overline{\left(\frac{\partial r'}{\partial x_1}\right)^2} \tag{6.18}$$

If furthermore we apply Taylor's hypothesis (see Woods, Chapter 15):

$$\frac{\partial}{\partial x_1} = \bar{u}_1^{-1}\frac{\partial}{\partial t} \tag{6.19}$$

then we may write

$$\varepsilon = \frac{15\nu}{\bar{u}_1^{-2}}\,\overline{\left(\frac{\partial u_1'}{\partial t}\right)^2}\;, \quad N_\theta = \frac{3\mu_\theta}{\bar{u}_1^{-2}}\,\overline{\left(\frac{\partial \theta'}{\partial t}\right)^2}\;,$$

$$\text{and} \quad N_r = \frac{3\mu_r}{\overline{u}_1^{-2}} \overline{\left(\frac{\partial r'}{\partial t}\right)^2}, \qquad (6.20)$$

which means that we may obtain the desired dissipation rates by measuring the variance of the time derivative of the fluctuations.

Dissipation rates for turbulent energy and temperature variance have been determined in this way (Stegen, et al., 1973; Schedvin, et al., 1973; Boston and Burling, 1972; Gibson, et al., 1970), but the method has not yet been applied to humidity as far as we know. The method requires instrumentation with extremely low electronic noise levels and very fast response (in general up to more than 1 KHz).

A more practical solution is to assume the existence of an inertial subrange of isotropic eddy motions, in which the one-dimensional wave number spectra may be written (Tennekes and Lumley, 1972)

$$\left.\begin{aligned} F_{11}(k) &= \alpha_1 \varepsilon^{2/3} k^{-5/3} \\ F_\theta(k) &= \beta_\theta \varepsilon^{-1/3} N_\theta k^{-5/3} \\ F_r(k) &= \beta_r \varepsilon^{-1/3} N_r k^{-5/3} \end{aligned}\right\} \qquad (6.21)$$

and where k is the wave number in radians per unit length, and the coefficients α_1, β_θ, and β_r are supposed to be universal constants, the so-called Kolmogorov constants. In the following, the spectra are defined so that

$$\int_0^\infty F(k)dk = \sigma^2, \qquad (6.22)$$

where σ^2 may be $\overline{2q}$, $\overline{\theta'^2}$, or $\overline{r'^2}$.

If the Kolmogorov constants are known, and the spectral densities F(k) can be determined for wave numbers in the inertial subrange, then the dissipation rates may be obtained through Eq. (6.21).

Conversely, the determination of α_1, β_θ, and β_r from spectral measurements in the inertial subrange requires that the dissipation rates can be evaluated either by the "direct" method implied in Eq. (6.18), or by assuming that they are equal to the production rates. The Kolmogorov constants can also be determined from the second and third order structure functions (Paquin and Pond, 1971).

Businger (1975), Wucknitz (1974), and Busch (1973a) have reviewed recent results concerning the values of the Kolmogorov constants. They agree that α_1 lies in the range from 0.48 to 0.6 with a reasonable average of about 0.55, although estimates as high as 0.65-0.69 have been obtained (Shei, et al., 1971; Gibson, et al., 1970). Typical standard deviations are 10-20%.

The constant β_θ for the temperature spectrum shows considerably more scatter from 0.6 (Garratt, 1972) to 1.6 (Boston, 1970) and 2.0 (Gibson, et al., 1970). Most of the available estimates vary from 0.6 to 0.9, so we suggest that the most appropriate value to use in the atmosphere is β_θ=0.80±10 to 20%.

Based on the structure function approach, Paquin and Pond obtained β_θ=0.83±0.13 and β_r=0.80±0.17, and concluded that there is no difference between the constants for temperature and humidity fluctuations. Smedman-Högström (1973) assumed that production of humidity variance equals dissipation and found β_r=0.58±0.2, a value which on the basis of the same assumption is supported by Leavitt and Paulson (1975).

Returning to Eqs. (6.9), (6.11), and (6.12), we bring the equations in non-dimensional form

$$\frac{\kappa x_3}{u_{*o}^3}\frac{\partial \bar{q}}{\partial t} = \phi_m - \frac{x_3}{L} - \frac{\kappa x_3 \varepsilon}{u_{*o}^3}; \tag{6.23}$$

$$\frac{\kappa x_3}{u_{*o}T_*^2}\,\overline{\frac{\partial \frac{1}{2}\theta'^2}{\partial t}} = \phi_\theta - \frac{\kappa x_3 N_\theta}{u_{*o}T_*^2}; \tag{6.24}$$

$$\frac{\kappa x_3}{u_{*o}r_*^2}\,\overline{\frac{\partial \frac{1}{2}r'^2}{\partial t}} = \phi_r - \frac{\kappa x_3 N_r}{u_{*o}r_*^2}; \tag{6.25}$$

where we still neglect the divergence terms, and where the non-dimensional gradients are defined through

$$\phi_m = \frac{\kappa x_3}{u_{*o}}\frac{\partial \bar{u}_1}{\partial x_3}, \quad \phi_\theta = \frac{\kappa x_3}{T_*}\frac{\partial \overline{T'}}{\partial x_3}, \quad \text{and } \phi_r = \frac{\kappa x_3}{r_*}\frac{\partial \bar{r}}{\partial x_3}. \tag{6.26}$$

The so-called Monin-Obukhov stability length

$$L = -\frac{u_{*o}^3 T_{Vo}}{\kappa g \overline{u_3'\theta_V'}} \tag{6.27}$$

is related through Eq. (6.16) to the temperature length scale L_θ and the humidity length scale L_r by

$$\frac{1}{L} = \frac{1}{L_\theta} + \frac{1}{L_r} \tag{6.28}$$

where

$$L_\theta = \frac{u_{*o}^2 T_o}{g\kappa T_*} \quad \text{and} \quad L_r = \frac{u_{*o}^2\, T_{Vo}}{0.61 g\, \kappa r_* \bar{T}}. \tag{6.29}$$

In most investigations L_r has been neglected; i.e., the influence of the humidity flux on the static stability has been neglected, which, over the sea, is often a poor approximation.

Equations (6.23)-(6.25) lead naturally to the stability parameter x_3/L and to the hypothesis that all the ϕ functions may be unique functions of x_3/L in a quasi-stationary, horizontally homogeneous surface layer.

The von Kármán constant κ is defined so that in neutral stratification ($x_3/L=0$) we have $\phi_m(0)=1$. Its customary value of 0.4 is subject to considerable discussion (Busch, 1973a). Businger, et al. (1971) and Högström (1974) find a value of 0.35. Tennekes (1973) has suggested that it may vary with the roughness of the underlying surface.

The forms of ϕ_m, ϕ_θ, and ϕ_r have recently been reviewed by Plate (1971), Monin and Yaglom (1971), and Högström (1974). Most atmospheric data are well represented by

$$\phi_m = \begin{cases} 1 + \beta_m x_3/L & \text{for } x_3/L \geq 0 \\ (1 - \gamma_m x_3/L)^{-1/4} & \text{for } x_3/L \leq 0 \end{cases} \tag{6.30}$$

and

$$\frac{\phi_\theta}{\phi_\theta(0)} = \frac{\phi_r}{\phi_r(0)} \begin{cases} 1 + \beta_\theta x_3/L & \text{for } x_3/L \geq 0 \\ (1 - \gamma_\theta x_3/L)^{-1/2} & \text{for } x_3/L \leq 0 \end{cases} \tag{6.31}$$

with $\beta_m \simeq 5$, $\gamma_m \simeq 15$, $\beta_\theta \simeq 6$, and $\gamma_\theta \simeq 9$.

The linear form for the ϕ functions in stable stratification is generally accepted, whereas different suggestions have been made for the unstable ϕ_m function. One of the often used forms is the "Keyps" equation

$$\phi_m^4 - \gamma \frac{x_3}{L} \phi_m^3 = 1, \tag{6.32}$$

where the estimate of γ ranges from 9 (Businger, 1973) to 18 (Panofsky, 1973). Carl, et al. (1973) have suggested a compromise

$$\phi_m = (1 - 15\, x_3/L)^{-1/3}. \tag{6.33}$$

Equations (6.32) and (6.33) have the advantage of being in accordance with the "classical" idea of "free convection" in the limit of large, negative x_3/L.

The values of $\phi_\theta(x_3/L=0)$ and $\phi_r(x_3/L=0)$ are also subject to some discussion. It is generally accepted that they are both equal to unity, but Businger, et al. (1971) find that $\phi_\theta(0)=0.74$ and Högström (1974) that

$\phi_\theta(0) \simeq 0.74$ and $\phi_r(0) = \phi_m(0)$, whereas ϕ_r approaches ϕ_θ as the stratification becomes strongly unstable. Leavitt and Paulson (1975) find strong similarity between momentum and humidity transfer and suggest $\phi_r \simeq \phi_m$ rather than $\phi_r \simeq \phi_\theta$. The value of ϕ_θ is difficult to determine in near-neutral conditions because of the smallness of temperature gradients and heat fluxes.

By use of Eq. (6.26) we reformulate Eqs. (6.13)-(6.15) and obtain

$$\left.\begin{aligned} u_{*0}^2 &= [\kappa x_3 \frac{\varepsilon}{\phi_m - x_3/L}]^{2/3} \\ |\overline{u_3'\theta'}| &= [N_\theta \kappa x_3 u_{*0}/\phi_\theta]^{1/2} \\ |\overline{u_3' r'}| &= [N_r \kappa x_3 u_{*0}/\phi_r]^{1/2} \end{aligned}\right\} \tag{6.34}$$

which in near-neutral stratification reduce to

$$\left.\begin{aligned} u_{*0}^2 &\simeq [\kappa x_3 \varepsilon]^{2/3} \\ |\overline{u_3'\theta'}| &\simeq [N_\theta \kappa x_3 u_{*0}]^{1/2} \\ |\overline{u_3' r'}| &\simeq [N_r \kappa x_3 u_{*0}]^{1/2} \end{aligned}\right\} \tag{6.35}$$

Since the stratification in the surface layer over the sea is generally near neutral, Eqs. (6.35) are often used instead of Eqs. (6.34), because in this case flux determination requires no measurements of the gradients of the mean flow. It is possible to use Eq. (6.34) without measurements of mean gradients if one assumes universal forms ϕ_m, ϕ_θ, and ϕ_r such as given in Eqs. (6.30) and (6.31). Initial estimates of fluxes are obtained by use of Eq. (6.35). These estimates are used to estimate L and to determine new estimates by use of Eq. (6.34). Iteration continues until flux estimates converge.

The dissipation method of obtaining the fluxes is attractive, in particular in conjunction with Eqs. (6.21) and (6.35), because the demands on precise spatial alignment of the sensors are much less stringent than those required for the direct eddy correlation method. However, it should be borne in mind that local isotropy and inertial subrange are required, that thermal stratification may invalidate Eq. (6.35), and that the assumption of negligible flux divergences may not always be valid, especially not close to the surface wave field.

The dissipation method has recently been applied by Wucknitz (1974), who also reviews some of the older measurements, and by Stegen, et al. (1973); Denman and Miyake (1973a); McBean, et al. (1973); and Pond, et al. (1971). Wucknitz reports that when applied to nearly isotropic conditions (see, e.g., Busch, 1973c), the dissipation method yields stresses which are about 20% smaller than the stresses obtained by the direct method. Stegen, et al. (1973) find stresses which are about 15% smaller. Wucknitz attributes this to the influence of the surface waves on the turbulent field above.

Wucknitz also finds that when the heat flux is not too small, the dissipation method yields results for the heat flux which are in reasonable agreement with results obtained by the direct method. For small heat fluxes the dissipation method yields fluxes which are far too large. Pond, et al. (1971) find that in some cases the method may not be applicable for sensible heat flux. Leavitt (1975); Phelps and Pond (1971); and Donelan and Miyake (1973) report striking differences between the behavior of the humidity and temperature fluxes. Pond, et al. partly attribute these differences to unexpected cold spikes in the temperature field (see also Friehe and Schmitt, 1976).

In general it must be said that the limitations to the applicability of the dissipation method need to be further investigated.

6.6 The profile or gradient method

Simple manipulation of Eq. (6.28) by use of Eqs. (6.26) and (6.29) leads to the relation

$$\frac{x_3}{L} \simeq Ri \frac{\phi_m^2}{\phi_\theta} \left(1 + \frac{0.61\, c_p T}{A_c B}\right), \tag{6.36}$$

where Ri is the gradient Richardson number:

$$Ri = \frac{g}{T_o} \frac{\partial \overline{T'}/\partial x_3}{(\partial \overline{u}_1/\partial x_3)^2}. \tag{6.37}$$

A_c is the latent heat of condensation, and B is the Bowen ratio.

$$B = \frac{c_p \overline{u_3'\theta'}}{A_c \overline{u_3' r'}}. \tag{6.38}$$

In the process of obtaining Eq. (6.36), we have neglected a factor $(1+0.61\,\overline{r})$.

Equation (6.36) illustrates that if the ϕ functions are known functions of x_3/L, then the ϕ functions are known functions of Ri and $\overline{T}/B$. This means that if the Bowen ratio can be estimated, then measurements of the mean velocity gradient and potential temperature gradient are sufficient to produce value of ϕ_m, ϕ_θ, and ϕ_r.

Rewriting Eq. (6.26) we obtain

$$u_{*o}^2 = \kappa^2 x_3^2 (\partial \overline{u}_1/\partial x_3)^2/\phi_m^2 ; \tag{6.39}$$

$$\overline{u_3'\theta'} = -\kappa^2 x_3^2 (\partial \overline{u}_1/\partial x_3)(\partial \overline{T'}/\partial x_3)/(\phi_m \phi_\theta); \tag{6.40}$$

$$\overline{u_3' r'} = -\kappa^2 x_3^2 (\partial \overline{u}_1/\partial x_3)(\partial \overline{r}/\partial x_3)/(\phi_m \phi_r); \tag{6.41}$$

which then readily lend themselves to the determination of fluxes by gradient measurement.

If the Bowen ratio which can now be computed is significantly different from that used at the beginning of the calculation in Eq. (6.36), one may then repeat the calculation with the new Bowen ratio and, if necessary, continue this iteration until it converges.

Equations (6.26) may be integrated to yield the profiles (Paulson, 1970)

$$\Delta\overline{u}_s = \overline{u}_1 - u_s = \frac{u_{*0}}{\kappa}\left[\ln\frac{x_3}{z_0} - \psi_m(x_3/L)\right]; \tag{6.42}$$

$$\Delta\overline{T} = \overline{T}' - \overline{T}'(z_0) = \frac{T_*}{\kappa}\left[\ln\frac{x_3}{z_0} - \psi_\theta(x_3/L)\right]; \tag{6.43}$$

$$\Delta\overline{r} = \overline{r} - \overline{r}(z_0) = \frac{r_*}{\kappa}\left[\ln\frac{x_3}{z_0} - \psi_r(x_3/L)\right]; \tag{6.44}$$

where u_s is the surface current (assumed to be in the direction of the surface stress) and

$$\psi_\alpha = \int_0^{x_3/L} \frac{1-\phi_\alpha(\xi)}{\xi}\,d\xi; \qquad \alpha = m,\ \theta,\ r. \tag{6.45}$$

We have assumed that the roughness length z_0 fulfills the condition $|z_0/x_3|<<1$, which is normally the case over the sea, and, for convenience, that $\phi_\alpha(0)=1$.

It should be pointed out that z_0, $\overline{T}\,(z_0)$, and $\overline{r}(z_0)$ are integration constants that can be found by curve fitting when profiles are measured. They are not necessarily connected in a simple way with the surface state (e.g., r.m.s. wave height), the sea surface temperature T_s, and the sea surface specific humidity r_s (the saturation value of r at temperature T_s). However, if we make the simple *ad hoc* assumptions that

$$\overline{u_3'\theta'} = a_\theta u_{*0}(T_s - \overline{T}'(z_0)) \text{ and } \overline{u_3'r'} = a_r u_{*0}(r_s - \overline{r}(z_0)), \tag{6.46}$$

where a_θ and a_r are constants, then we obtain

$$\Delta\overline{T}_s = T_s - \overline{T}' = -\frac{T_*}{\kappa}\left[\ln\frac{x_3}{z_{0\theta}} - \psi_\theta(x_3/L)\right],$$

and (6.47)

$$\Delta\overline{r}_s = r_s - \overline{r} = -\frac{r_*}{\kappa}\left[\ln\frac{x_3}{z_{0r}} - \psi_r(x_3/L)\right],$$

with $z_{0\theta}=z_0\exp(-\kappa/a_\theta)$ and $z_{0r}=z_0\exp(-\kappa/a_r)$.

Equations (6.39)-(6.47) with suitable choices for the ϕ functions have been used by a rather large number of research workers, recently by Krügermeyer (1975); Hsu (1974b); Pond, et al. (1974); Dunckel, et al. (1974); Badgley, et al.

(1972), and Paulson, et al. (1972).

Due to the typically rather small mean gradients over the sea, great care must be taken in order to obtain usable profile measurements. Measurements from buoys, ships, and other moving or bulky platforms should be carefully examined for possible effects of the motion and aerodynamic interference between structure and flow.

6.7 Other methods

The Monin-Obukhov similarity theory predicts that

$$\sigma_w/u_{*o} = f(x_3/L), \tag{6.48}$$

where $\sigma_w = (\overline{u_3'^2})^{1/2}$, and $f(y)$ is a unique function of the argument. Merry and Panofsky (1976) suggest that

$$\sigma_w/u_{*o} = 1.3[\phi_m - 2.5\, x_3/L]^{1/3}. \tag{6.49}$$

Hence, if σ_w of the vertical velocity component is measured, which requires an instrument with a relatively fast response but only moderate sampling rates, then the stress may be computed from Eq. (6.49).

Along the same lines, we may assume that the vertical velocity spectrum $S_{33}(\omega)$ obeys

$$\frac{\omega S_{33}(\omega)}{u_{*o}^2} = g_1\left(\frac{\omega x_3}{\overline{u}_1}, \frac{x_3}{L}\right), \tag{6.50}$$

where ω is the frequency and $g_1(x,y)$ is a unique function of the arguments. In general, $\omega S_{33}(\omega)$ will show a relatively broad maximum, the value of which we shall denote $\hat{S}$. It follows that

$$\frac{\hat{S}}{u_{*o}^2} = g_2\left(\frac{x_3}{L}\right). \tag{6.51}$$

Moravek, et al. (1976) suggest

$$\frac{\hat{S}}{u_{*o}^2} = 0.43[\phi_m - 2x_3/L]^{2/3}. \tag{6.52}$$

The latter method is in some ways attractive because the frequency $\hat{\omega}$ at which $\omega S_{33}(\omega)$ has its maximum will normally lie in the range

$$\frac{0.2\, \overline{u}_1}{x_3} < \hat{\omega} < \frac{\overline{u}_1}{x_3}. \tag{6.53}$$

The determination of $\hat{S}$ thus involves far fewer demands on response time and sampling rate than do the direct and the dissipation method.

However, the scatter displayed in the two papers mentioned above indicates that the stress can hardly be evaluated with greater accuracy than ±50%, say, if one assumes that the directly measured stresses are correct.

6.8 Bulk parameterization

A great deal of effort has gone into attempts to parameterize the surface fluxes in terms of more easily obtained quantities. We shall concern ourselves only with the so-called bulk aerodynamic approach, in which the fluxes are related to mean flow quantities at standard height (normally 10 m) and surface characteristics by means of bulk exchange coefficients C_m, C_θ and C_r:

$$\left.\begin{aligned} u_{*o}^2 &= C_m(\overline{u}_1 - u_s)^2 \\ \overline{u_3'\theta'} &= C_\theta(\overline{u}_1 - u_s)(T_s - \overline{T}') \\ \overline{u_3'r'} &= C_r(\overline{u}_1 - u_s)(r_s - \overline{r}) \end{aligned}\right\} \qquad (6.54)$$

By use of Eqs. (6.42) and (6.47) we obtain

$$C_\alpha = \frac{\kappa^2}{[\ln \frac{x_3}{z_{0\alpha}} - \psi_\alpha]\,[\ln \frac{x_3}{z_0} - \psi_m]}, \qquad (6.55)$$

which shows three things: 1) that if $z_{0\alpha}$ and z_0 depend on the sea state, then we should expect the exchange coefficients to depend on wind speed (stress) and the characteristics of the surface wave spectrum (e.g., r.m.s. wave height, dominant wave phase velocity and wave length); 2) that the exchange coefficients should be influenced by hydrostatic stability stratification; 3) that if $\phi_\theta=\phi_r$, as suggested earlier, then $C_\theta \approx C_r$ as we would expect $z_{0\theta} \approx z_{0r}$.

The assumptions leading to Eq. (6.55) are many, and the physical mechanisms which provide the transfer of momentum, heat and moisture in the layer closest to the surface are only partly understood, so Eqs. (6.54) should probably be regarded as dimensionally correct proposals which are to be tested experimentally. However, the prediction of the variation of C_α with stability given by Eq. (6.55), where ψ_α are determined from Eqs. (6.30) and (6.31), are not likely to be greatly in error. For typical stabilities encountered over the ocean, the variation of C_α with stability is of order ±10% (Deardorff, 1968; Paulson, 1969).

Over the last decade a great number of investigations have been carried out, and we shall not attempt a review of the work. Quite adequate collections of results can be found in Businger (1975), Friehe and Schmitt (1976), Pond (1972), Kraus (1972), and Kitaigorodskii (1970). Instead, we shall briefly mention a few typical results and some of the possible reasons for differences in the values of the exchange coefficients obtained.

a. Momentum flux

Reliable estimates of C_m vary generally from 1.1×10^{-3} to 1.7×10^{-3} with scatter which is typically $\pm 0.3 \times 10^{-3}$. In most cases there is no clear indication of a variation of C_m with wind speed (Krügermeyer, 1975; Dunckel, et al., 1974; Brocks and Krügermeyer, 1970). This may be due to the fact that most investigations cover very limited wind speed ranges (typically 3-10 m s^{-1}).

The same comment is pertinent in connection with a variation of C_m with stability. The range of stabilities in which measurements have been made has been narrow and near neutral. Krügermeyer (1975) analyzes profiles, "corrects" them for stability, and arrives consistently at $C_m = 1.34 \times 10^{-3}$, when the drag coefficient is reduced to neutral. Hedegaard and Busch (unpublished) have analyzed profiles measured along a tower of 25 m of water in the Kattegat and find a reduction of C_m of typically 60% in stable conditions, as compared to neutral and no clear variation with stability on the unstable side.

High wind speed results are few. Smith and Banke (1975) present results for wind speeds ranging from 7 to 21 m s^{-1} over breaking waves on the beach of Sable Island. They also review a number of earlier investigations and conclude that

$$C_m = (0.63 + 0.066\, \overline{u}_{10}) \times 10^{-3} \pm 0.23 \times 10^{-3}, \qquad (6.56)$$

with $\overline{u}_{10}$ in m s^{-1}, describes the results well. They point out that

$$C_m = \kappa^2 [\ln \frac{x_3}{z_0}]^{-2} \qquad (6.57)$$

fits the data as well as Eq. (6.56), if

$$z_0 = a u_{*0}^2 g^{-1}, \qquad (6.58)$$

$\kappa=0.4$, and $a=1.44 \times 10^{-2}$, which is very close to Charnock's (1958) suggestion of $a=1.23 \times 10^{-2}$ (see Kitaigorodskii, 1970 for a summary of values).

The analysis by Hedegaard and Busch of the Kattegat profiles yields

$$C_m = (0.64 + 0.14\, \overline{u}_{10}) \times 10^{-3} \pm 0.29 \times 10^{-3}, \qquad (6.59)$$

corresponding to $a=.217$ in Eq. (6.58) with $\kappa=0.35$. In the light of the results of others this is an inexplicably strong dependence of C_m on wind speed. If it is correct, it may be due to the long, unobstructed fetch of 135 km and the absence of swell.

Kondo (1975) proposes that the aerodynamic roughness of the sea is closely related to the roughness Reynolds number $h_p u_{*0}/\nu$, where h_p is a representative dimension of the sea surface irregularities associated with the high frequency ocean waves. On the basis of Eq. (6.55) he presents an analysis of the influence of density stratification on the bulk exchange coefficients, and suggests that in neutral stratification the drag coefficient is well approximated by

$$C_m = (0.87 + 0.067\,\overline{u}_{10}) \times 10^{-3} \text{ for } 2 < \overline{u}_{10}\ 8 \text{ m s}^{-1};$$

$$C_m = (1.2 + 0.025\,\overline{u}_{10}) \times 10^{-3} \text{ for } 8 < \overline{u}_{10} < 25 \text{ m s}^{-1}; \qquad (6.60)$$

$$C_m = 0.073\,\overline{u}_{10} \times 10^{-3} \text{ for } \overline{u}_{10} > 25 \text{ m s}^{-1}.$$

Davidson (1974) analyzes data from Bomex and finds

$$C_m = \kappa^2[\ln \frac{x_3}{z_0} + 6.44\,\frac{x_3}{L}]^{-2}, \qquad (6.61)$$

with $z_0 = z_0' \exp(-0.13\ c/u_{*0})$, where $z_0' \simeq 7$ mm and c is the phase velocity of the dominant wave.

Kitaigorodskii, et al. (1973) find that z_0 is a strongly decreasing function of c/u_{*0} and thus that the sea becomes very much smoother as the sea waves change from wind waves to swell. Hsu (1974a) arrived at $z_0 = (2\pi)^{-1}\,\zeta\,(c/u_{*0})^{-2}$, where ζ is the wave height. Stewart (1974) suggests that a in Eq. (6.58) should be a function of c/u_{*0}. Baines (1974) presents a theoretical analysis indicating that C_m for shallow water waves should increase with wind speed to a maximum and then decrease, which theory is in agreement with the observations by Hicks, et al. (1974).

Thus, we find that there are conflicting theories and conflicting data. There is little doubt that in order to properly account for scatter and variation of C_m, the surface wave characteristics must be included more comprehensively in the analyses and to a much larger extent than has been the case hitherto.

b. Humidity and heat flux

Friehe and Schmitt (1976) have compiled data sets from several sources and conclude that the moisture flux or latent heat flux is adequately parameterized by

$$C_r = 1.32 \times 10^{-3} \pm 0.07 \times 10^{-3}. \qquad (6.62)$$

They suggest that the exchange coefficient for humidity transfer is larger than the coefficient for sensible heat.

Kondo (1975) finds $C_r \simeq C_\theta$ and that in near-neutral stability conditions the value increases from around 1.08×10^{-3} for $\overline{u}_{10}=2$ m s^{-1} to a constant value of about 1.26×10^{-3} for $\overline{u}_{10}>8$ m s^{-1}. In case of light winds, the effect of stability is found to be considerable even when the sea-air temperature difference is small.

Pond, et al. (1974) suggest that $C_r \simeq C_\theta$ with a great deal of scatter and that the value is about 1.5×10^{-3}. In their data collection C_r varies from

1.23×10^{-3} to 1.47×10^{-3} so Eq. (6.62) appears to be appropriate.

The exchange coefficient C_θ for sensible heat flux shows considerably more scatter and is in various ways problematic.

In strong wind, Smith and Banke (1975) obtain

$$C_\theta = 1.46 \times 10^{-3} - \frac{0.4 \pm 14}{\Delta \overline{T}_s \overline{u}_{10}} \times 10^{-3}, \tag{6.63}$$

with $\Delta \overline{T}_s \overline{u}_{10}$ in K m s^{-1}. Wucknitz (1974) finds for $|\Delta \overline{T}_s \overline{u}_{10}| \geq 4.5$ K m s^{-1} that $C_\theta \simeq 1.5 \times 10^{-3}$, in good agreement with Krügermeyer (1975), who by profile analysis obtains an average of $(1.42 \pm 0.28) \times 10^{-3}$ for neutral conditions. The latter author also finds that when $|\Delta \overline{T}_s \overline{u}_{10}| \leq 4.5$ K m s^{-1} then the fluxes obtained by the profile method are distinctly smaller than those obtained by the direct mehtod, and parameterization is not possible. In this case Wucknitz finds wide discrepancy between the direct and the dissipation method, which discrepancy may be caused by cold spikes such as those mentioned in Section 6.5.

Pond, et al. (1974) report C_θ values that vary from 10^{-3} to 5.7×10^{-3}. The high values are probably due to anomalous cold spikes, which are discussed in some detail by Friehe and Schmitt (1976), who summarize their analysis of 152 determinations of sensible heat fluxes by

$$\overline{u_3'\theta'} = A + C_\theta \overline{u}_{10} \Delta \overline{T}_s \qquad [\text{K m s}^{-1}], \tag{6.64}$$

with $A = 2.6 \times 10^{-3}$ and $C_\theta = 0.86 \times 10^{-3}$ for $\Delta \overline{T}_s \overline{u}_{10} < 0$;
$A = 2 \times 10^{-3}$ and $C_\theta = 0.97 \times 10^{-3}$ for $0 < \overline{T}_s u_{10} < 25$;
$A \simeq 0$ and $C_0 = 1.46 \times 10^{-3}$ for $\Delta \overline{T}_s \overline{u}_{10} > 25$;
$A = 1.2 \times 10^{-3}$ and $C_\theta = 1.41 \times 10^{-3}$ (all data).

The small positive value of the heat flux for $\Delta \overline{T}_s \overline{u}_{10} = 0$ has also been observed by Dunckel, et al. (1974), Müller-Glewe and Hinzpeter (1974), and Hasse (1971).

An additional source of uncertainty in the bulk parameterization of heat and humidity fluxes is that the sea surface temperature T_s in Eq. (6.54) is not easily measured. Instead, the bucket temperature is used or some other temperature representative of the well mixed layer below the surface. The suppression of turbulent transfer at the air sea interface may cause strong temperature gradients at the interface, and hence a difference between the bucket temperature and the appropriate skin temperature (Saunders, 1973a; Paulson and Parker, 1972; Hasse, 1970).

6.9 Concluding remarks

Other parameterization schemes have been used, such as the geostrophic drag and exchange coefficients (Clarke and Hess, 1975; Hasse and Dunckel, 1974; Hasse and Wagner, 1971). However, parameterization by the bulk aerodynamic method appears to be the most promising and a quite adequate method in

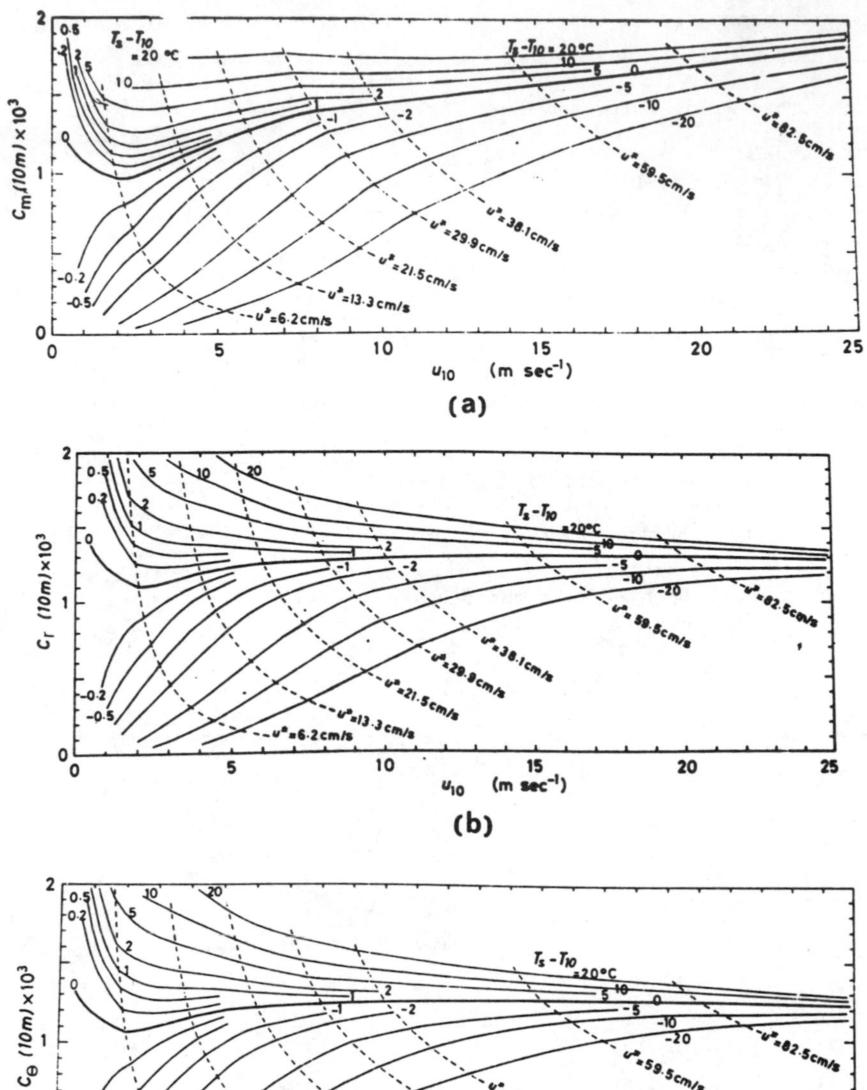

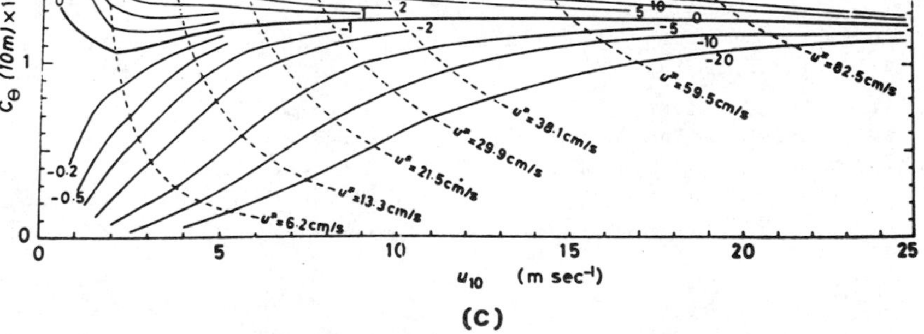

Fig. 6.1 - Bulk transfer coefficients at 10 m reference height for (a) momentum, (b) latent heat, and (c) sensible heat for varying winds and stabilities. From Kondo (1975).

most cases. Figure 6.1 (from Kondo, 1975) shows typical variation of the bulk coefficients with respect to wind speed, surface stress, and surface temperature difference.

Problems remain. It is still uncertain to what extent and by what mechanism the exchange coefficients vary with wind speed, especially in very high winds; neither is it clear how and to what extent extreme stability conditions affect the exchange. The heat exchange is rather poorly modelled at numerically small $\Delta\overline{T}_s\overline{u}_{10}$. What is needed is partly more information about the heat transfer immediately across the interface, so that the bucket temperature which is normally used can be replaced by a more appropriate surface temperature, and partly better understanding of the interaction between the surface waves and the turbulent field above them, and of tbe radiative and evaporative processes in the lowest layer of the atmosphere.

Acknowledgments: I am indebted to Professor C. A. Paulson of Oregon State University for his help and valuable suggestions during the preparation of the manuscript. I wish to thank Dr. S. D. Smith of Bedford Institute of Oceanography for his useful comments.

Chapter 7
ENTRAINMENT

O. M. Phillips

7.1 Introduction

Entrainment is a general term describing the set of processes involved when turbulence spreads into adjacent non-turbulent fluid. It is a characteristic property of virtually all free turbulent flows - ones not contained by boundaries - and is involved in the spread of turbulent wakes, of patches of clear air turbulence at high altitude, and in the deepening of the upper mixed layer of the ocean under the continued influence of wind or convective stirring. In some situations, the density of the fluid inside and outside the turbulent region is uniform, but in the ocean or the atmosphere, the mean density or mean buoyancy in the turbulent region is often different from that of the fluid outside. When an oceanic mixed layer develops in a region in which the ambient fluid is initially statically stable, a pycnocline develops at the bottom of the mixed layer and entrainment must proceed against the ever increasing density difference between the mixed layer itself and the underlying water.

The processes of entrainment provide an important element of irreversibility in the nutrient supply to the upper layer associated with upwelling. Colder, denser, nutrient rich water is brought towards the surface by the dynamical processes described elsewhere in this book (see O'Brien, et al., Chapter 11). If there were no entrainment associated with upwelling, the deep water would subside when, say, the longshore wind disappeared and there would be no net entrainment of nutrient to the surface water. In fact, however, the elevation of the pycnocline will be shown to enhance entrainment, so that the biological consequences of upwelling are the result of both the vertical movement and the enhanced entrainment. In order to understand these phenomena as well as to predict the upper ocean structure far from boundaries, the nature of the entrainment process must first be understood in order that a realistic parametrization be introduced into the analytical and numerical models being developed.

7.2 Entrainment in a homogeneous fluid

As long ago as 1943, it was realized by Corrsin that in a homogeneous fluid, the interface between turbulent and non-turbulent fluid, though massively convoluted, is remarkably sharp. A probe, placed near the outside edge of a turbulent wake, records intermittent regions of high turbulent activity as the billows move past the observation point, interspersed with intermittent intervals of much lower activity, the relatively abrupt transition between the two occurring as the turbulent interface moves past the probe. The volume of turbulent fluid grows continuously as the turbulence encroaches upon the neighboring fluid. A distinction must be drawn between two velocities associated with this entrainment process. The first, w_e , the entrainment velocity, is defined as the average flow of fluid per unit area normal to the mean position of the interface, or the rate at which the <u>mean</u> position of the interface advances with time into the non-turbulent fluid. But the interface is extensively convoluted and each element of area on the surface is itself advancing into the non-turbulent region. The velocity of advance, u_A, is defined as the velocity with which this element of area on the surface is advancing, normal to itself, into the adjacent, non-turbulent fluid. If the instantaneous position of the interface is specified by $\zeta(x,y,t)$, then w_e in the z-direction and u_A are related by

$$w_e = u_A[1+(\nabla_H\zeta)^2]^{1/2}. \tag{7.1}$$

Now, how does an element of fluid just outside the turbulent interface become turbulent? Turbulence is characterized by a random distribution of vorticity and, as Corrsin pointed out, vorticity can be acquired in the first instance only as a result of viscous diffusion of vorticity from neighboring vortical fluid elements. Once vorticity is present, however, it can be amplified by the straining set up as a result of the neighboring turbulence. The speed of advance u_A must therefore be determined by a balance between these two processes - the viscous diffusion of vorticity and its subsequent amplification as a result of straining, and must therefore, on dimensional grounds, depend upon the fluid viscosity ν and the mean-square rate of strain close to the interface. As the latter is related directly to the dissipation ε_0, the speed of advance u_A must therefore be proportional to the Kolmogorov microscale $(\nu\varepsilon_0)^{1/4}$. Similarly, the thickness of the interface must be proportional to the Kolmogorov length scale $(\nu^3/\varepsilon_0)^{1/4}$. On the other hand, in homogeneous flows, the entrainment velocity w_e is found to scale with the length scale of the energy containing eddies and this is considerably larger than u_A; in order that these findings be consistent, Eq. (7.1) shows that necessarily the surface must be extensively convoluted on a large scale.

Some aspects of the kinematics of the advancing turbulent front can be demonstrated by supposing that the large-scale eddies, responsible for the massive convolutions, are statistically independent of the small-scale motions near the interface which determine the statistics of the local strained field in this vicinity. If the large-scale velocity field $\underset{\rightarrow}{u}$ is supposed given and the advance of the interface is parameterized by u_A, proportional to the Kolmogorov microscale, then the location of the interface is specified by

$$\frac{d\zeta}{dt} = u_\zeta + w_e = u_\zeta + u_A[1+(\nabla_H\zeta)^2]^{1/2} \tag{7.2}$$

where u_ζ represents the fluid velocity normal to the mean position of the interface, at $z=\zeta$, as given by Phillips (1972). It does not seem possible to find a general solution to this equation for a given $u_\zeta(\underset{\rightarrow}{x}, t)$, but several useful conclusion have emerged from consideration of particular examples. It is found that indeed the structure of the interface is self adjusting, in which the extent of the convolutions develops until the balance described by Eq. (7.1) obtains. It is also found that the billow structure that develops is associated with the larger eddies which move (relative to the outside, non-turbulent fluid) with a velocity less than u_A. Only the relatively slowly moving eddies distort the interface substantially into billows; faster moving eddies, imbedded deeper in the turbulence, produce only a transient perturbation of the surface.

If the ocean were of homogeneous density and initially at rest, the development of a surface mixed layer as a result of wind action would involve entrainment of this kind. If the wind stress is τ_0, then the scale velocity for the turbulence in the water is $u_* = (\tau_0/\rho)^{1/2}$. This friction velocity determines the velocity scale for the energy containing eddies of the motion and so for the entrainment rate w_e. With a constant wind stress, the thickness of the layer which is being stirred would increase linearly with time until Coriolis effects became important or until this mixed layer attained the depth of any pre-existing homogeneous region.

If, however, the initial state is statically stable, as the entrainment proceeds, the density distribution in the mixed layer is made more nearly homogeneous by the turbulent mixing and an appreciable density difference between the turbulent and non-turbulent regions develops. As soon as this happens, some new effects arise. A sharp pycnocline or thermocline develops in which the density distribution is statically stable. However, if there is sufficient mean shear that the local Richardson number drops below 1/4, dynamical instability can lead to the formation of rolls or billows, whose internal mixing is enhanced by the overturned density distribution. Beautiful and revealing experiments on these processes have been made by Thorpe (1971). On the other hand, the larger eddies of the turbulence above, being most strongly influenced by the mean buoyancy field, are less able to produce massive convolutions of the interface as the density difference increases. Internal waves may be generated on the interface; if the fluid below has sufficient stratification, energy may be radiated downwards by bodily internal waves. Finally, if the density difference is sufficiently great that the kinetic energy of the turbulent eddies is unable to lift more than an occasional wisp of dense fluid into the less dense fluid above, the rate of entrainment virtually ceases. This variety of new physical phenomena is very difficult to untangle in field observations and has not yet yielded to theoretical study, but it has been the subject of laboratory experimentation.

7.3 Stirring experiments

A number of interesting experiments of two types have been made to clarify the processes of entrainment against a density jump. The first kind, pioneered by Rouse and Dodu (1955) and developed greatly by Turner (1968) and Thompson (1969) involves the mechanical stirring of the upper part of a container that is initially stratified, either continuously or with two miscible fluids. A sharp interface develops between the turbulent upper layer and the more quiescent fluid below and gradually moves downwards as the entrainment proceeds, and the entrainment velocity w_e is measured simply by the speed of advance of this front. Turner's authoritative monograph (1973) discusses the results of these experiments in some detail. In the later experiments of this kind, hot film techniques were used to measure the root-mean-square value u_1 of the horizontal component of the turbulent velocity near the surface and the integral length scale l; if the entrainment velocity w_e is a function of these turbulent parameters, as well as the buoyancy difference $g\Delta\rho/\rho_r$ across the interface, then on dimensional grounds,

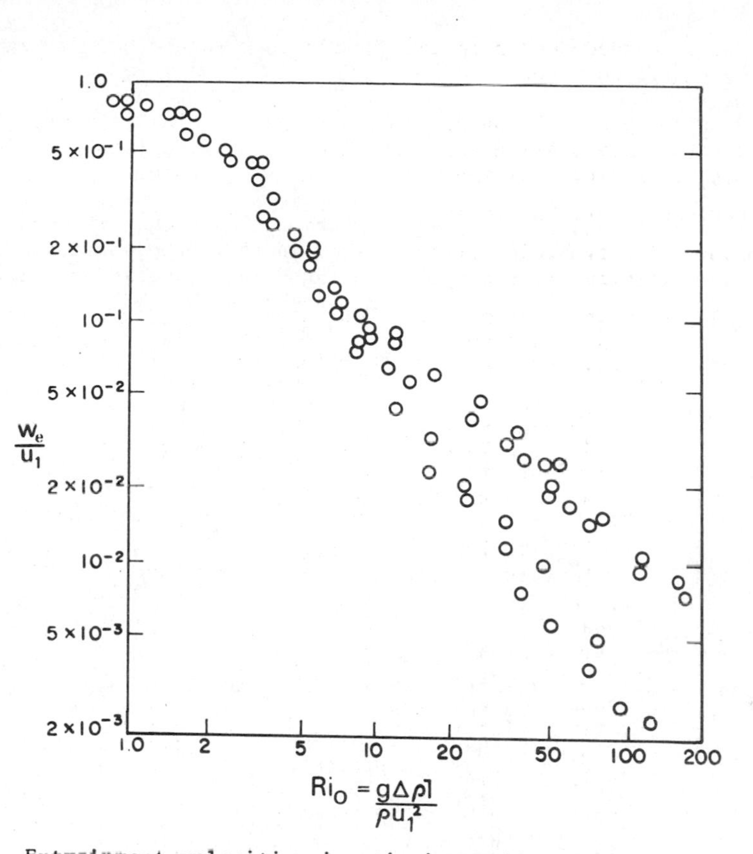

Fig. 7.1 - Entrainment velocities in stirring grid experiments, after Turner (1973). The solid points represent results when the density difference is the result of temperature differences and for the open circles, salinity differences.

$$\frac{w_e}{u_1} = f\,(Ri_o), \tag{7.3}$$

where

$$Ri_o = \frac{g\Delta\rho l}{\rho_r u_1^2} \tag{7.4}$$

is an overall Richardson number. Some of the results obtained by Turner and Thompson are illustrated in Fig. 7.1. In experiments in which the buoyancy difference was maintained by temperature differences between the upper and the lower fluid, the entrainment rate decreases approximately as $(Ri_o)^{-1}$, but when they are the result of salinity differences, the decrease at large Richardson numbers is more rapid, approaching $(Ri_o)^{-3/2}$. The differences suggest that the relation (7.3) is incomplete, that molecular processes in the experiments are important, and that

$$\frac{w_e}{u_1} = f\,(Ri_o, Pe) \qquad (7.5)$$

where Pe is a Péclet number, $Pe=u_1 l/\mu$, μ being the diffusivity of either heat or salt in water.

Visual observations of these experiments indicate that the massive surface convolutions are generally supressed and that the entrainment is the result of wisps or thin sheets of fluid being drawn into the turbulent region as a result of the scouring of the turbulent eddies above. In such experiments there is, of course, no difference in mean velocity across the interface as one might expect in the ocean when the turbulence in the upper layer is generated by wind - it will appear later that shear across the interface greatly augments the entrainment rate. It is likely that the results of these experiments are most pertiment to the oceanographic situation in which turbulent (convective) motions are produced in the surface layer by cooling rather than by wind action. In seeking to apply the results of these experiments to field situations, however, there does remain the difficulty of relating the intensity and length scales of the turbulent motion to the overall parameters of the flow, mainly the layer depth and the rate of buoyancy flux.

7.4 Applied stress experiments

A second type of experiment (Kato and Phillips, 1969; Kantha, 1975; Kantha and Phillips, 1976) seeks to model rather more closely the oceanic entrainment that occurs as a result of surface wind stress. The surface stress, communicated to the water, generates both a mean flow and a distribution of turbulent intensity. If the surface stress is applied to a rectangular laboratory tank, no matter how long, the distribution of mean velocity and turbulent intensity is modified by the return flow required by the presence of the ends of the tank. To avoid this, the experiments make use of an annular tank filled with stratified fluid initially at rest, the surface stress being applied by means of a rotating screen. In the Kato-Phillips experiment, the Brunt-Väisälä frequency N was initially constant. The impulsively started screen initially generates billows of the kind observed by Thorpe (1971), but within a second or two these break down, the upper layer is fully turbulent and the entrainment rate is measured as the experiment proceeds. If the entrainment velocity w_e is independent of molecular processes, then in an ideal experiment, w_e must depend on the external parameters specifying the flow, namely, the friction velocity, the layer depth h and the buoyancy difference across the interface $g\Delta\rho/\rho_r$. Consequently

$$\frac{w_e}{u_*} = f\,\left(\frac{g\Delta\rho h}{\rho_r u_*^2}\right) = f(Ri), \qquad (7.6)$$

where Ri is a different overall Richardson number based upon the friction velocity and the layer depth. In an actual experiment, not only molecular processes but also the influence of aspect ratio (the ratio of layer depth to width, which is zero in the ocean) and possibly variations in inertia (neglected in the Boussinesq approximation) must be considered. The Kantha-Phillips experiment involves a simple but significant improvement. In it, the tank initially contains two homogeneous layers with a density difference $(\Delta\rho)_0$. As the entrainment proceeds, conservation of mass requires that $\Delta\rho \cdot h$ is constant so that in a given experiment, and with u_* held fixed, the overall Richardson number is constant. In the absence of extraneous effects, the entrainment velocity w_e is also constant. The experiment is then self checking, departures from linearity in $h(t)$ as a function of time indicating the onset of unwanted side effects.

The structure of the interface in these experiments is seen most clearly by adding dye to the upper layer. In an experiment in which Ri is relatively small (of order 50) the interface is still very convoluted as it is in homogeneous entrainment, with not only the turbulent eddies scouring the lower fluid, but also with occasional bursts of more regular vortices indicative of a coherent shear instability near the interface even in the presence of turbulence. At a larger overall Richardson number (500 or so) the appearance of the interface is quite different - the convolutions and eddies have gone and it looks more like an air-water interface ruffled (in slow motion) as if by the wind, but with occasional wisps of the lower fluid being drawn up. The entrainment ratio measured as a function of Richardson number in the Kantha-Phillips experiment is shown in Fig. 7.2. It is evident that the function f in Eq. (7.6) is not a simple power law over an extensive range as is found in the experiments with oscillating grids. The Kato-Phillips measurements, with a fair scatter and limited to $15 < Ri < 250$ had suggested that $w_e/u_* \propto Ri^{-1}$, which is the approximate slope in this range in Fig. 7.2, but it is evident from the later experiments that at higher Richardson numbers, the decrease is more rapid.

The earlier Kato-Phillips results also lie rather below those shown in Fig. 7.2. One possible reason for this is that, in the older experiments, the underlying fluid was stably stratified so that energy could be radiated down from the surface as internal waves, while in the later experiments, with two homogeneous layers, this is not possible.

An Ri^{-1} law for turbulent entrainment has considerable appeal on theoretical grounds. The entrainment and subsequent mixing of denser fluid increases the potential energy of the system. With a relation of this kind it can be shown simply that the rate of increase of potential energy is proportional to $\rho_r u_*^3$, and so proportional to the rate of energy input from the wind. This point was made by Turner (1968) while Long (1975), in a critical discussion of various sets of experiments, argues plausibly for a relation of this kind. However, it clearly cannot hold as $Ri \to 0$ and Fig. 7.2 shows that it does not hold either as $Ri \to \infty$.

There is, indeed, little reason to expect a simple power law form for Eq. (7.6) in view of the apparently different physical processes involved at zero, low and high Richardson numbers. If the buoyancy difference across the surface is sufficiently large, the interface is virtually flat. Wisps of fluid can be entrained by the turbulence only if the kinetic energy of the eddies, of order $\rho_r u_*^2$ is sufficient to supply the potential energy needed to lift fluid elements through the interface. If δ is the interfacial thickness, then, for fully turbulent entrainment

$$\rho_r u_*^2 \gtrsim g\Delta\rho\delta,$$

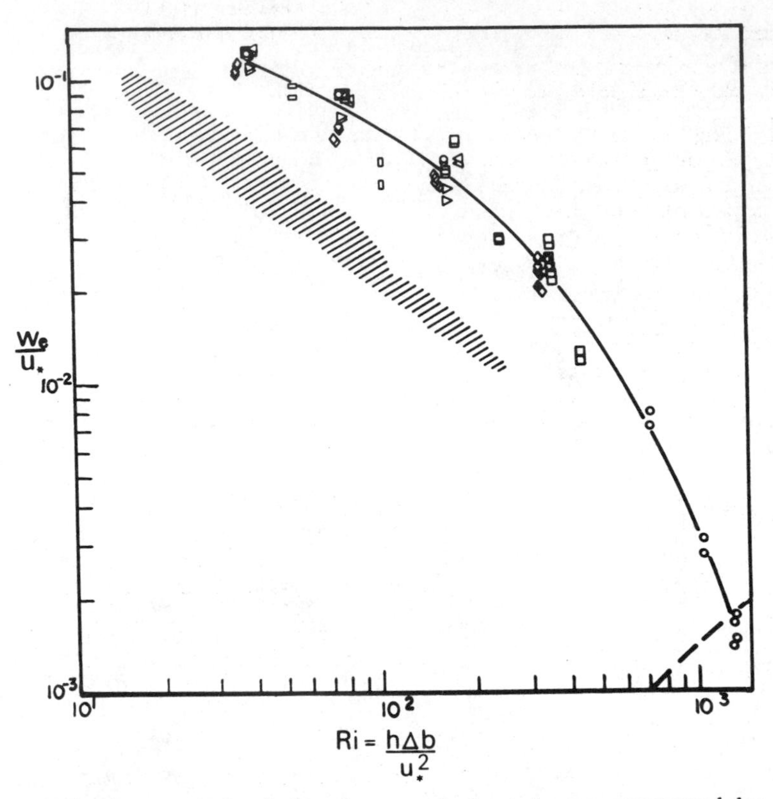

Fig. 7.2 - Entrainment velocities in a turbulent layer generated by an applied stress. The experimental points are from Kantha (1975) and cover a considerable range of applied stress, initial density difference and molecular diffusivity. The shaded area represents the region in which the Kato-Phillips results lay.

or

$$Ri = \frac{g\Delta\rho h}{\rho_r u_*^2} < \frac{h}{\delta} . \qquad (7.7)$$

When Ri is large and the interface is flat, a lower limit to δ is the diffusion thickness μ/w_e, where μ is the diffusivity and w_e the speed of advance of the interface, so that

$$Ri \leqslant \left(\frac{w_e}{u_*}\right)\left(\frac{u_* h}{\mu}\right) . \qquad (7.8)$$

In the Kantha-Phillips experiments $u_* h/\mu \sim 7 \times 10^5$. The precipitous decline in the curve f(Ri) as the broken line is approached probably reflects the increasing rarity of turbulent detachment of the underlying fluid. In the ocean, h is characteristically 50m or 10^3 times that in the experiment, so that fully turbulent entrainment can be expected up to values of the overall Richardson number of order 10^8 if the stratification is the result of salinity variations, or 10^7 if it results from temperature. Characteristic values in the ocean are perhaps 10^3 to 10^5.

7.5 An application to the oceanic mixed layer

Although a fuller appreciation of the detailed properties of the oceanic mixed layer demands both more detailed experiments and analysis, the results of the above experiments can be used to provide a simple parameterization for a slab model of the kind offered by Pollard, Rhines, and Thompson (1973). With the assumption that, in the upper mixed layer, vertical momentum exchange is sufficiently vigorous that the velocity is independent of depth, the horizontal components of the momentum equation can be integrated throughout the layer. If the lower level of integration is taken at a fixed depth below the layer $h'>h(t)$, then $u=0$ when $|z|>h(t)$ and

$$\int_{-h'}^{0} \frac{\partial u}{\partial t}\, dz = \frac{\partial}{\partial t}\int_{-h'}^{0} u\, dz = \frac{\partial}{\partial t}(uh).$$

Consequently

$$\left.\begin{aligned} \frac{\partial(uh)}{\partial t} - fvh &= \rho_r^{-1}\tau^x \\ \frac{\partial(vh)}{\partial t} + fuh &= \rho_r^{-1}\tau^y \end{aligned}\right\} \tag{7.9}$$

where $f=2\Omega \sin \phi$ is the Coriolis parameter. Note that the Reynolds stress at the base of the mixed layer is non-zero but serves to accelerate the entrained fluid to the speed u, and

$$\rho_r^{-1}(\tau^x)_{-h} = u\,\frac{\partial h}{\partial t}\,.$$

With a prescribed surface stress, these two equations do not themselves specify $u(t)$, $v(t)$ and $h(t)$; a further statement concerning the rate of entrainment is required and this the experiments can provide. Pollard, Rhines, and Thompson assume that a different overall Richardson number, defined as

$$Ri' = \frac{g\Delta\rho h}{\rho_r(u^2+v^2)}\,,$$

always has the value unity, viewing the layer as a whole as marginally dynamically unstable. However, the experiments indicate that the upper layer is completely turbulent, any marginal stability being local and concerned with the mean thickness of the transition region between the turbulent and non-turbulent flow, rather than the layer depth as a whole. If, in place of this assumption, one interprets u_* in Eq. (7.6) as parameterizing the velocity difference across the interface, the quantity relevant to entrainment, then in the ocean

$$\dot{h} \equiv w_e = (u^2+v^2)^{1/2} f(Ri'), \tag{7.10}$$

where Ri' is as given above. In fully turbulent entrainment, the Kantha-Phillips experiment can be fitted approximately by

$$\frac{w_e}{u_*} \cong 6\left[\frac{g\Delta\rho h}{\rho_r u_*^2}\right]^{-1},$$

and the mean velocity difference across the interface was about 10 u_*. Consequently,

$$\dot{h} \sim 6\cdot 10^{-3} \frac{\rho_r}{g\Delta\rho} \frac{(u^2+v^2)^{3/2}}{h}, \tag{7.11}$$

which is the relation required to close the system of Eqs. (7.9).

If a constant wind stress $\underset{\rightarrow}{\tau}=(\tau, 0)$ is applied from time $t=0$, then, from (7.9),

$$uh = \frac{\tau}{\rho_r f} \sin ft,$$

$$vh = \frac{\tau}{\rho_r f} (\cos ft - 1),$$

and

$$h^2(u^2+v^2) = 2\left(\frac{\tau}{\rho f}\right)^2 (1 - \cos ft). \tag{7.12}$$

If, further, the stratification is initially uniform with Brunt-Väisälä frequency N, the buoyancy difference when the mixing has proceeded to a depth h is

$$\frac{g\Delta\rho}{\rho_r} = \frac{1}{2} N^2 h,$$

so that from Eqs. (7.11) and (7.12)

$$\dot{h} \sim 12\cdot 10^{-3} \frac{(u^2+v^2)^{3/2}}{N^2h^2} = 10^{-1} \frac{\tau^3|\sin^3\frac{1}{2}ft|}{\rho_r{}^3 f^3 N^2 h^5}. \tag{7.13}$$

This equation can be integrated readily, leading to

$$h(t) = 0.9 \frac{u_*}{f^{2/3}N^{1/3}} \{2 - 3\cos(\tfrac{ft}{2}) + \cos^3(\tfrac{ft}{2})\}^{1/6}, \tag{7.14}$$

when $0 \leq ft \leq 2\pi$, and since $\tau = \rho_r u_*^2$. After one inertial period when $ft=2\pi$,

$$h = 1.1\, u_*(f^{2/3}N^{1/3})^{-1}. \tag{7.15}$$

If the turbulence in the upper layer continues to be maintained by the wind and the wind induced shear, the depth of the mixed layer continues to grow, but it does so much more slowly in view of the h^{-5} factor in Eq. (7.13). The ratio $u_*/f^{2/3}N^{1/3}$ can be taken as a convenient depth scale for the mixed layer. Pollard, Rhines, and Thompson's analysis, with its assumption that Ri'=const. gives $u_*/(fN)^{1/2}$ as this scale; although the physical basis for this is questionable it may be difficult to distinguish between the two on purely observational grounds.

In order to determine the distribution of mean velocity, density and perhaps turbulent intensity in the mixed layer, more elaborate models are necessary involving closure assumptions to relate the Reynolds stress to the properties of the mean flow, the turbulent intensity and the degree of stratification. These models are all rather *ad hoc* and may involve a choice of a large number of arbitrary constants. Some examples of this genre are to be found elsewhere in this book.

It should be emphasized that the use of Eq. (7.11) assumes that the entrainment is a local process depending on the velocity difference across the region at the base of the mixed layer. A more realistic simulation of oceanic entrainment may require the use of a large scale apparatus which itself rotates, so that Coriolis forces can be scaled properly. Possibly the only existing facility in which this could be done is the large (14m) rotating platform at the Institut de Mécanique of the Université Scientifique et Médicale de Grenoble and it is to be hoped that an appropriate experiment might soon be designed and performed.

Acknowledgement: It is a pleasure to acknowledge the support of the U. S. Office of Naval Research under Contract No. N00014-67-A-0163-0009 during the performance of this research and the preparation of this contribution.

Chapter 8

OBSERVATIONS AND MODELS OF THE STRUCTURE OF THE UPPER OCEAN

R. T. Pollard

8.1 Introduction

This book is primarily concerned with models of the upper ocean, but models and observations cannot be separated. Observations stimulate models. Models suggest observational tests to verify or disprove them. Progress in a field can be held up by a lack of either ingredient. I believe that a lack of observations is what is primarily hampering progress in upper ocean modelling, so that a survey of what is known observationally is particularly appropriate.

This paper is a combination of three talks given at Urbino on aspects of the structure of the upper ocean. The emphasis is on the mixing layer, though the seasonal thermocline is also mentioned in the section on inertial oscillations. My aim has been to describe what is known observationally, with close attention to the needs of those interested in modelling. Accordingly, the observations described in each section are complemented with a survey of models that attempt to duplicate the particular physical process under discussion.

8.2 Ekman theory and eddy viscosity

Ekman (1905) used the concept of a constant eddy viscosity K_ν to predict the Ekman spiral. He defined an Ekman depth $D_e=\pi/a$, where a equals $(\frac{f}{2K_\nu})^{1/2}$, as the depth at which the current has rotated through 180° and diminished in amplitude by a factor $\exp(\pi)$. Variations on Ekman's model incorporating different boundary conditions or viscosity as a function of depth have been derived by various authors. For example, Gonella (1970), in a variation applicable to the mixing layer, assumed a frictionless interface at a depth h (usually less than D_e) which he identified with the mixing layer/thermocline interface. As $h/D_e \to \infty$, the classic Ekman spiral is recovered, and as $h/D_e \to 0$ the motion becomes independent of depth and at 90° *cum sole* to the wind direction.

While rotation must be taken into account in considering diffusion of momentum, it can be ignored when considering diffusion of heat. By solving the diffusion equation with a fixed periodicity (usually the diurnal variation of temperature), "hodographs" similar to Ekman spirals can be found for the changes in temperature oscillation with depth (Fig. 8.1), where the angular differences represent the time lags between the oscillations at different depths. Several authors, seeking approximations to the above models in nature, have used such hodographs to estimate K_ν (Table 8.1).

a. Ekman spirals

Well defined Ekman spirals are very rarely observed in the ocean surface layer. Many factors mitigate against such observations. It is unlikely that K_ν is constant with depth or time, and it is usually large enough that $D_e>h$. The time-dependent problem contains inertial oscillations (Ekman, 1905) which dominate the mean at most depths. It is difficult to make reliable current measurements near the ocean surface (Saunders, 1976).

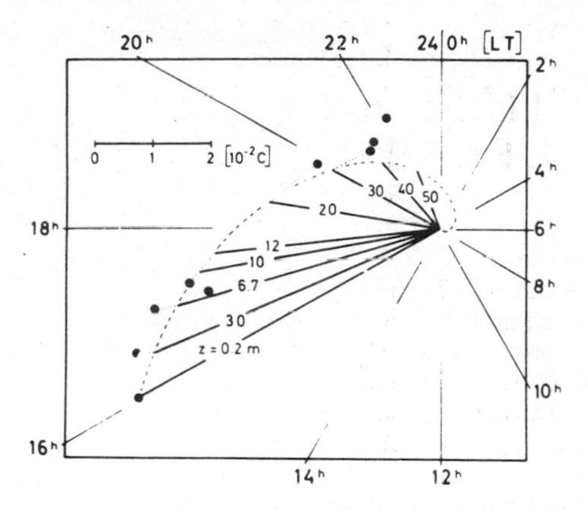

Fig. 8.1 - Harmonic dial of the diurnal variation of water temperature. Results of ATEX (dots) and solution vectors of the Ekman spiral with K_ν = 320 cm^2s^{-1} as a function of depth. (Numbers denote levels of measurements in meters.) (From Prümm, 1974).

TABLE 8.1

Estimates of Eddy Viscosity From Observations

ESTIMATES OF EDDY VISCOSITY FROM OBSERVATIONS

AUTHOR	EDDY VISCOSITY (cm^2/sec)	COMMENTS
	USING EKMAN THEORY	
Brennecke (1921)	160	sparse data, near South Pole under ice
Hunkins (1966)	24	under ice near North Pole
Swallow and Bruce (1966)	300 ± 200	near equator, after J. C. Swallow (personal communication)
Assaf et al.(1971)	150 and 225	
Gonella (1970)	120	July } graphic methods, using modified Ekman spiral theory
	500	October } graphic methods, using modified Ekman spiral theory
	USING DIURNAL OSCILLATION	
Bowden, Howe and Tait (1970)	146	0 - 4 m
	96	4 - 8 m
	47	8 - 12 m
Hoeber (1972)	420	upper 12 m, large variations (150 to 10,000) correlated with wind variations
Prümm (1974)	480	0 - 12 m
	265	20 - 50 m
	(320)	mean
Ostapoff and Worthem (1974)	20 - 120	

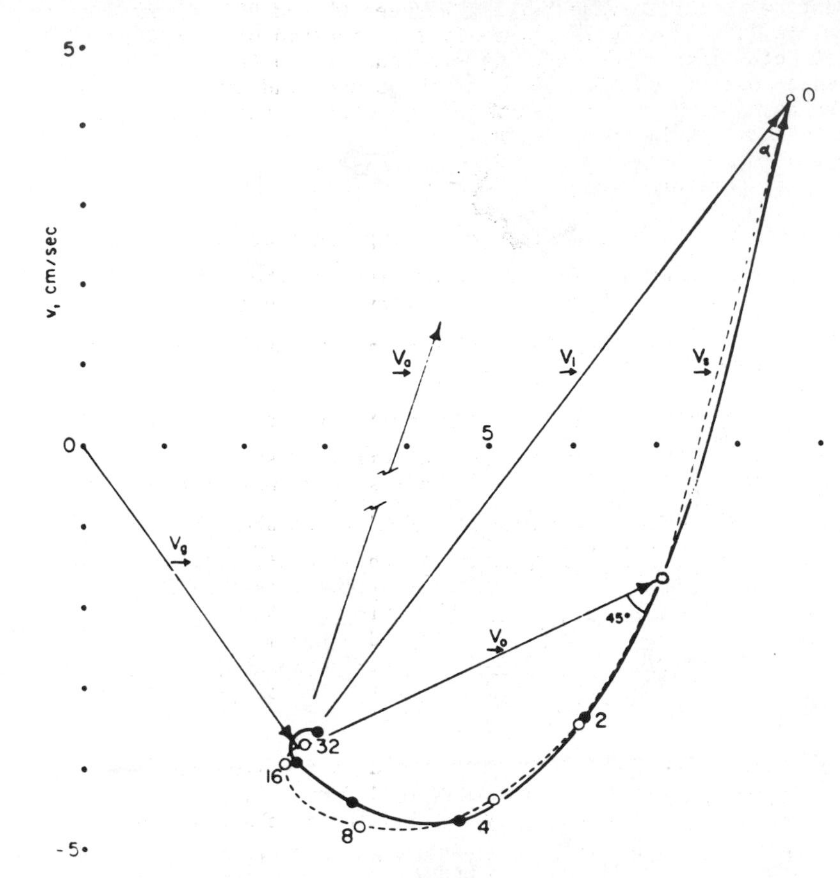

Fig. 8.2 - Comparison between mean observed hodograph for periods 8-13 (solid line and circles) and theoretical curve for surface boundary layer and Ekman layer with D_e=18 m (dotted line and open circles). Numbers refer to depths in meters below the base of the ice. (From Hunkins, 1966).

Although Brennecke's (1921) observations under the Antarctic ice have been interpreted to contain an Ekman spiral, the first and only convincing spiral was observed by Hunkins (1966) under the Arctic ice (Fig. 8.2). In the open ocean, Swallow and Bruce (1966) observed spiral structures in the steady trade wind situation off east Africa between 4° and 8°N. Apart from these three sets of observations, all at rather anomalous locations, the only other data I am aware of that show spiral-like structures are visual observations of dye patches made by scientists at Lamont Observatory (Katz, et al., 1965; Ichiye, 1967; Gordon, 1970; Assaf, et al., 1971) which show a tendency to a *cum sole* deflection with depth.

The lack of observed spirals is not, of course, a negation of Ekman theory. Ekman's conclusion that the total wind driven momentum flux is at right angles to the wind does not depend on any assumptions about the eddy viscosity. Further, any profile of $K_\nu(z)$ will lead to a spiral structure of some sort in the current, with the surface current between 0° and 90° *cum sole* to the wind. Thus, Durst's (1924) analysis of nearly two thousand observations of ship's drift from both hemispheres gives general support to Ekman theory, with its finding that, over a wide range of latitudes and wind conditions, mean near surface currents lie between 5° and 55° *cum sole* to the wind. But such analyses tell us little about $K_\nu(z)$.

b. Diurnal temperature cycles

Estimates of K_ν from the diurnal heating cycle are averages over one or preferably several diurnal cycles. In particular, they average daytime and nighttime situations, in which K_ν may be very different if strong heating is damping turbulent transfers during the day, while convection may be penetrating to the layer depth at night.

Bowden, Howe, and Tait (1970) computed K_ν over a seven day period in the North Atlantic by this technique during a period of strong heating (July) with a mean wind speed of 8 m s^{-1}. They found a strong dependence of K_ν on depth (Table 8.1), which implies time scales ($T=D_e^2/K_\nu$) of about 5, 12, and 25 minutes for the heat to penetrate from the surface to 4, 8 and 12 m respectively. Prümm (1974), using data from the Tropical Atlantic, also found a decay of K_ν with depth, obtaining the best fit with a value of 480 cm^2s^{-1} in the top 12 m, and 265 between 20 and 50 m. Ostapoff and Worthem (1974), using a free-drifting, sensitive (0.001°C) thermistor chain, were able to trace diurnal variations down to 30 m. They found a wide scatter of estimates of K_ν, but of order 100 cm^2s^{-1}.

Hoeber (1972), over an eleven day period in the Tropical Atlantic, found a mean eddy viscosity between 0 and 12 m of 420 cm^2s^{-1}, but noted that on shorter times K_ν varied enormously with wind speed, changing from 400 to 4000 cm^2s^{-1} for an increase of only 1 m s^{-1} in wind speed from 7 to 8 m s^{-1}.

TABLE 8.2

Time scale $T=L^2K_\nu^{-1}$ for ranges of L and K_ν

K_ν (cm^2/s) \ L (m)	10	20	30	40
50	6 hours	1 day	2 days	4 days
100	3 hours	12 hours	1 day	2 days
200	1.5 hours	6 hours	12 hours	1 day
500	.5 hours	2 hours	5 hours	10 hours

c. Are eddy viscosity observations useful?

While the above review is far from comprehensive, it is sufficient to show the weakness of eddy viscosity estimates. Nearly all estimates of K_ν (Table 8.1) lie in the range 50-500 cm^2s^{-1}. All we can conclude from this (Table 8.2) is that, on average, turbulence can stir the mixing layer down to a few tens of meters on time scales between a few hours and a few days. We learn nothing of the structure of the turbulence, or of the physical processes (e.g., convection, wave breaking, winds, current shear) that drive it. It is clear that K_ν varies greatly as the physical processes change. It tends to decrease with depth, and there is a suggestion from Hoeber's findings that vertical transfer processes can be speeded up greatly in high winds. While numerical modellers may be forced to use eddy viscosity parameterizations for want of better, the experimenter needs much more detailed measurements if he is to understand how the structure of the upper ocean is created and maintained.

8.3 Inertial oscillations in the upper ocean

It has been known for some years that intermittent oscillations at about the inertial frequency [$f = \sin(\text{latitude})/12\ \text{hr}^{-1}$] often dominate current records in many parts of the ocean. Webster (1968) gave a complete review of observations up to that date. More recently, it has become clear that the largest inertial oscillations occur in the mixed layer (Pollard, 1970a) and are predominantly wind-driven (Day and Webster, 1965; Hunkins, 1967; Gonella, 1971; Pollard and Millard, 1970).

a. Dispersion

Any transient forcing mechanism will generate a whole spectrum of internal waves, with frequencies between f and the Brunt-Väisälä frequency $N(z)$. The clue to the dominance of inertial and near inertial frequencies lies in their small energy propagation velocities relative to higher frequencies. For a wave of frequency ω and wave number $\underset{\rightarrow}{\kappa}=(k, l, m)$, the dispersion relation is

$$\omega^2 = \frac{k^2+l^2}{\underset{\rightarrow}{\kappa}^2} N^2 + \frac{m^2}{\underset{\rightarrow}{\kappa}^2} f^2 ; \tag{8.1}$$

$$\omega^2 = N^2 \sin^2\theta + f^2 \cos^2\theta ; \tag{8.2}$$

where θ is the angle between $\underset{\rightarrow}{\kappa}$ and the upward vertical. The group velocity vector is accordingly:

$$\underset{\rightarrow}{c}_g = \nabla_{\underset{\rightarrow}{\kappa}} \omega = \frac{N^2-f^2}{\underset{\rightarrow}{\kappa}^4 \omega} m^2 \{k, l, \frac{-(k^2+l^2)}{m}\} ; \tag{8.3}$$

$$\underset{\rightarrow}{c}_g \cong \frac{N^2 \sin^2\theta \cos^2\theta}{\omega (k^2+l^2)^{1/2}} \{1, -\tan\theta\} . \tag{8.4}$$

The terms in brackets represent the vector components in the x,y,z directions in Eq. (8.3); in Eq. (8.4) the first term is the unit vector in the horizontal direction (k, l) and the second term is again the vertical component. I have assumed $N^2 \gg f^2$, which is well satisfied in the seasonal thermocline (taking $f^2=10^{-8}\ s^{-2}$, $N^2=10^{-4}\ s^{-2}$).

It is clear that $\underset{\rightarrow}{c}_g \rightarrow 0$ as $\omega \rightarrow f$, so that all waves except those with frequencies close to inertial quickly disperse from a surface forced region. The question is, "how close and how quickly?" Note first that if $\omega \leqslant 2f$, say, then $\theta \leqslant 1°$, so inertial oscillations have near vertical wave numbers and near horizontal group velocities. If we minimize the group velocity, by choosing a large inertial period (24 hours) and small horizontal wavelength (100 km), we find, for $\omega=2f$, $\underset{\rightarrow}{c}_g \simeq (150\ \text{km day}^{-1},\ 2.5\ \text{km day}^{-1})$. The vertical group velocity is still fairly large, and a wave of frequency 2f will disperse through the seasonal thermocline in well under a day. Reducing ω to $\omega=1.1f$ reduces the group velocity to $\underset{\rightarrow}{c}_g=(19\ \text{km day}^{-1},\ 88\ \text{m day}^{-1})$, which is still on the large side. Hence internal waves are highly dispersive, and if a spectrum of waves is forced at the surface, only those with frequencies within a few per cent of inertial remain in the mixing layer or seasonal thermocline for several inertial periods. It is these near inertial frequencies that cause the inertial oscillations often observed for several days after a storm, and that cause the inertial peak to dominate in some near surface current spectra.

b. Local generation

In spite of their dispersive nature, Pollard (1970a) showed that inertial oscillations driven by storms could absorb a significant fraction of the energy input $\underset{\rightarrow}{\tau}\cdot\underset{\rightarrow}{v}$. Oscillations are most strongly generated by features in the wind field that have time scales less than an inertial period. Slower changes in wind are not efficient at generating internal waves, driving mostly geostrophic adjustment velocities through the curl of the wind stress, and of course a mean Ekman flux while the wind lasts.

The fact that only waves of very near inertial frequency remain in the surface layer for more than a day allowed Pollard and Millard (1970) to simplify Pollard's model, making the motion independent of the horizontal coordinates x and y. By continuity, this forces the vertical velocity to be identically zero so that stratification drops out of the problem and the forced surface layer is decoupled from the thermocline. The angle θ is thus set to zero so that $\omega=f$ (Eq. 8.1), while $\underset{\rightarrow}{c}_g=0$ (Eq. 8.4). In this form, two major assumptions have been introduced, that spatial variations in the wind field can be ignored, and that there is no dispersion. The second assumption is a poor one, so the effect of dispersion, to remove energy, was modelled in Pollard and Millard's model by a damping term based on Pollard's computations.

The omission of spatial variations in the wind field can be defended using the group velocity arguments developed above. For frequencies within a few per cent of inertial, the horizontal component of the group velocity is about 0.06 L_H km/inertial period ($\omega=1.03f$), or up to 0.2 L_H km/inertial period ($\omega=1.1f$), where L_H is the dominant horizontal scale of the wind field. Thus it would take many inertial periods for energy to propagate a distance L_H from where it was generated, in which time it would have dispersed through the seasonal thermocline.

Begis and Crepon (1975), using a two-layer ocean, considered the effect of a horizontal boundary on the oscillations at a point some way away from it. The group velocity is again the determining factor, and Begis and Crepon found that the current would depend only on the local wind if the distance from the shore to the observation point were much greater than the internal radius of deformation (the distance energy in the baroclinic mode travels in $(2\pi)^{-1}$ inertial periods).

We conclude that only the local wind is effective at generating inertial oscillations in the mixed layer. Pollard (1971) was able to verify this using wind and current data from moorings set 50 km apart (Pollard and Tarbell, 1975). A factor of two difference between the amplitudes of the inertial oscillations at the two moorings at one time was reproduced in the currents generated by

Pollard and Millard's model, using the local winds. The difference was attributable to a depression which passed the moorings at different times and with different time lags between the passages of the warm and cold fronts.

c. Modelling the forcing function

Although wind is agreed to be the driving mechanism for inertial oscillations, the way in which the wind stress is parameterized has not proved critical when modelling the oscillations. Pollard (1970a), arguing that turbulence would distribute momentum input by the wind through the mixing layer in a time short compared to an inertial period, modelled the wind stress by a body force evenly distributed through the layer. Hollan (1969) employed a similar parameterization. Pollard (1969) showed that other distributions, such as a body force falling off with depth, could equally well be used. The resultant oscillations have a very similar depth dependence to the forcing function.

Crepon (1969), Gonella (1971), and Krauss (1972), following Ekman (1905), used an eddy viscosity parameterization in the surface layer. Ursell (1950) and Hasselmann (1970) suggested that inertial oscillations could be driven by the Stokes drift associated with surface waves, but Pollard (1970b) doubted that the amplitude of the Stokes drift was large enough to account for observations. It may well be that much of the wind driven momentum passes initially into surface waves (e.g., Dobson, 1971), but dissipation must certainly be invoked to transfer surface wave momentum into current momentum.

d. Oscillations in the thermocline and coherence

Pollard and Milliard's (1970) simple model applies only to the mixing layer. In the thermocline, inertial motions are considerably more complex, because of the complexity of the dispersion pattern (Eq. 8.4), particularly when horizontal and vertical variations of N are present. Energy propagates at a very small angle to the horizontal (θ) which is inversely proportional to N (for ω constant and of order f, Eq. 8.2) so that oscillations only a few meters below the base of the mixing layer are likely to have been generated many days earlier and some kilometers away. The complexity of the dispersion pattern is shown, for example, by Pollard (1970a) and Krauss (1972). Even with constant N and rather simple forcing functions, both found that the interaction of oscillations with slightly different frequencies (modes) caused beats below the surface layer.

It is clear from the above that a forecast of oscillations in the thermocline on the lines of Pollard and Millard could only be achieved with unrealistically detailed information about the wind field and stratification over a large area of space and time. One must fall back therefore on statistical descriptions of the inertial wave field. For the dispersion pattern envisaged, one would expect the highest coherence between records horizontally separated in the mixing layer. Horizontal coherence should fall off rapidly with depth because of spatial and temporal variations in N and the wind.

Pollard (1971) analyzed data from three moorings 50 km apart, each carrying wind recorders and current meters at 10, 30, 50, and 70 m (Pollard and Tarbell, 1975). He found that the stratification $N(\vec{x},t)$ was significantly different at the three moorings, and changed markedly over a period of six weeks. The inertial oscillations were correspondingly complex, but the horizontal coherence across 50-70 km was 0.83 at 10 m (which was in the mixing layer most of the time), dropping to 0.72 at 30 m, and 0.36 and 0.25 at 50 and 70 m. Instruments separated vertically were mostly coherent (0.45-0.75) except for the largest separations (60 m). The most significant feature of the vertical comparisons was that the phases of the near inertial oscillations at any instrument invariable lagged those at any deeper instrument. This showed that the vertical component of the wave number was directed upwards ($m>0$), confirming that the vertical component of the group velocity was directed downwards (from Eq. 8.3), and the oscillations were indeed propagating downwards from a forcing region near the surface.

Schott (1971) analyzed data from four moorings at the center and corners of a 4.5 km triangle in the northern North Sea. The water depth was 82 m and there was two-layer stratification, i.e., almost homogeneous top and bottom layers, separated by a strong thermocline between 28 and 38 m. In such shallow water, modes will form even at frequencies close to inertial, and Schott found that the first mode dominated, with the oscillations in antiphase across the thermocline. Coherences were correspondingly high (0.7-0.9) but there was some coherence loss in the thermocline where (*i*) the oscillations are affected by changes in stratification, and (*ii*) the first mode is less dominant, so that the phase mixing of higher modes can reduce the coherence (Hasselmann, 1970).

e. Importance of inertial oscillations

There are two main reasons why internal and inertial waves are of interest to the upper ocean modeller.

i. Transfer of energy and momentum out of the surface layer

I have already discussed the downward dispersion of internal waves in some detail. Other authors who have drawn attention to its importance are Frankignoul and Strait (1971) and Linden (1975). Frankignoul and Strait developed a general correlation technique to investigate the flow of internal wave energy in the ocean. They considered the whole internal wave band including inertial frequencies and looked at records spanning the main thermocline. They concluded that overall there was a downward propagation of energy which could reasonably be explained in terms of travelling internal waves. Linden estimated that sufficient energy could be radiated away from the surface to reduce significantly the energy available to mix and deepen the surface layer.

ii. Contribution to shear and shear instability

Estimates of shears caused by inertial oscillations are summarized in Table 8.3

In the thermocline: Shear instability in the seasonal thermocline (e.g., Woods, 1968) causes intermittent turbulent patches and hence diffuses the thermocline. Orlanski and Bryan (1969) suggested that gravitational instability of large amplitude internal waves would contribute to mixing, but Garrett and Munk (1972) concluded that shear instability is very much more likely than overturning because much of the internal wave energy is near the inertial frequency where the shear to slope ratio is largest. It is the small group velocity that accounts for large shears at inertial frequencies. The smaller the dispersion the slower can shears be reduced.

Garrett and Munk also found that the mean Richardson number was inversely

TABLE 8.3

Shears Caused by Inertial Oscillations

	Shear (cm/sec/m) (i.e. $10^{-2}s^{-1}$)	Instrument Separation (m)	Instrument Depths (m)
	(a) in the thermocline		
Pollard (1971)	0.1-0.7	20	32 and 52
	(b) across mixed layer/thermocline interface		
Gonella (1970)	2	15	15 and 30
Schott (1971)	1	14	18 and 32
Pollard (1971)	0.5-2	20	12 and 32
	(c) in the mixing layer		
Gonella (1970)	1	10	5 and 15
	0.1	20	5 and 25
Pollard, Rhines and Thompson (1973)	0.25	40	13 and 53

proportional to N, so that shear instability is most likely in the seasonal thermocline, where N is largest. It can be seen from Eqs. (8.1) and (8.2) that for constant ω and k^2+l^2, m is largest (θ smallest) where N is largest, i.e., vertical scales are smallest in the thermocline, so shears are largest.

Pollard (1971) found a maximum inertial shear of 7×10^{-3} s^{-1} across the thermocline, a factor of three too small to cause shear instability if $N=10^{-2}$ s^{-1}. Thus, instabilities are likely to be intermittent on a time scale much less than inertial, as they can only form when shears associated with high frequencies, added to the mean and inertial shears, are sufficient to cause instability. Nevertheless, the inertial shear may be a major contribution to the total.

Across the mixed layer/thermocline interface: The largest inertial shears are recorded when one instrument is in the mixing layer, the other in the thermocline, which is consistent with models (e.g., Pollard, 1970a) which forecast a discontinuity in velocity at the base of the forced layer. Shears of 2×10^{-2} s^{-1} have been observed, and might well have been higher if the instrument spacing had been smaller. Inertial shears of this magnitude are sufficient to cause instability without invoking any other source of shear, and led Pollard, Rhines, and Thompson (1973) to model deepening of the mixed layer driven by shear instability at the interface. They found that rapid deepening would take place while the shear acts (a fraction of an inertial period). (See Niiler and Kraus, Chapter 10, for a detailed discussion).

Within the mixing layer: Models frequently assume slab-like behavior in the mixing layer, which is reasonable if shears across the layer are smaller than shears at the interface or if it is only the integrated momentum of the surface layer that is forecast. However, the contribution of shears in the mixing layer to the turbulent energy budget should not be ignored (Chapter 10). For example, if the stress is 1 dyne cm^{-2}, a shear of $10^{-2}s^{-1}$ will cause a shear production term of 10^{-2} (c.g.s.), which is of the same order of magnitude as estimates of dissipation (Stewart and Grant, 1962).

It should be noted that in addition to inertial shears, recent investigators have found large mean shears across the mixing layer. Gonella (1970), Halpern (1975), Thorpe (Loch Ness data, unpublished) and Pollard (JASIN 1972 data, 1973, and unpublished) have all reported mean shears of order 10^{-2} s^{-1} in the direction of the wind, and approximately independent of depth across the mixing layer (typically between 5 m and 20 m).

8.4 Langmuir circulations

Since Langmuir (1938) first produced evidence of organized circulations in the surface layers of lakes and oceans, numerous investigators have confirmed their existence, and produced theories of their generation. As yet, no theory has received unanimous acceptance, so I shall here first summarize much of what is known observationally about Langmuir circulations before summarizing the theories that have been proposed.

a. Observations

The cellular structure (Fig. 8.3) that Langmuir described is well supported by all subsequent observations, in both oceans and lakes (Table 8.4). Langmuir circulations consist of alternate left and right handed helical roll vortices, aligned more or less along wind, with along wind surface velocities strongest in the convergence zones. The lines of convergence are often made visible by particulate matter or oil films that collect there, but they can also exist in the absence of surface pollutants, as has been shown by scattering cards, pieces of paper, or dye on the surface (Assaf, et al., 1971; Faller, 1964; Harris and Lott, 1973; Ichiye, 1967; Katz, et al., 1965).

The most variable reported feature of Langmuir circulations is the row spacing, ranging from 2-25 m in lakes, and from 2-300 m in the ocean. Cells of several different scales can exist together. The factors which control the circulation scale have not been unambiguously determined. Langmuir found larger (15-25 m) spacings in October and November than in May and June (5-10 m), suggesting a correlation with the depth of the seasonal thermocline, but other authors (Faller and Woodcock, 1964) have found this correlation to be not significant. Scott, et al. (1969) found that Langmuir circulations in Lake George did not usually mix the diurnal heat input right down to the seasonal thermocline (at about 10 m) but formed a secondary thermocline a few meters above it.

There is unanimous agreement that the surface velocity in the direction of the wind is larger in the convergence or streak zones than out of them, and velocity differences of up to 17 cm s^{-1} have been reported. There have been several attempts to measure the downwelling velocity in the convergence zones using dye, drag plate current meters, and even the vertical velocity necessary to submerge the pelagic Sargassum that accumulate in streaks (Woodcock, 1950). Sutcliffe, et al. (1963) have reviewed the data on downwelling, correlating the vertical velocity w with wind speed u_a by the line $w=0.85x10^{-2}\ u_a$, or about

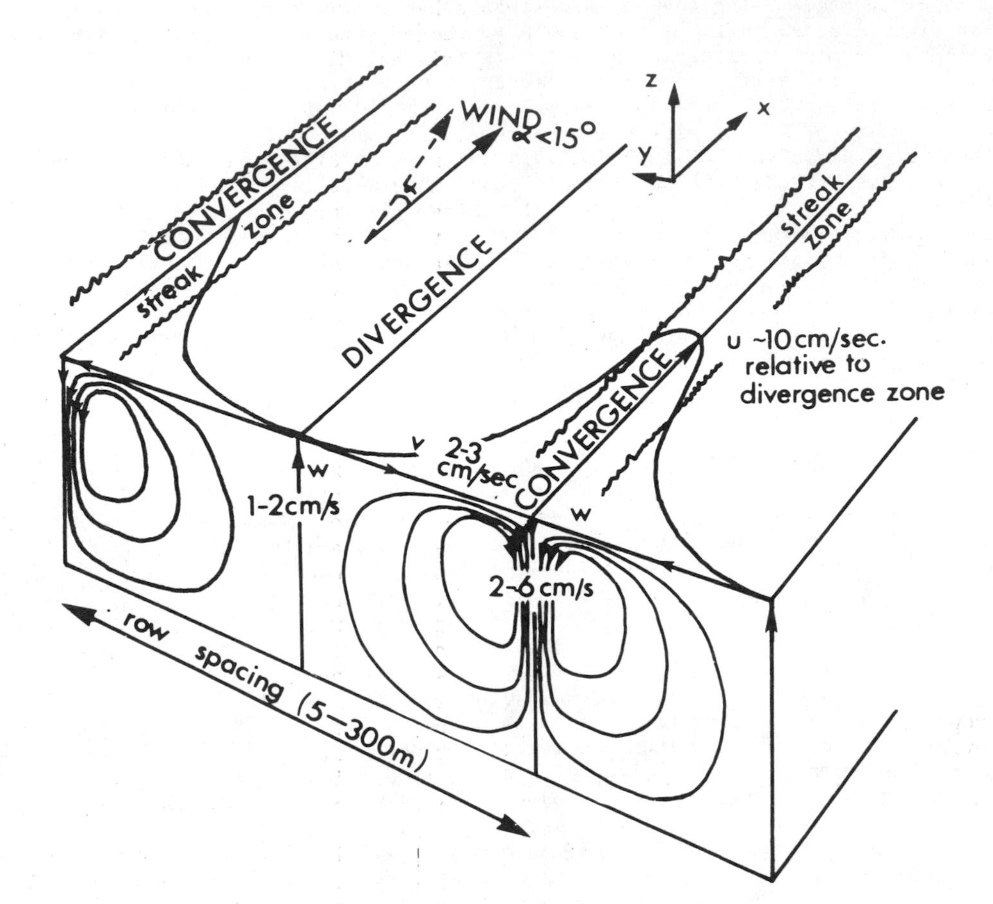

Fig. 8.3 - Diagrammatic representation of the main features of Langmuir circulations.

1 cm s^{-1} per 1 m s^{-1} wind. Although horizontal velocities perpendicular to the wind can be inferred from the convergence of material into streaks, only Langmuir has estimated their amplitude. By tracking surface debris he estimated a sideways velocity of 2-3 cm s^{-1}.

Langmuir circulations do not normally exist in winds less than 3 m s^{-1} (Faller and Woodcock; Harris and Lott; Ichiye; Langmuir; Scott, et al.; Welander, 1963) though Ichiye remarked that circulations can exist even in calm seas when there are pronounced swells. After the onset of larger winds however, circulations, or at least surface streaks, appear within a few tens of minutes. Langmuir, for example, recorded that lines of seaweed realigned within 20 minutes when the wind shifted by 90°. It is not clear how rapidly the circulations penetrate downwards. Welander suggested that the reorientation is initially confined to a very thin surface layer. He found that 10 minutes after a shift in a 9 m s^{-1} wind, surface streaks had rearranged themselves in the new wind direction and surface floats converged into the streaks but floats at 1 and 2 m depth continued to move in the original direction.

b. Theories

Although Langmuir gave a thorough description of the circulations now named after him, he gave no account of how they were generated. He clearly believed they were wind driven, and explained the larger forward velocities in the convergence zones by pointing out that the water there had been on the surface since it rose in the divergence zone and had therefore been accelerated by the wind for longest. The lack of a comprehensive explanation has taxed modellers ever since, and a large number of theories have been proposed, most of which have several points against them. The theories are tabulated chronologically in Table 8.5.

i. Thermal convection

The knowledge that convective plumes in the atmosphere can be aligned into rows by wind shear (see review by Kuettner, 1971) naturally led investigators to try to attribute oceanic circulations to the same mechanism. However, evidence that thermal convection is not a primary mechanism in the ocean is overwhelming. Stommel (1951), Faller and Woodcock, Csanady (1965), Ichiye, and Scott, et al. have all observed Langmuir circulations in conditions of stable stratification. Scott, et al. observed tongues of warm surface water downwelling under surface streaks on several occasions (Fig. 8.4). Several of the above authors calculated surface heat fluxes, but found no correlation between heat fluxes and slick spacing or mixed layer depth. Baylor and Sutcliffe (reported by Faller, 1969) observed that the surface temperature in lines of surface convergence could be relatively warm, indicating a downward flux of heat. Harris and Lott compared downwelling velocities in stable and unstable surface heat flux conditions but found no significant difference.

All of the above is negative evidence. There is no proof that surface cooling cannot assist cell formation, and my attempt to estimate growth rates from expressions given by Kuettner for convective instabilities gave values of the same order as observed growth rates. Convective rolls exist in the atmosphere and many experimenters have generated them in the laboratory. However, Faller (1969) found that thermally driven rolls could be quickly swamped by surface wave driven circulations which were qualitatively much stronger.

Paradoxically, the only example of naturally occurring thermally driven rolls in the ocean seems to be that of Owen (1966), in calm heating conditions. Convection can occur in this case because the radiative heat input is distributed through a few meters, while cooling takes place right at the surface. However, the rolls Owen observed (Table 8.4) were significantly smaller than most Langmuir circulations and confined to the top 60 cm. The streaks seen by Stommel are possibly due to the same cause.

ii. Atmospheric rolls and surface films

The possibility that oceanic roll vortices are coupled to atmospheric rolls can be quickly disposed of, since the patterns in the air move too fast across the surface of the water to be coupled to it. Welander's (1963) hypothesis of a feedback mechanism involving the modification of the surface wind over streaks can be similarly discounted. Kraus (1967) suggested the first of

TABLE 8.4

Observations of Langmuir Circulations

AUTHOR	OCEAN OR LAKE	ROW SPACING (R) (M)	COMMENTS	VERTICAL VELOCITY ($CM\ S^{-1}$)	DOWN OR UP	FORWARD VELOCITY OF CONVERGENCE ZONE ($CM\ S^{-1}$)	REORIENTATION TIME (MIN)	OTHER COMMENTS
ASSAF ET AL. (1971)	OCEAN	3-6 5-12 30-50 90-300	CALM HIERARCHY IN 5-15 M/S WIND			10		
FALLER & WOODCOCK (1964)	OCEAN	20-50						WIND W AND ROW SPACING R CORRELATED. R=WX 4.8 SEC
GORDON (1970)	OCEAN	5		1.5	UP	1-3		
HARRIS & LOTT (1973)	LAKE ONTARIO	3-4		2-9	DOWN	≤6		
ICHIYE (1967)	OCEAN	20-200	ACTUAL QUOTE "TENS TO HUNDREDS OF METERS"			17		
KATZ ET AL. (1965)	OCEAN	2-10 10-45	SURFACE 0-7 M DEEP			>0	30	WIND AND R CORRELATED R= W x4 SEC
LANGMUIR (1938)	OCEAN LAKE	100-200 5-10 15-25	 MAY-JUNE OCT-NOV	1-2 2-3 1-1.5	DOWN (AVERAGE 0-6 M) DOWN (AT 2M) UP	>0	20 20	
OWEN (1966)	OCEAN	1.5						CALM, THERMALLY DRIVEN, AND ONLY 60 CM DEEP
STOMMEL (1951)	PONDS ON CAPE COD						1-2	LESS THAN 30 CM DEEP
SUTCLIFFE ET AL. (1963)	OCEAN			3-6	DOWN			
SCOTT ET AL. (1969)	LAKE GEORGE			4-7	DOWN		"ALMOST INSTANTANEOUS"	R AND DEPTH TO FIRST STABLE LAYER CORRELATED
WELANDER (1963)	BALTIC SEA						10	
WOODCOCK (1950)	OCEAN			>3	DOWN			

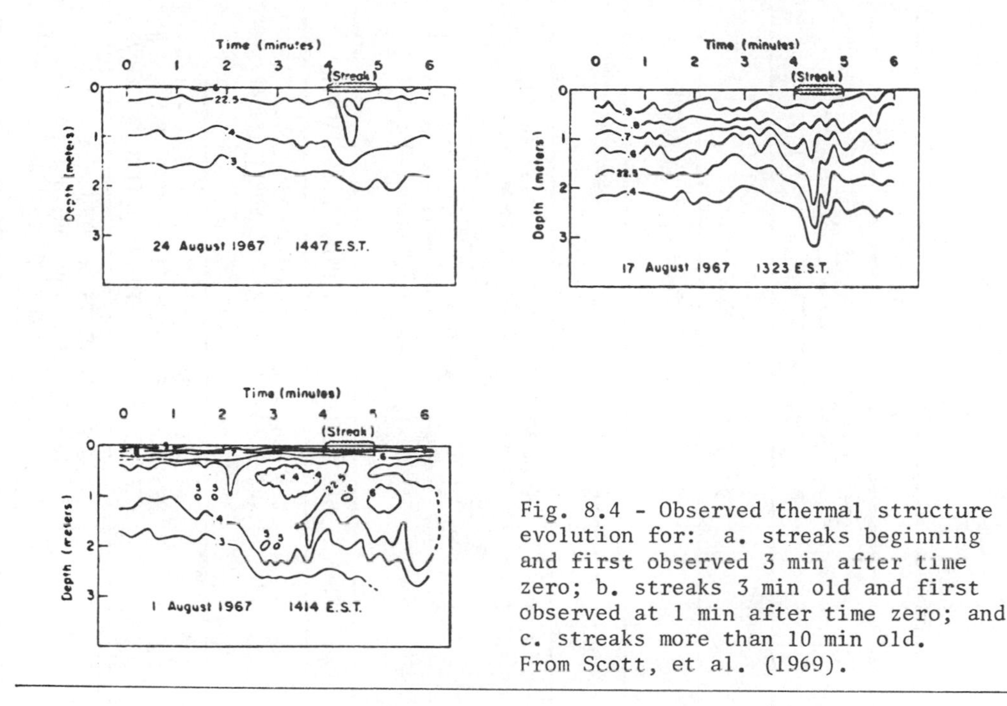

Fig. 8.4 - Observed thermal structure evolution for: a. streaks beginning and first observed 3 min after time zero; b. streaks 3 min old and first observed at 1 min after time zero; and c. streaks more than 10 min old. From Scott, et al. (1969).

several wave interaction mechanisms, that the damping of capillary waves as they approach a slick generates a radiation stress that will enhance the slick. Kraus' hypothesis is a forerunner of a recent theory by Garrett (1976), but can be discounted in the form given by Kraus for several reasons. As with convection, the main reason is the negative one that while slicks may be caused by Langmuir circulations, surface contamination is not a pre-requisite for generation of the circulations (Section 8.4a).

iii. Instability of sheared rotating flow

Faller (1964) suggested that instability of the Ekman spiral (Faller, 1963; Faller and Kaylor, 1966) could be the cause of Langmuir circulations. Faller supported his hypothesis with observations that showed windrows lying at about 15° to the right of the wind. The main objection to this theory has been the long time scales (of order f^{-1}) for growth of the instability (compare growth times in Table 8.4), but recently Gammelsrød (1975) has found a time scale in a sheared flow of $(f\frac{du}{dz})^{-1/2}$ where $\frac{du}{dz}$ is the mean shear. This can be as short as 20 minutes for the shears estimated in Section 8.3e.

iv. Winds and waves

All the remaining theories that have been advanced rely on winds or wind driven waves as the driving force for Langmuir circulations. This is consistent with all observations, including Langmuir's own, and it is certain that winds are the primary mechanism. It is not at all clear, however, how the

TABLE 8.5

Theories of Langmuir Circulations

THEORY	ORIGIN	PRESENT STATUS
CONVECTIVE INSTABILITY, ROLLS ALIGNED BY WIND	ANALOGY WITH ATMOSPHERIC BOUNDARY LAYER, SEE REVIEW BY KUETTNER (1971)	NOT A PRIMARY MECHANISM, AS CELLS OFTEN OBSERVED TO GROW IN STABLE CONDITIONS AND BREAK DOWN STABLE STRATIFICATION
COUPLING WITH ATMOSPHERIC ROLLS	UNKNOWN, MENTIONED BY STOMMEL (1951)	DISCOUNTED, ATMOSPHERIC VORTICES MOVE TOO FAST OVER OCEAN SURFACE
MODIFICATION OF WIND OVER SURFACE SLICKS	WELANDER (1963)	DISCOUNTED, FOR REASONS GIVEN ABOVE
INSTABILITY OF THE EKMAN SPIRAL	FALLER (1964)	NOT A PRIMARY MECHANISM, CANNOT ACCOUNT FOR OBSERVED GROWTH RATES
DAMPING OF CAPILLARY WAVES IN SLICKS PROVIDES RADIATION STRESS TO DRIVE ROLLS	KRAUS (1967); A FORERUNNER OF GARRETT'S (1976) THEORY (BELOW)	DISCOUNTED AS CELLS MAY EXIST IN THE ABSENCE OF SURFACE CONTAMINANTS. ALSO ENERGY SUPPLY TOO SMALL TO EXPLAIN OBSERVED GROWTH RATES
INTERACTION OF TWO LINEAR WAVE TRAINS	STEWART AND SCHMITT (1968)	DISCOUNTED, CANNOT PROVIDE VORTICITY
"EDDY PRESSURE" OF SURFACE WAVES	FALLER (1969)	DISCOUNTED, CANNOT PROVIDE VORTICITY
INTERACTION OF PAIRS OF INVISCID WAVE TRAINS IN A SHEAR FLOW	CRAIK (1970)	DISCOUNTED BY LEIBOVICH AND ULRICH (1972), INVISCID THEORY CREATES VORTICITY OF WRONG SIGN
INTERACTION OF WAVES AND SURFACE CURRENT, WITH WAVE DISSIPATION	GARRETT (1976)	UNTESTED, BUT CAN EXPLAIN ALL OBSERVED FEATURES OF CIRCULATIONS WITH REASONABLE GUESSES FOR MAGNITUDES OF FORCING TERMS
INSTABILITY OF COUETTE FLOW IN A ROTATING INVISCID FLUID	GAMMELSRØD (1975)	UNTESTED, PREDICTS SLANTED CELLS, AND MAXIMUM VERTICAL AND HORIZONTAL VELOCITIES AT HALF THE CELL DEPTH

winds can set up the circulation. Shear instability theories with rotation implicitly invoke the wind to set up the shear flow. Most of the theories so far suggested which do not include rotation can be shown to be inadequate.

The inviscid wave theories of Stewart and Schmitt (1968) and Faller (1969) cannot account for the vorticity of the circulations. The vorticity in a similar theory by Craik (1970) involving the interaction of two wave trains, was supplied by the vertical shear of a mean current, but Leibovich and Ulrich (1972) point out that the circulation generated would have maximum forward velocity in the upwelling zones, which is contrary to all observations (Section 8.4a). Leibovich and Ulrich conclude that no inviscid mechanism involving production of longitudinal vorticity by distortion of cross stream vorticity will work.

The recent theory by Garrett (1976) seems the most plausible to date. Garrett points out that the presence of a current will modify the wave field passing through it in order to conserve wave action (Bretherton and Garrett, 1968; see also Phillips, Chapter 12). If small variations in surface current exist, the waves will enhance them by (a) causing convergence towards the line where the horizontal current is maximum, (b) strengthening the horizontal current where it is maximum by preferential wave breaking there. Thus a feedback loop exists, and the cells are generated by an instability mechanism. Garrett's mechanism is supported by Myer's (1971) observation that wave amplitudes are largest in the convergence zones. Ichiye's observation that circulations can exist even in calm seas when there are pronounced swells supports mechanisms involving waves rather than direct wind action, as do Faller's (1969) observations in a wave tank.

c. Mixing due to Langmuir circulations

Reynolds stresses associated with Langmuir circulations are potentially very efficient at redistributing momentum through the mixing layer from the thin surface layer in which it is initially deposited by surface wave breaking. This is the primary reason why such circulations are important to mixed layer dynamics. It is easily seen from Fig. 8.3 that u and w variations along a horizontal line perpendicular to the cell axes are in phase, so that the Reynolds stress $\overline{uw}$ averaged along that line is nonzero, and transfers u momentum downwards. For example (from Fig. 8.3) if one assumes sinusoidal variations with a difference between u in the convergence and divergence zones of 6 cm s^{-1} and between down and upwelling velocities of 4 cm s^{-1} at a depth of 2 m, say, then $\overline{uw}=-3$ cm^{2}s^{-2}, corresponding to a downward stress of 3 dynes cm^{-2}. Gordon (1970) in a similar calculation, estimated the stress as 2-3 dynes cm^{-2}, which was more than enough to transport downwards all the momentum input by the surface stress.

Stewart (1970) includes several references to stress calculations, and himself estimated that the stress necessary to overcome the near surface thermal stability on Lake George was 0.6 dyne cm^{-2}. He suggested that the wind speed needed to supply this stress could be the explanation of the threshold speed of 3-4 m s^{-1} below which Langmuir circulations are not observed. Langmuir circulations appear rapidly after the onset of winds above this minimum, making it likely that they, rather than small scale turbulence generated by breaking waves, control the downward diffusion and redistribution of wind generated momentum through the surface layers. In the terminology of Section 8.2, Langmuir circulations, when they exist, can easily account for the largest eddy viscosities that have been observed.

Chapter 9
PARAMETERIZATION OF UNRESOLVED MOTIONS

J. D. Woods

9.1 Introduction

a. Turbulent eddies

The Reynolds number of the ocean is large so it is always and everywhere turbulent, with eddy sizes covering an immensely broad range bounded at the large end by the physical dimensions of the ocean and at the small end by molecular viscosity. The ratio between the horizontal dimensions of the smallest and largest eddies is approximately 1 cm : 5000 km or 1 : 5×10^8. The corresponding ratio for the vertical dimensions is three orders of magnitude smaller because the ocean is shallow. In general small eddies have short lifetimes and large eddies have long lifetimes, and, anticipating theoretical predictions to be presented below, the ratio of the lifetimes of the smallest and largest eddies is approximately 1 second : 1 year or 1 : 5×10^7. The distribution of turbulent kinetic energy within this broad spectral range is still the subject of speculation. Some recent studies suggest that there will be a strong spectral peak of energy at the Rossby radius of deformation, where potential energy enters the turbulent kinetic energy spectrum through the action of baroclinic instability; and a second spectral peak has been identified at the Ozmidov scale (see below), where internal wave energy enters the turbulent kinetic energy spectrum through the action of Kelvin-Helmholtz instability. The energy level is much lower between these peaks, but there is no reason to suppose that no eddies exist in these spectral valleys. On the contrary, experience of other turbulent flows suggests that turbulent kinetic energy flows from the source regions at spectral peaks to smaller and larger scale eddies, either continuously or in a series of jumps, to give what is normally called the turbulent energy cascade. While direct evidence of the existence of eddies on all possible scales in the ocean is incomplete, the most plausible hypothesis is that eddies of any size in the permitted range can and do exist, but not necessarily with all sizes occurring everywhere at any instant. Furthermore, it seems reasonable to suppose that eddies occurring at the same geographical location at the same time will interact strongly in a way that contributes to the overall cascade of energy through the spectrum. The ultimate sink of turbulent kinetic energy passing down the cascade occurs in very small eddies, whose kinetic energy is converted to heat by molecular motion. However, larger eddies do work against gravity, thereby losing a tiny fraction of their kinetic energy to be temporarily* stored as potential energy.

b. Waves

Propagating through these turbulent eddies are internal waves, which are often so intense that over a broad band they make a larger contribution to the kinetic energy density in one dimensional spectra of velocity fluctuations

---

*The storage period is long compared with the eddy lifetime, but eventually the potential energy stored in this way will be recycled through the turbulent kinetic energy cascade, assuming there is no tendency for an overall increase in potential energy of the oceans.

in the ocean (e.g., Webster, 1968; Wunsch, 1972). It is still not clear from the available field observations, whether these waves should be treated as a quasi-homogeneous field of motion or as an ensemble of discrete wave packets which demand rather different approaches in the study of interactions between the waves and the eddies they pass through. On the one hand, there are satellite observations of apparently long lived internal wave packets (Apel, Byrne, Proni and Charnell, 1975) implying that energy passes only slowly from the waves to the eddies, while on the other hand, recent mooring data (IWEX: see, e.g., Briscoe, 1975) have been interpreted as showing rapid degradation of individual wave packets (e.g., Olbers, 1975). These two approaches lead to rather different conclusions concerning the role of internal waves in transporting momentum through the ocean and in stimulating the diffusion of scalars by energizing eddies, which achieve the mixing that the waves cannot perform (see, e.g., Muller, 1974).

A classic problem of oceanography has been how to discriminate between motion classed as eddies and motion classed as internal waves, as detected by time series from current meters. Some authors have implied that it is impossible to devise an experiment to discriminate between waves and turbulence and this may well be true if wave packets rapidly lose their identity because of interactions with eddies. But if the wave-eddy interactions are weaker, so that wave packets persist for long enough to permit measurement of their characteristic and dispersion ratios, then, provided these are different from the equivalent ratios for eddies (and the theory given below suggests this is true), it is in principle possible to tell waves from eddies. Some tentative steps have recently been made in this direction (Woods, ed., 1975).

c. Computer integration of the primitive equations

There would be no need to parameterize the transport properties of these eddies and waves if all the motion could be computed by numerical integration of the primitive equations. If this were possible, all the eddies and waves, large and small, would be described in sufficient detail to permit computer predictions of the transports of momentum, heat, etc., as required. But this is not possible at present and not even the most sanguine extrapolation of contemporary computing techniques suggests that it will become possible in the future. If the spectral extent of natural variability in the ocean, expressed in terms of the logarithms of the ratios of the maximum and minimum scales in the zonal, meridional, vertical and temporal dimensions, is greater than (8x8x5x7), then the corresponding expression for the spectral window of a typical present day computer model is typically (2x2x1x3). A computer model with more than 10^{16} times greater resolution than this would be needed to follow all the motions deterministically, and as no foreseeable increase in computing power is likely to make that possible, there is no point in considering the problems of initiation and error growth in such a comprehensive model.

In practice, therefore, primitive equation models cover a narrow spectral window imbedded somewhere in the vast band of natural variability. As shown in Fig. 9.1, the choice of where to locate the spectral window rests with the designer of the computer model. Some, such as those described by Holland (Chapter 2) and by Somerville (Chapter 3), locate the spectral window at the low wave number end of the natural band, so that the largest eddies and waves are resolved, and all neglected motions have scales smaller than the resolution of the model. These are the sub-grid scale motions. Other primitive equation models have spectral windows floating inside the natural band, so

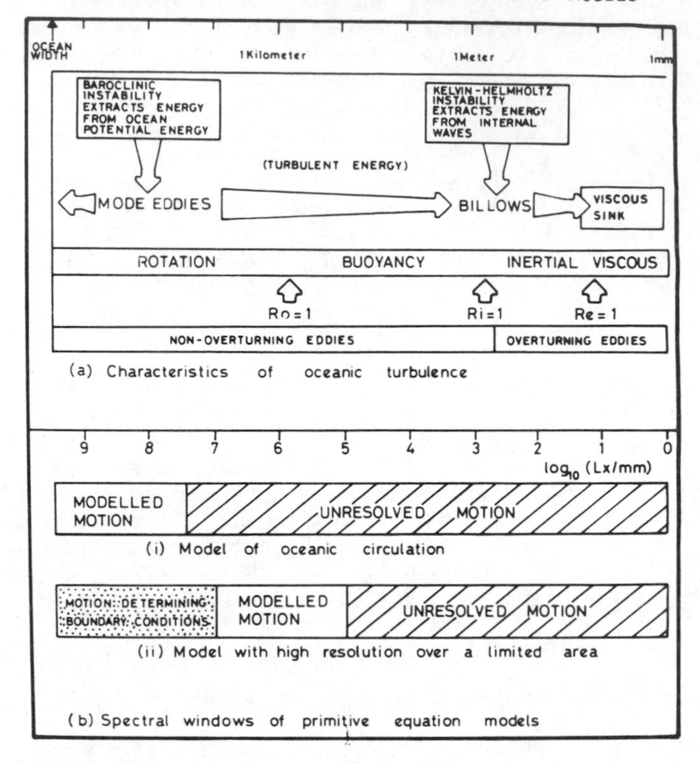

Fig. 9.1 - A comparison of the scales of turbulence eddies following the classification given in Section 9.3 and the spectral windows of primitive equation computer models given in Section 9.1.

that the intermediate scales of motion are described deterministically to the neglect of larger scale motions, whose action is treated in terms of boundary conditions for the model, and the sub-grid scale motions are incorporated parametrically into the model equations.

d. Parameterization of unresolved motions

The small eddies and waves too small to be resolved in the model usually produce significant transport of momentum, heat, and dissolved and particulate constituents of seawater. They also interact dynamically with the larger eddies described deterministically in the model, causing changes which, if neglected, progressively degrade the accuracy of the model predictions. Finally, the unresolved motions extract energy from the larger eddies described deterministically in the model. Some attempt can be made to minimize the errors arising from inadequate representation of unresolved motions by adding terms to the model equations which describe economically the diffusive, dissipative and interactive effects.

The process of representing these net effects of the unresolved motions is called parameterization. Ideally, parameterization should be based on a detailed analysis of the properties of the small eddies described deterministically, so that account can be taken of local variations due to the distributions of velocity, density, etc. produced by the larger motions resolved by the model. For example, in one of the parameterizations of sub-grid scale transport to be described later, the eddy transport coefficients vary with the Richardson number, whose magnitude is determined by the vertical current shear and density gradient resolved by the model.

An important assumption made in parameterization is that the magnitudes of the rate parameters appearing in the model equations (e.g., eddy viscosity, etc.) are controlled by the largest unresolved motion present at that locality. For a detailed discussion of this assumption the reader is referred to Townsend's (1976) monograph. This can be justified in terms of simple dimensional arguments for turbulent flows with red spectra (in which energy density increases monotonically with eddy/wave size) and physically similar motions for all eddy or wave sizes. The assumption may not be justified when the Nyquist boundary of the model lies close to a boundary between physically different flow regimes (e.g., between the different turbulent ranges described below, or between different classes of waves), or close to a spectral peak in the unresolved motions (e.g., at the MODE eddy size, see Chapter 15).

In order to minimize the errors arising from interactions between the motions resolved by the model and the unresolved eddies and waves, it is best to locate the Nyquist boundary of the model in a gap or, at least, a valley in the eddy and wave spectra. The interaction errors are likely to be worst if the Nyquist boundary lies close to a spectral peak, from which energy is likely to cascade strongly both down and up the spectrum. It would therefore be bad practice to design a model with Nyquist boundary close to the MODE eddy peak (horizontal scale $L_H \simeq 100$ km), or the Kelvin-Helmholtz billow peak ($L_H \simeq 1$ m; vertical scale $L_V \simeq 20$ cm). If the local distribution of eddies and wave packets is concentrated into an ensemble of narrow spectral oases separated by spectral deserts, as postulated by Woods (1975) (see below), then the ideal solution would be to adjust the Nyquist boundary locally to fall in the desert, rather than pass through one of the oases. As well as minimizing cross boundary interactions, this technique also provides a clear separation between motions treated deterministically and those parameterized.

Flexible Nyquist boundaries may become a refinement in future models. Meanwhile, all too many models incorporate diffusion terms in which the rate parameters (eddy viscosity, etc.) are constant in space and time and have a magnitude chosen arbitrarily or to satisfy the needs of computational stability, rather than as a representation of eddy and wave motions whose physics are understood, but which are not resolved in the model.

9.2 Observations of eddies and internal waves

a. Statistical properties of temperature microstructure

Ensemble mean spectra of horizontal variance of sea surface temperature generally exhibit a k^{-p} trend, where k is the wave number and the power p is usually in the range $p=2\pm0.3$ (Fig. 9.2). Such ensemble spectra contain no obvious evidence of a climatological spectral gap or valley corresponding to that believed to occur in the meteorological mesoscale. The magnitude of the variance density in the horizontal wave band 10 cm to 10 km is modulated diurnally by up to two orders of magnitude (Moen, 1973), probably by the vertical heat transport mechanism described in Section 9.4d below.

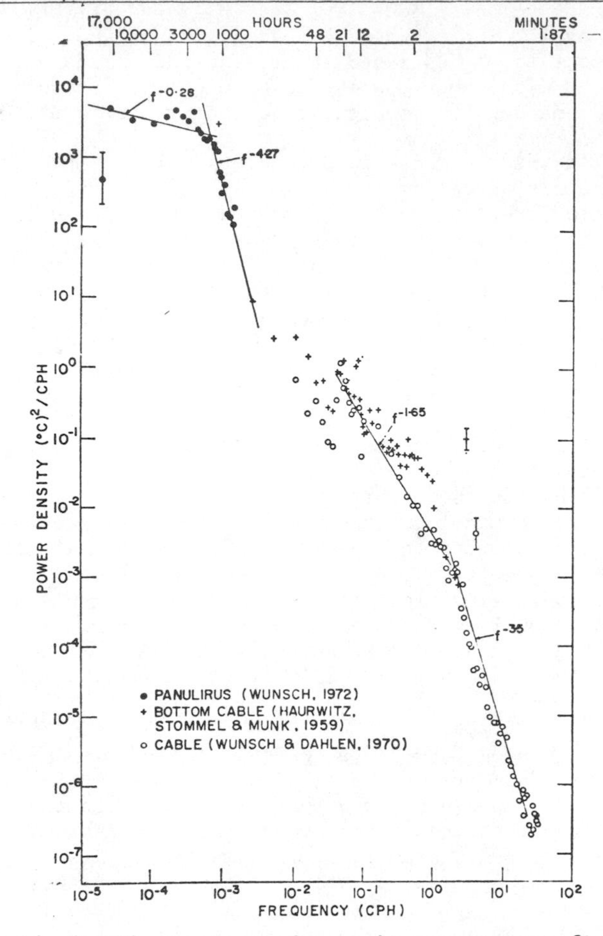

Fig. 9.2 - Wunsch's (1972) spectrum of oceanic temperature fluctuations.

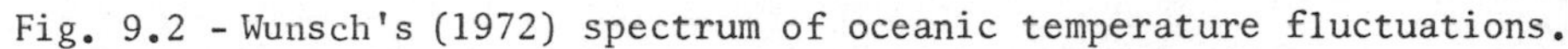

Local spectra exhibit wide fluctuations in spectral slope and maximum resolution spectra (see Chapter 15) show that the temperature variance is concentrated into hills and ridges in the (x,k)-plane. The kurtosis of horizontal temperature gradient ranges up to 40, confirming that the temperature variations are distributed very intermittently.

These observations in one dimension cover a broad spectral band. Two-dimensional statistics of sea surface temperature (e.g., Morrice, 1974) or of temperature distribution on potential density surfaces inside the thermocline (Fig. 9.3) cover much narrower spectral bands, but tend to confirm the very patchy distribution of temperature variance. The latter are particularly interesting because, by definition, they reveal the patterns associated with eddies, uncontaminated by internal waves, which will be discussed later.

Vertical profiles of temperature microstructure and vertical sections show the same kurtosis of temperature gradient and the same intermittency.

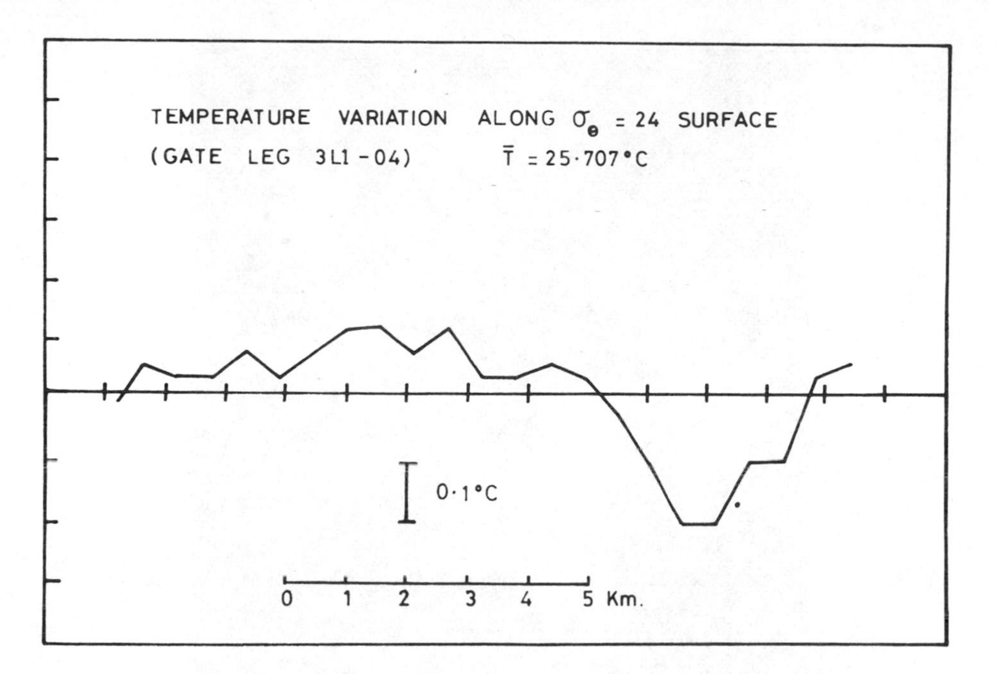

Fig. 9.3 - Variation of temperature along σ_θ=24.0 potential density surface measured during GATE (Woods and Minnett, unpublished).

b. The condensed variance model of turbulence

The picture of the distribution of temperature variance as a function of physical and Fourier coordinates*, $E_T(\underset{\rightarrow}{x}\ ;\ \underset{\rightarrow}{\kappa})$, that best fits these observations is a cloud of condensed droplets, rather than the continuum picture that emerges from ensemble spectra. In any narrow wave band, the variance "droplets" occupy a small fraction of physical space, typically about 10%. In addition to the droplets in the "condensed variance" model, which correspond to eddies, there are ridges, corresponding to sharp discontinuities at fronts.

There are no corresponding velocity fluctuation measurements† but the temperature microstructure must be created by velocity fluctuations that fit the same model. So, if the condensed variance model of temperature microstructure is accepted, the velocity variance distribution would also consist of condensed droplets corresponding to eddies, plus frontal ridges. This leads to a picture of turbulent motion characterized by discrete eddies, separated locally by spectral gaps and concentrated so sparsely that only a relatively small selection of eddies is present at a given geographical location at any instant. A motion film of $E(\underset{\rightarrow}{x}\ ;\ \underset{\rightarrow}{\kappa})$ would show the selection of

*This concept is discussed in more depth in Chapter 15.

†Moored current meter data are excluded because internal waves mask the turbulent motions of interest here.

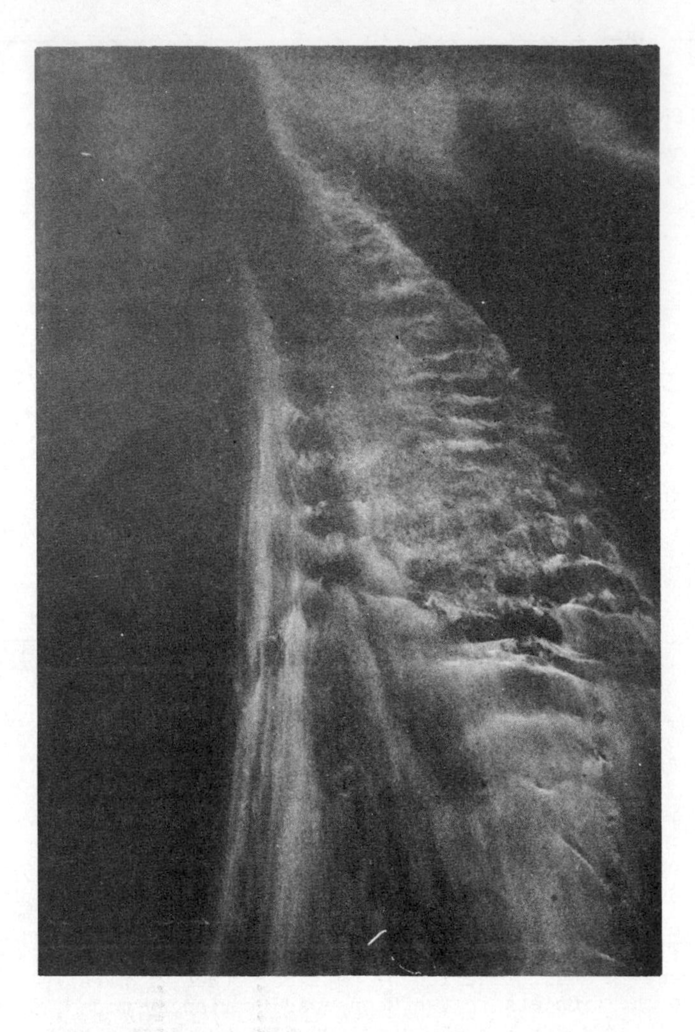

Fig. 9.4a - A patch of billow turbulence revealed by dye tracers: general view from above (wave length ≃ 75 cm).

eddies at each location changing with time, as eddies interact one with another.

c. Observations of individual eddies

If the condensed variance model is a good representation of distributions in the turbulent upper ocean, it should be possible to identify individual eddies in field observations, in the same way that a meteorologist identifies cyclones. The criterion for such identification is that the eddy should be a variance concentration separated by a clear spectral gap from eddies of different sizes at the same location and by a clear physical gap from the nearest eddy of the same size. Very few observations in the upper ocean cover a sufficiently broad volume in physical and Fourier space to satisfy this criterion. As is usual in fluid dynamics the best evidence comes from flow visualization and remote sensing rather than measurements with _in situ_ instruments.

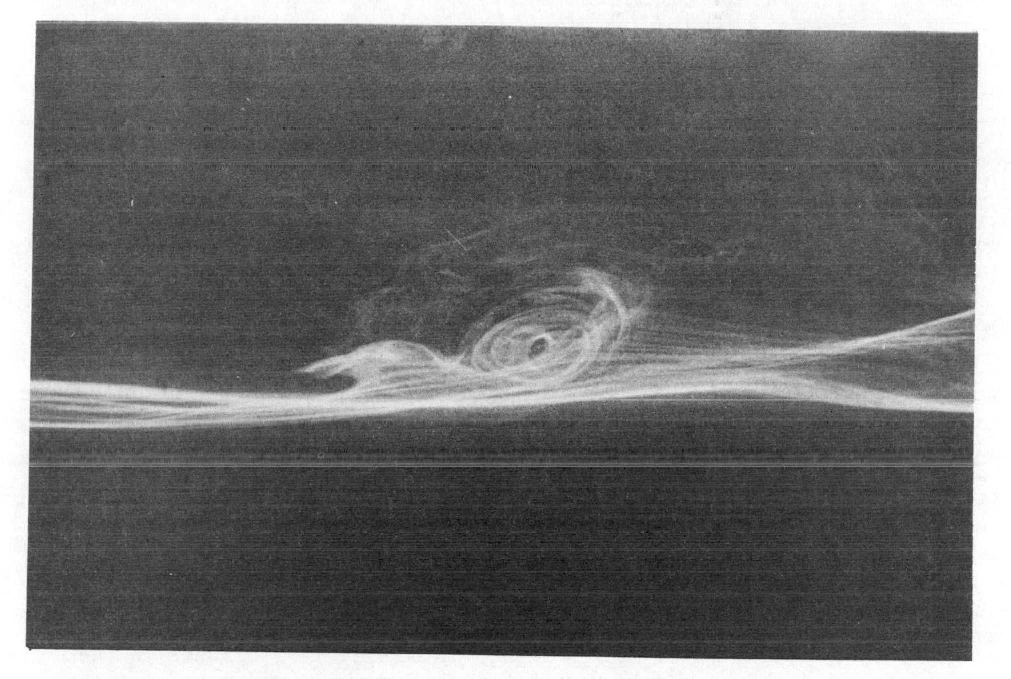

Fig. 9.4b - A patch of billow turbulence revealed by dye tracers close up from the side of a billow (20 cm high) with a secondary eddy.

In the upper ocean, flow visualization experiments by Woods (1968) have revealed the presence of patches of billow turbulence, equivalent to the meteorologist's clear air turbulence (Fig. 9.4). These eddies, which are a few decimeters high inside the thermocline, are the largest motions revealed by flow visualization that involve overturning of density surfaces. Larger motions, up to the largest scale detected in flow visualization (about 30 meters depending on water clarity) have the appearance of layers, with sharply defined upper and lower surfaces characterized by enhanced vertical density gradients and shear, on which the billow turbulence events occur. These layers are identified with the rotation and buoyancy range eddies to be described later (Section 9.3).

At much larger scales, there is some indication in satellite photographs of the sea surface for discrete eddies with horizontal dimensions of order 100 km and it is likely that this will prove an increasingly important source of information in the future when resolution increases. Such eddies have also been identified in STD sections made near islands (Patzert, 1969), and the MODE and POLYGON data sets have been interpreted as showing discrete eddies of similar size, despite their narrow experimental bandwidths. At intermediate scales (2-20 km), the GATE data set has shown preliminary indications of discrete eddies, but the full analysis has not yet been concluded.

d. Internal waves

The story is much the same for internal waves. Ensemble spectra of velocity variance from moored current meters shows a remarkably consistent k^{-p} form, where $p \simeq 2$. The local spatial distribution has not yet been measured, but there is evidence of intermittency, at least at a wave band close to the inertial frequency.

Individual wave packets have been identified in satellite photographs, airborne radar photographs and sonargraphs from towed transducers recording echoes from layers within the thermocline. They are also a common feature of divers' flow visualization studies where trains of waves with $\lambda \simeq 10$ m have been observed to trigger Kelvin-Helmholtz instability leading to billow turbulence.

e. Fronts

The existence of lines of strong sea surface temperature and salinity gradient, with associated surface current convergence and shear, has been known for a long time. These have now been identified as the surface outcrops of fronts, the equivalent of those found in mid-latitude atmospheric depressions, in which the surface transition zone extends down into the thermocline with locally enhanced horizontal gradients centered on an interface sloping at 1 part in about 300. Observations of the detailed structure of these fronts have shown large amplitude waves with wave lengths of about 10 km with an associated switchback modulation of the frontal jet, giving vertical displacements of up to 50 meters above and below the mean level. These observations form the basis for the parameterization in Section 9.4d. There are as yet no reliable estimates of the frequency of occurrence of fronts in the upper ocean; theoretical considerations lead to an upper estimate of one every 100 km, which may be modulated according to local conditions by a factor of 10.

9.3 Turbulence theory

The aim of turbulence theory is to replace the primitive equations, which cannot be integrated over the broad spectral range of natural variability, with simpler equations in which prediction of events has been abandoned in favor of statistics of the motions dispersed over a broad band of scales. The specification for a statistical theory of turbulence depends upon the application; here the parameterization of unresolved motions in a form suitable for computer models of the upper ocean is sought. It is hoped that parameterization can be largely based on the properties of the largest sub-grid scale eddies but, as was pointed out in Section 9.1d, this depends upon the absence of nearby spectral peaks or boundaries between physically dissimilar eddy regimes. An appropriate first aim of turbulence theory is, therefore, to consider the distribution of turbulent kinetic energy across the whole range of natural eddy sizes and to identify regions of physical similarity. It is convenient to start with the second of these aims.

a. Ranges of physical similarity

i. Eddies that do not overturn

The largest eddies, those with Rossby number ($Ro=Uf^{-1}L_H^{-1}$, where f is the Coriolis frequency and U is a velocity scale) less than unity, are strongly influenced by the earth's rotation, so that the flow is quasi-geostrophic; these lie in the rotation range. Next down the spectrum are those eddies which have Rossby number and Richardson number ($L_V^2N^2U^{-2}$, where N is the Brunt-Väisälä, or buoyancy, frequency; L_V is the length scale, usually a vertical scale for Richardson numbers) both greater than unity; these lie in the buoyancy range. Eddies in the rotation and buoyancy ranges contribute fluctuations to the turbulent motion which are two-dimensionally isotropic, with the orthogonal components lying in the plane of the density surface being statistically indistinguishable from one another, but both far stronger than the normal component to the density surface. It is important to note that the normal component, although far weaker, may play a crucial role in the eddy kinematics (and hence in parameterization) and that it is seldom correct to neglect the small local inclination of the density surface to the horizontal, since the vertical component of the motion along an inclined density surface is often much larger than the vertical component of the normal component. Eddies in the rotation and buoyancy ranges cannot therefore be classified as "two-dimensional" nor can the predictions of "two-dimensional turbulence" theory necessarily be applied to them.

ii. Eddies that overturn

Eddies with Richardson number much less than unity behave quite differently from those in the buoyancy and rotation ranges. They involve overturning of density surfaces by motions which are statistically close to three-dimensional isotropy at $Ri \simeq 1$ and rapidly lose any residual anisotropy as the eddy size and therefore the Richardson number decreases.

This class of eddies is divided into three main ranges. Firstly, the Kelvin-Helmholtz billows which effect the transition from non-overturning to overturning flow; these consist of a series of parallel rotors aligned with axes in the plane of the density surface at right angles to the direction of shear. Secondly, a class of smaller eddies with Richardson numbers much less than unity, but Reynolds number ($Re=LU/\nu$, where ν is the kinematic viscosity) greater than unity; these correspond to Kolmogorof's (1941) inertial subrange. And, thirdly, the final viscous subrange of eddies, for which the Reynolds number is less than unity.

b. Eddy lifetimes

Assuming a model in which the turbulent energy cascade consists of a succession of little leaps from large eddies to successively smaller eddies, then the lifetime of a particular eddy, spectrally separate from its neighbors, is a measureable quantity τ, whose magnitude will on average increase with eddy size. Thus an eddy somewhere in the midst of the spectrum will live a shorter life than the larger more powerful eddy which gives it life in a single transfer of energy, and a longer life than the multitude of smaller less energetic eddies which cause its progressive decline. Turbulent cascade theories based on eddy interaction dynamics (see, for example, Leslie, 1974) are not yet sufficiently well developed to give a quantitative account of this process, but provided there is some simple monotonic relationship between eddy size and lifetime, then the ratio of eddy lifetimes between donor and recipient in any interaction passing energy down the cascade will depend on the ratio of eddy sizes or, alternatively, upon the width of the spectral leap of the energy exchange. In an extreme example, taken from the atmosphere but probably paralleled in the upper ocean, for a rotation range eddy (a cyclone) losing energy to billows by Kelvin-Helmholtz instability along its fronts, the ratio of eddy lifetimes is a few days for the cyclone to a few minutes for the billows, or about 2000 : 1. More usually energy cascades between eddies located rather close together in the spectrum and with correspondingly close lifetimes.

When relating theoretical models of the turbulent cascade to experimental data it is important to remember that eddy lifetimes are usually much longer than encounter periods recorded as they are advected through moored instruments. Lifetime is Lagrangian, encounter time Eulerian.

c. Classification by eddy lifetime

The classification of geophysical turbulence into ranges given above (Section 9.3a) was based on dimensionless numbers incorporating the Coriolis and Brunt-Väisälä frequencies and kinematic viscosity. It is now possible to interpret this classification in terms of eddy lifetime τ. Eddies in the rotation range (Ro<1) spin up to a quasi-geostrophic balance, so their lifetimes must exceed f^{-1}. Eddies in the buoyancy range (Ro,Ri>1) do not have time to spin up and do not overturn, so their lifetimes must be in the range, $f^{-1}<\tau<N^{-1}$, and those in the billow turbulence range (Ri<1) have lifetimes shorter than N^{-1}. By introducing a further assumption about the turbulent cascade, namely that there is no flux divergence of turbulent kinetic energy in Fourier space in the inertial subrange, Kolmogorov (1941) derived a time scale for the start of the viscous subrange (Re=1), namely $\tau_\nu=(\varepsilon/\nu)^{-1/2}$, where ε is the energy dissipation rate.

d. Relationship between horizontal and vertical eddy dimensions

The dimensional quantities defined above can be rearranged to obtain the following relationship between eddy dimensions

$$L_V/L_H = (f/N)\ Ro\ Ri^{1/2}.$$

For rotation range eddies Ro $Ri^{1/2} \simeq 1$, so that the ratio of vertical to horizontal eddy dimensions is approximately (f/N) or 1/500. An eddy 1 km wide will be only a couple of meters thick.

This ratio is progressively decreased in the buoyancy subrange until L_V/L_H is approximately 1/4 for Kelvin-Helmholtz billows. There is evidence that individual eddies in the inertial and viscous subranges are highly anisotropic (Townsend, 1976), but they are randomly orientated, so that $\overline{(L_V/L_H)}=1$.

e. Relationship between eddy size and lifetime

Although eddies do not have a natural time scale like the frequency of a wave, and cannot therefore have anything equivalent to a dispersion relationship, it seems reasonable to suppose that the probability distribution for the lifetimes of eddies of a given size will only extend over a relatively narrow band width centered on the mean lifetime for that size. The observation that variance is patchily distributed in Fourier space leads inevitably to the conclusion that the turbulent cascade must proceed in fits and starts, depending on just which eddies happen to be present at a given location to interact and pass on energy down the spectrum, so the rate of energy flux through Fourier space, ε, will not be uniform in physical space-time, i.e., $\varepsilon=\varepsilon(\underset{\rightarrow}{x},t)$. Supposing that the mean lifetime of an eddy of a given size is uniquely determined by the mean value of the turbulent kinetic energy flux, $\bar{\varepsilon}$, then the probability distribution of lifetimes for that size will be related to the probability distribution of ε, and these two distributions might be classified in terms of depth, season, weather, etc.

The first step is to consider the mean lifetime as a function of eddy size. In order to do this it is necessary to make some assumption about the spectral form of the mean energy flux rate, i.e., of $\bar{\varepsilon}(\underset{\rightarrow}{\kappa})$. Kolmogorov (1941) introduced the famous hypothesis that $\bar{\varepsilon}$ is independent of spectral location within the inertial subrange, where there are no sources or sinks of turbulent kinetic energy and therefore the flux divergence of $\bar{\varepsilon}$ must be zero*. This leads to the identification of eddy size and lifetime at which the Reynolds number is unity and the viscous subrange begins; these are $L_\nu=(\varepsilon/\nu)^{1/4}$ and $\tau_\nu=(\varepsilon/\nu)^{-1/2}$. For all the eddies in the inertial subrange $L=(\varepsilon\tau^3)^{1/2}$. Ozmidov (1965) extrapolated to the end of the inertial subrange where $\tau=N^{-1}$, where the eddy size is $L_N=(\varepsilon/N^3)^{1/2}$, sometimes referred to as the Ozmidov length scale. Panchev (1971, p. 322) extended the argument to include eddies in the buoyancy range, terminating at a horizontal scale $L_f=(\varepsilon/f^3)^{1/2}$ and to a new range, the curvature range, in which eddy dynamics are dominated by the earth's curvature, parameterized in terms of $\beta=(df/d\phi)$. Panchev gives the boundary scale between rotation and curvature ranges as $L_\beta=(\varepsilon\beta^{-3})^{1/5}$ and the time scale as $\tau_\beta=(\varepsilon\beta^2)^{-1/5}$.

Panchev's hypothesis is, therefore, that the relationship between eddy lifetime and size in the rotation and buoyancy ranges, like the inertial subrange, is independent of the magnitude of f or N. This hypothesis has not been tested experimentally or fully explored theoretically.

f. Energetics

i. Energy sources

Recent experimental evidence suggests that the primary source of turbulent kinetic energy in the mid-ocean (i.e., away from the bottom, surface, and coastal boundary layers) is the generation of "MODE" eddies at a scale close to the Rossby radius of deformation by baroclinic instability. There is some indication that the intensity of this source may have strong regional

---

*This assumes a Lagrangian frame so that there is no flux divergence due to advection.

variation (Robinson, 1975), but there have been no reports concerning temporal variation. The spectral width of this peak has not yet been resolved by experimental observation or predicted by theory, but it is presumably small compared with the whole range of oceanic eddy sizes. The MODE eddies derive their kinetic energy from the potential energy of the ocean, the density of kinetic energy in them is much lower than the density of available potential energy. It is assumed that they lose energy to smaller scale eddies though the surface and bottom boundary layers and by internal dynamical interactions, thereby powering the cascade of turbulent kinetic energy through the spectrum.

The second source of turbulent kinetic energy is the generation of billows by Kelvin-Helmholtz instability. The shear required to initiate Kelvin-Helmholtz instability by reducing the Richardson number below the critical value of 1/4 comes partly from eddies in the rotation and buoyancy ranges, and partly from internal waves. Thus the energy for billow turbulence comes partly from larger eddies as a part of the energy cascade originating primarily in MODE eddies, and partly from internal waves, which are powered mainly by the wind, and to a lesser extent by tidal and other currents. Near the sea surface, in the seasonal thermocline, the wave energy density is strongly correlated with the local wind, and Nasmyth (1970) has shown that the intermittency factor for billow turbulence is similarly correlated, ranging from 5% during prolonged periods of calm weather up to 30% or more in rough weather. The strength of the turbulent kinetic energy source at the Ozmidov scale therefore appears to be dependent on the local wind strength, and may be negligible away from boundaries when the weather is calm.

Other sources of turbulent kinetic energy include internal wave breaking (Orlanski and Bryan, 1969), the strength of which will also probably follow the internal wave variation with local wind strength, as do the eddies associated with convective and forced circulations in the surface wind mixed layer (e.g., Langmuir cells, see Pollard, Chapter 8). A rather different kind of source of turbulent kinetic energy is the spin up of density anomalies generated at the sea surface by spatial inhomogeneities of air-sea interaction, such as rain showers and atmospheric eddies in the lee of oceanic mountains (Patzert, 1969).

To sum up, the relative strengths of the various sources of turbulent kinetic energy are not known, nor are their spatial and temporal variations, but it seems likely that the source at the Rossby radius of deformation ($L_H \approx 100$ km) is more steady than the others which depend more or less directly on short term local variations of air-sea interaction, and may become negligible away from the coast during extended periods of calm weather.

ii. Energy sinks

Energy is lost from eddies to heat by the action of viscosity at small scales ($L<1$ cm), to internal waves and to potential energy (e.g., by transporting heat downwards from near the surface where it is absorbed).

Measurements with hot film flow meters have established that the local energy dissipation rate due to viscosity is normally in the range 10^{-4} - 10^{-3} erg $gm^{-1}s^{-1}$, in the seasonal thermocline, with values as high as 10^{-2} erg gm^{-1} s^{-1} reported from the equatorial undercurrent. These instrument records also show that viscous dissipation occurs in patches, which are assumed to correspond with the patches of billow turbulence revealed by flow visualization studies. The intermittency factor associated with these patches ranges from 5% in calm weather to more than 30% in rough weather, giving spatially averaged dissipation rates of $\bar{\varepsilon}$=5x10^{-6}-5x10^{-5} erg gm^{-1}s^{-1} (calm) and $\bar{\varepsilon}$=3x10^{-5}-3x10^{-4} erg gm^{-1}s^{-1} (rough) in the seasonal thermocline away from major currents.

The rate of loss of turbulent kinetic energy to internal waves has not yet been estimated, but some attention has been given to the possible interaction between eddies in the surface mixed layer and internal waves in the thermocline. At present there is no evidence to suggest that this loss is significant in comparison with viscous dissipation and it will be neglected in the energy cascade.

The rate of loss of turbulent kinetic energy to potential energy can be estimated on a seasonal basis by comparing changes in monthly mean density profiles. The rate of increase in potential energy due to the working of eddies against gravity is about 1% of the rate of energy dissipation by viscosity so this energy sink will also be ignored in the discussion of flux divergence in the energy cascade.

iii. The turbulent energy cascade

If the only significant energy sink is viscous dissipation and if the input of energy from internal waves can be neglected (at least in calm weather) in comparison with the input from potential energy, then the average turbulent energy cascade may be described approximately as having zero flux divergence of $\bar{\varepsilon}$ between a single source at L_H≈100 km to a single sink at L_H<1 cm.

Nothing is known of the corresponding flux of energy from the source at L_H≈100 km to larger scales. One might guess that energy flows equally to larger and smaller eddies from the MODE eddies so that the magnitude of $\bar{\varepsilon}$ is the same at scales both larger and smaller than L_H.

The magnitude of the ensemble mean flux $\langle\bar{\varepsilon}\rangle$ in this idealized cascade will be somewhere in the range of values of $\bar{\varepsilon}$ measured at small scales, i.e., $\langle\bar{\varepsilon}\rangle$≈10^{-5} erg gm^{-1}s^{-1}. In calm weather, the width of the probability distribution for $\bar{\varepsilon}$ may be estimated from the observed distribution of $\bar{\varepsilon}$ values as being about a factor of 2 either way, i.e., for ε[erg gm^{-1}s^{-1}], $\log_{10}\langle\bar{\varepsilon}\rangle$ = -5±0.3. The range of values of ε, the energy flux rate inside a patch of eddies is related to the spatially averaged value $\bar{\varepsilon}$ by the intermittency factor, which may depend on eddy size.

g. Space-time diagrams

i. Eddies

Substituting the estimated mean value for the energy flux $\langle\bar{\varepsilon}\rangle$ into the relationships between eddy lifetime and size given in Section 9.3e gives the space-time diagram (Fig. 9.5). Assuming that the width of the probability distribution is $\log_{10}(\Delta\bar{\varepsilon})$=±0.3, then all eddies will lie in a relatively narrow band through Fourier space-time (Fig. 9.5).

ii. Internal waves

Also shown in Fig. 9.5 is the Garrett and Munk (1971) dispersion

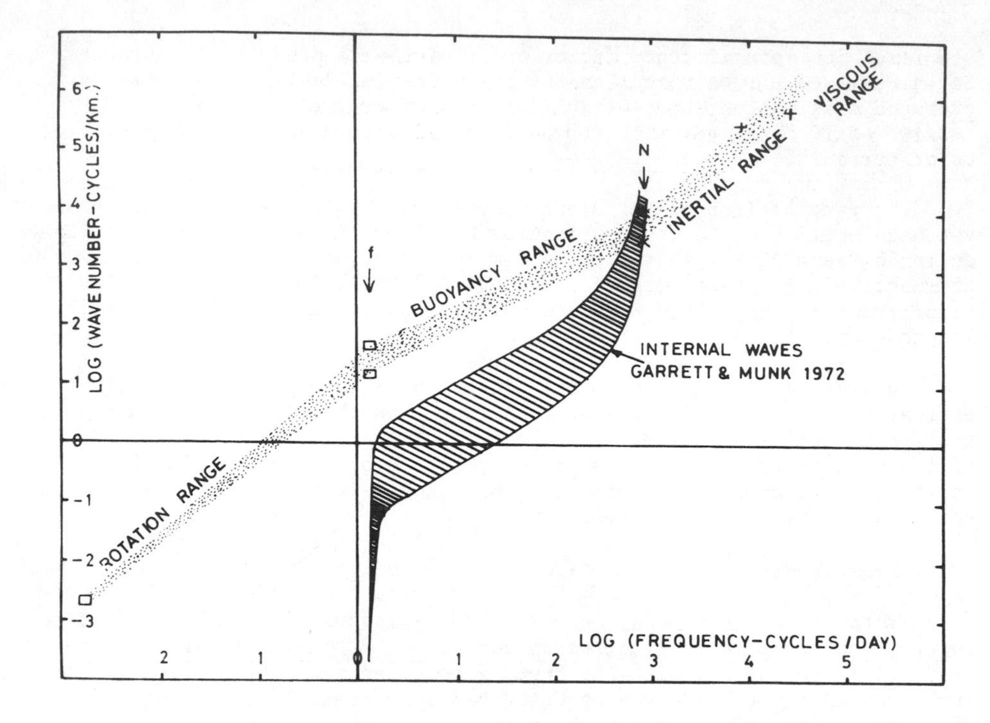

Fig. 9.5 - Space-time diagram for eddies and internal waves in the seasonal thermocline (Woods 1974a).

relation linking the wave length and frequency of internal waves. It is clear that the bands do not overlap except at wave frequencies close to N, so it should be possible in principle to discriminate between waves and eddies on the basis of the time scale for changes in structure of a given horizontal size. In such a test, the comparison is between wave period on the one hand and eddy lifetime on the other, both measured relative to the moving water rather than a fixed (Eulerian) frame of reference. It is not necessary to consider the lifetimes of wave packets in this context (see Section 9.1b above), although it is interesting to speculate on the form of the Fourier band relating the dominant length and mean lifetime of a wave packet.

9.4 Examples of parameterization schemes

a. Turbulent diffusion

The oldest and still the most popular method parameterizing sub-grid scale motions is to sweep aside arguments about individual eddies and variations in their behavior with spectral position, in favor of a uniform (or at best smoothly varying) property of the fluid akin to molecular viscosity and diffusivity. The latter works well because of the broad spectral gap between the smallest continuum motion and the mean free path of the molecules, but similar gaps do not exist at larger scales as ephemeral spaces between eddies. The underlying assumption of this micrometeorological approach is that useful progress can be made by ignoring the absence of a spectral gap and proceeding to parameterize the net effect of all sub-grid scale motions in terms of pseudo-Fickian transport coefficients. Having adopted this simplifying assumption, parameterization becomes a matter of devising formulae for the variation of the magnitudes of the transport coefficients in terms of the velocity and density distribution described deterministically by the model.

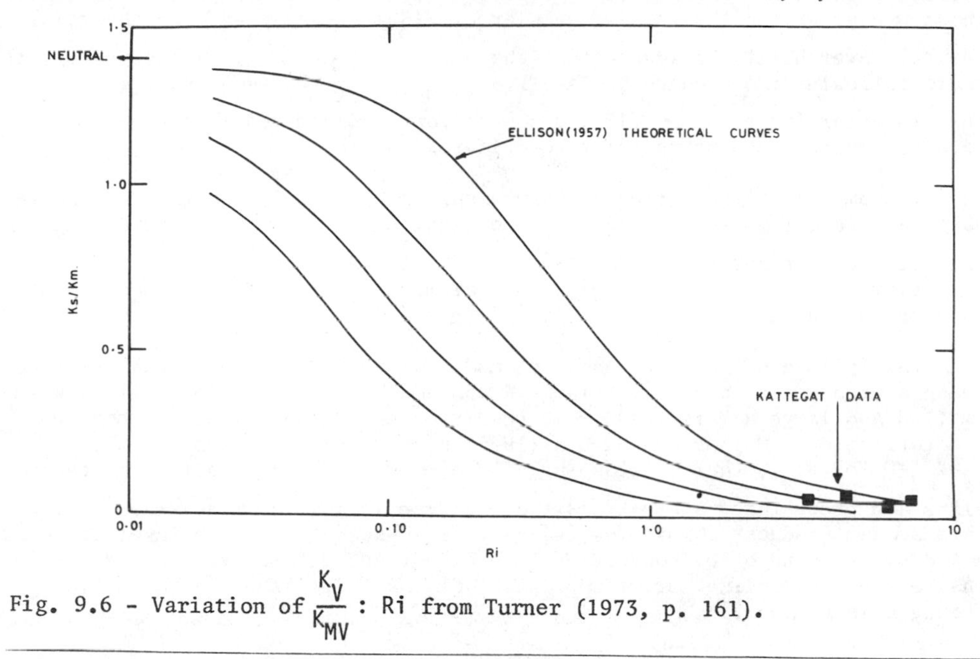

Fig. 9.6 - Variation of $\frac{K_V}{K_{MV}}$: Ri from Turner (1973, p. 161).

In a paper which pioneered the adoption of this approach in upper ocean models, Munk and Anderson (1948) used the empirical relationships between the transport coefficients for momentum and bouyancy and the Richardson number derived by Taylor (1931) from field measurements of Jacobsen (1913). The most important feature of these formulae (Fig. 9.6) is the decrease in eddy viscosity and the still faster decrease in eddy diffusivity with increase in Richardson number, but finite values of both for Ri as large as 1000.

The first objection to these and similar empirical formulae for the vertical transport coefficients K_{MV} (momentum) and K_V (heat, salt) as functions of Ri is that they are based on limited data sets, which may be unrepresentative of the conditions of the model in which they appear as parameters. Jacobsen's measurements were collected some sixty years ago with what would today be considered primitive apparatus at a site which encouraged a one-dimensional interpretation, but which differs greatly from the open ocean. But more important than the errors introduced by non-representativeness of region, is the error introduced by neglecting changes of Ri, K_{MV} and K_V with scale. Jacobsen's data were for eight levels at 2.5 m intervals (see Proudman, 1953, p. 112), which comfortably resolved the mean profile in the Kattegat halocline, but which is significantly larger than the typical Ozmidov length scale of 20 cm observed in the open ocean thermocline. It might be expected therefore that the vertical transport rate at L_V=2.5 m would not be controlled by an ensemble of billow turbulence events, within each of which the eddies are approximately isotropic, but by buoyancy/rotation range eddies, which have horizontal dimensions of order one kilometer plus possibly a vertical momentum transport by internal waves. The Kattegat site, being shallow and narrow, is unlikely to exhibit a sample of either kilometer wide eddies or internal waves, that is representative of the deep open ocean, even in calm weather, so the measurements at L_V=2.5 m are suspect. Furthermore, there is overwhelming evidence that the horizontal diffusivity increases with scale following the Richardson law (i.e., $K_H \propto L_H^{4/3}$) from scales of 10 cm to 1000 km according to Okubo (1971), so the formulae derived from the 2.5 m data are unlikely to be relevant to different scales.

To summarize, the Kattegat measurements are unlikely to provide a reliable data base from which to derive empirical expressions relating K_V, K_{MV} and Ri; not even for the vertical scale of the measurements. If this line of parameterization is to be pursued, the parameterization formulae must be based on a broader data base.

The failure of the Munk-Anderson model to predict a sharp transition between a wind mixed layer, with large K and small Ri, and a thermocline, with small K and large Ri, is the direct result of using the Taylor-Jacobsen parameterization. Mellor and Durbin (1975) have overcome this by putting $K_V=K_{MV}=0$ for Ri>1, thereby stopping all vertical transport inside the thermocline and producing a sharp transition between it and the wind mixed layer. This scheme produces a more realistic profile when the surface layer is being mixed by the wind or by convection, but cannot account for vertical diffusion inside the thermocline, or disappearance of the sharp elbow in the curve during calm weather.

i. Justification in terms of isotropic turbulence

The assumption that K=0 for Ri>1 is not consistent with Jacobsen's data. Attempts to justify it on the basis that isotropic turbulence cannot persist in regions where Ri>1, which is true, are based on the false premise that when the Richardson number calculated by finite-differences at the grid scale of the model exceeds unity, there are no sub-grid scale regions where Ri<1. On the contrary, the characteristic feature of turbulent fluids such as the upper ocean is a complex microstructure, with the average* value of Ri much lower than average on the same scale and at the same depth.

The Mellor-Durbin parameterization scheme reflects a generally held belief (e.g., Turner, 1973) that the vertical diffusion rates may be correlated with the mean energy density of isotropic turbulence in the upper ocean, which can be related by laboratory experiments to the local Richardson number. If this were true, it would be necessary to take account of the observation that isotropic turbulence in the thermocline occurs intermittently as small relatively intense patches, in otherwise laminar** flow, by developing relationships between the frequency of occurrence of such events and the bulk Richardson number at the model grid scale. Since the billow turbulence events occur

---

* This is based on ensemble average of the stability frequency and shear $\langle Ri \rangle = \{\langle N \rangle / \langle du/dz \rangle\}^2$.

**Laminar flow is defined as containing no overturning motion, such as occurs in billow turbulence.

when the local Richardson number falls below 1/4, it is argued that all one has to do is to arrive at a relationship between the bulk Richardson number and the spatial concentration of regions in which the local Richardson number is less than 1/4. A number of attempts are currently being made to derive empirical relationships of this type from measurements of velocity and density microstructure in the ocean (e.g., Simpson, 1972; Osborn, 1974) and in Loch Ness (e.g., Thorpe, 1976) so far without conclusion. If these efforts are rewarded with a satisfactory relationship between bulk Richardson number and the mean energy density of isotropic turbulence, the next step will be to relate the latter to the vertical diffusion rates on the (much larger) scale of the model, thereby closing the parameterization.

Let us review the pros and cons of this approach. It is attractive because it concentrates upon the small scale turbulence for which overturning is possible and assumes that the larger eddies achieve a negligible vertical transport because they cannot overturn. On the other hand, isotropic turbulence occurs in the thermocline on scales much smaller than the grid scale of most models; the neglect of eddies whose size lies intermediate between the resolution of the model and the motions parameterized leaves one a little apprehensive, since in most situations the <u>largest</u> unresolved eddies control the diffusion rate. The assumption that no overturning equals no vertical mixing is clearly correct, but no vertical mixing does not necessarily mean no vertical transport, as will be shown in a later example.

Remembering that the aim is to relate eddy transport coefficients to bulk Richardson number, perhaps the best plan is to cut away superfluous (and usually spurious) justifications in terms of isotropic turbulence. After all, it really does not matter, for the micrometeorological method of parameterization, whether the vertical transport is effected by two-dimensionally isotropic eddies with billow turbulence playing the subsidiary role of smearing out the strong gradients they produce (as proposed by Woods and Wiley, 1972) or whether the former achieve nothing and the latter everything.

ii. Objections to the micrometeorological approach

The major objection to the micrometeorological approach is that it treats sub-grid scale motions, which occur as rare events patchily distributed through the ocean, in terms of a smooth continuous diffusion, even though there is no spectral gap between them and the motions described deterministically by the model. This failure to take account of local variations in sub-grid scale transport introduces an error into the model at every integration step. Because the diffusion term is non-linear, parameterization error leads to errors in the model predictions, including those at space-time scales which are spectrally well separated from those of the sub-grid scale motions. This error propagation to large scales has been widely discussed in the context of computer models of the atmosphere (see, e.g., GARP, 1972; Somerville, Chapter 3).

The sensitivity of models to arbitrary changes in eddy viscosity has been demonstrated by a number of authors (e.g., Bryan and Cox, 1970; Holland, Chapter 2), but there are as yet no detailed assessments of the magnitude of the errors introduced by existing micrometeorological parameterization schemes. These will depend upon three main factors; firstly, the natural intermitency of eddies and wave packets even under conditions where they are present in the same mix as the data base from which the parameterization formulae were derived; secondly, the local departures from this representative mix of waves and eddy types due to the effects of weather, season, time of day, topography; and, thirdly, exclusion of different eddies and waves from parameterization as the resolution of the model changes. Locally, the eddy transport coefficients range from zero to two or more orders of magnitude greater than the mean value. The ultimate aim of the micrometeorological method must be to elaborate the parameterization formulae by the inclusion of factors which modulate the transport coefficients with the factors listed above, until they are capable of accurately predicting fluxes calculated from field data sets collected under the widest possible range of conditions.

The major obstacle to progress is the lack of suitable field data, and it is unlikely that there will be much improvement in micrometeorological parameterization schemes in the foreseeable future.

b. Transport due to waves

Internal waves propagating through the upper ocean carry with them latent momentum, which produces radiation pressure when they encounter a solid obstacle, and is released to the fluid as a jet if they break. Thus it is possible to conceive of momentum leaving the flow at one location, where waves are generated (e.g., by flow over bottom topography), to be later released at another level where the waves break. Furthermore, some of the wave energy lost in breaking goes into overturning and mixing the water in a short lived patch of isotropic turbulence, which produces local vertical fluxes of heat, salt and the other constituents of the water.

Parameterization of this process requires assumptions to be made about the following:

1. The rate at which energy flows into the wave field as a function of geographical location, weather, etc., and how this source is distributed across the wave spectrum.

2. The rate at which energy and momentum is lost to eddies without wave breaking.

3. The rate at which energy and momentum are transferred through the spectrum by wave-wave interaction.

4. The rate at which energy is lost to wave breaking as a function of wave length and environmental conditions and hence when and where wave momentum and energy is released to the flow.

Experimental and theoretical investigations of these processes have not yet yielded reliable data to justify constructing a complete parameterization of wave transport, but Olbers (1975) has shown how this can be done, on the basis of assumptions about the wave field. His scheme suggests that the vertical fluxes due to internal waves might be comparable with those derived from diffusion experiments.

c. Eddy mixing

The previous schemes have parameterized the transport of heat and momentum by sub-grid scale motions. In this third scheme, an attempt is made to describe in addition the transport through Fourier space of the variances of heat and momentum, with a view to retaining some predictability of the microstructure of temperature and velocity created by the unresolved processes. The unresolved motions are assumed to comprise a finite number of discrete eddies as described in Section 9.2. The mixing action of each eddy is parameterized in terms of a transfer of variance from the spectral location of the eddy to a neighboring location. In due course, smaller eddies pass the variance step by step down the spectrum until eventually molecular motion dissipates it.

Modelling the cascade of variance down the spectrum in this way requires a model of the turbulence, in which eddies are formed and decay according to prescribed rules, which obey general statistical constraints rather than the dynamical constraints of the deterministic model. Offsetting this extra computational complexity is the possibility of incorporating more physically realistic treatment of eddies of different sizes and variations in their spatial concentration. At present, a number of groups are working on parameterization schemes with spectral resolution, but so far none has reached a state of development that would justify its incorporation as a subroutine in models of the upper ocean.

d. Frontal upwelling

It was pointed out above (Section 9.4a) that while the absence of overturning excludes vertical mixing it does not necessarily exclude vertical fluxes of heat and other constituents of sea water. The relatively small vertical component of the motion of two-dimensionally isotropic eddies in the buoyancy and rotation ranges can achieve significant vertical transport under certain circumstances, which are common in the upper ocean. These special circumstances may be divided into two categories, which will now be summarized.

i. Inclined density surfaces

The motion of rotation and byoyancy range eddies lies very largely in the plane of density surfaces. Where these are inclined to the horizontal (i.e., where there is a horizontal density gradient) this isentropic motion has a vertical component. If, as usually happens, the horizontal density gradient is the net result of opposed contributions from horizontal salinity and temperature gradients, so that the T-S relationship is different on either side of the sloping density surface, then the motion associated with rotation and buoyancy range eddies mixing the fluid along the density surface produces corresponding fluxes of heat and salt along the density surface. And because the surface is inclined to the horizontal, these isentropic fluxes have horizontal and vertical components, in the ratio of the slope of the surface. Woods (1975) has pointed out that for the rotation range eddies, assumed to be in geostrophic balance, the slope $S=fN^{-1}Ri^{-1/2}$ and the ratio of the horizontal and vertical diffusivities must therefore be given by

$$\frac{K_H}{K_V} = \frac{N^2}{f^2} Ri$$

ii. Vertical gradient in distributed source or sink

Although isentropic motions of the kind just described do not achieve irreversible mixing in the cross-isentropic direction, they do have convergence and divergence patterns which produce a cross-isentropic component of motion. Such motions will persist only for the duration of the eddy that generates them: they do not produce irreversible* displacements of the density surfaces in the vertical. Nevertheless, while temporarily displaced, they raise deep water nearer to the surface, and therefore to a level more brightly illuminated by the sun. This increased illumination of temporarily raised water achieves, firstly, a greater heating by short wave absorption and, secondly, a greater rate of consumption of nutrients by phytoplankton. Thus, water squeezed upward by convergence along density surfaces becomes warmer and more depleted of nutrients than the water that was not. In due course the water relaxes back to its mean level (when the eddy decays), bringing down with it the extra heat and nutrient deficiency. Many such events combine to produce a net downward heat flux and a net upward nutrient flux.

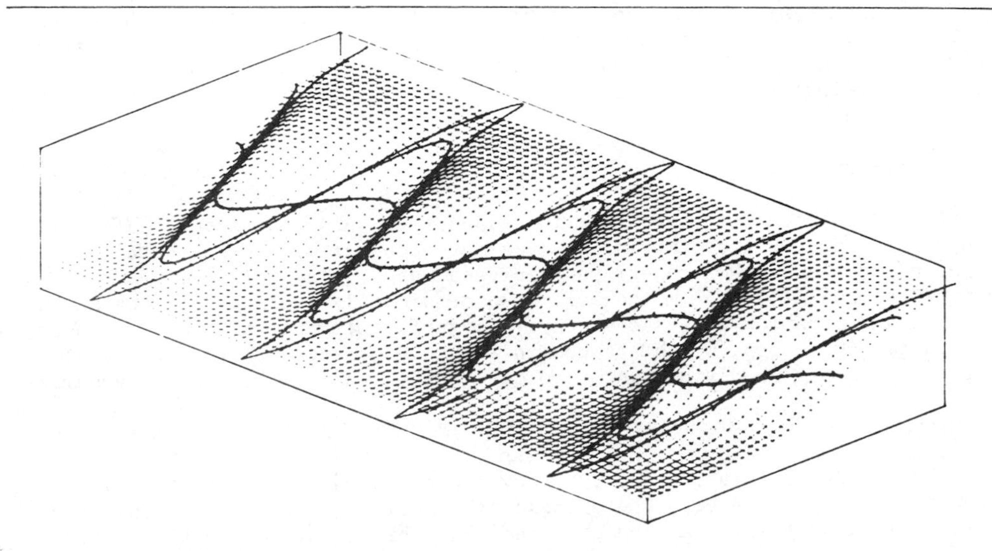

Fig. 9.7 - Particle trajectories along a front showing vertical undulations related to waves ($\lambda \simeq 10$ km). The grid lies on a density surface sloping at the front (from Woods, 1975).

iii. Frontal upwelling

The rates of vertical transport are achieved by non-overturning eddies in the two cases described above depends upon the slope of the density surfaces and the speed of isentropic motion and its convergence. These three factors all have maximum values at fronts (Fig. 9.7), and Woods (1975) has argued that vertical heat and nutrient fluxes inside the seasonal thermocline may depend almost entirely upon the existence of the vertical circulation at fronts. As an example, he parameterized the vertical heat flux in the central Mediterranean in terms of observed properties of fronts in that region, deducing an equivalent vertical diffusivity $K_V = 2\ cm^2 s^{-1}$ which is an order of magnitude larger than that consistent with the seasonal changes inside the thermocline.

---

*Irreversible displacements can only be achieved by cross-isentropic mixing, by billow turbulence, or by wave breaking.

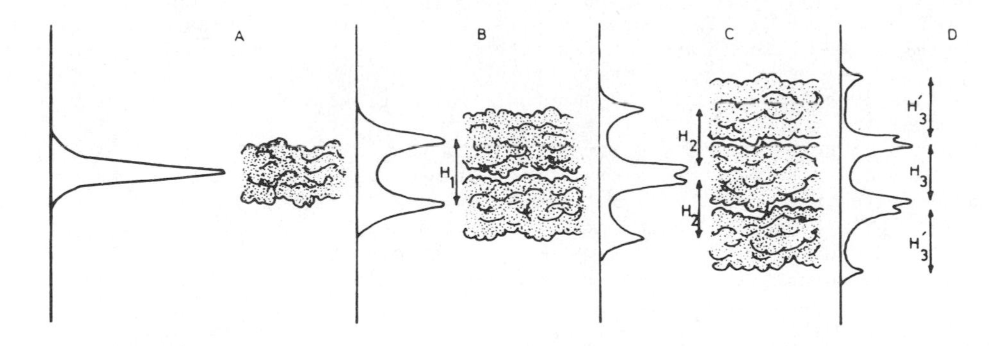

Fig. 9.8 - The breakup of a density sheet due to successive billow turbulence events in the model proposed by Woods and Wiley (1972).

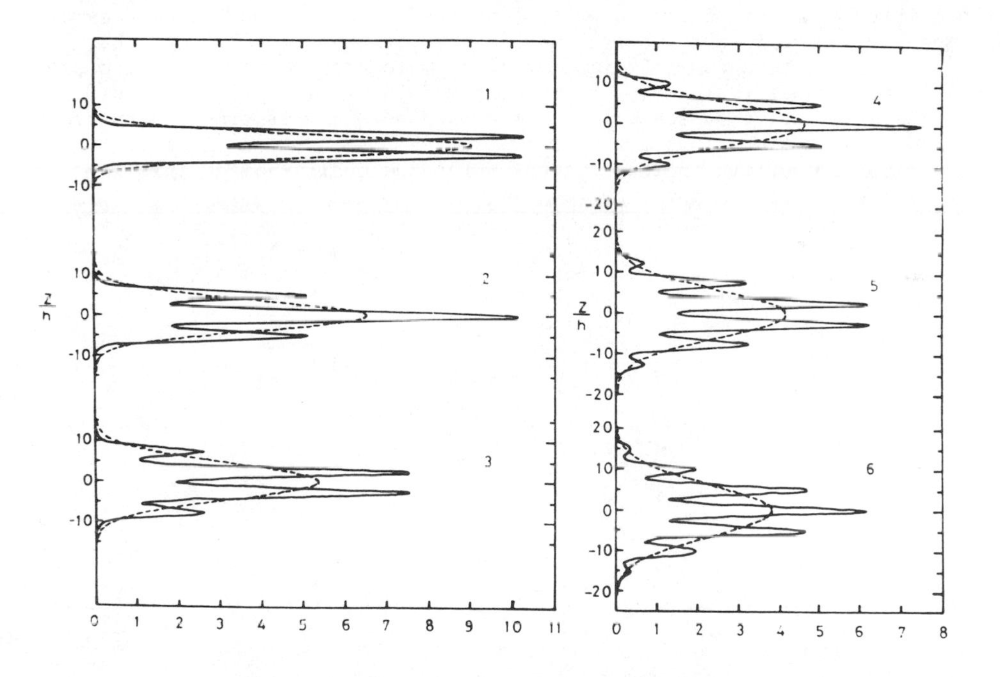

Fig. 9.9 - Parameterization of billow turbulence diffusion shown in Fig. 9.8, showing the increasing thickness of a smooth profile (dotted lines) fitted to the fragmented sheet profiles (continuous lines).

e. Billow turbulence

For computer models whose resolution is close to the Ozmidov length scale, sub-grid scale transport is dominated by billow turbulence events. Woods and Wiley (1972) proposed a physical model in which sheets (thin interfacial zones between layers a few meters thick) are progressively fragmented by a succession of events involving increasing numbers of vertically stacked patches of billows (Fig. 9.8), so that the originally sharp interface becomes thicker and thicker in a series of quantum jumps, each taking about five minutes to complete and separated from the next by several hours on average. The authors described the overall spreading of the interface in terms of an eddy diffusivity, which was related to (*i*) a form factor derived from observation of microstructure and (*ii*) the mean interval between successive events in the diffusion sequence estimated from the observations of the duration of individual events and the spatial concentration of events (Fig. 9.9).

9.5 Conclusion

It has been demonstrated that quite small arbitrary changes in the diffusion coefficients incorporated in primitive equation models as a method of parameterizing unresolved motions produce large changes in the model predictions. Observations of these unresolved motions show that they are intermittent and attempts to parameterize selected aspects of the motion, such as frontal upwelling lead to diffusion coefficients which fluctuate in space and time from zero to a hundred times larger than the seasonal mean value. These fluctuations, which are much larger than those found to produce large effects in primitive equation models, are not separated from the spectral window of the model by a persistent spectral gap. It is, therefore, unlikely that it will be possible to adequately improve the simple micrometeorological parameterization schemes that are found in most models at present. The most promising alternative, which will involve much more computation, is to treat the unresolved motions in terms of a statistical model based on closure assumptions appropriate to the internal waves and eddies responsible for vertical fluxes through the upper ocean.

Part III

PHYSICAL MODELS OF THE UPPER OCEAN

Chapter 10

ONE-DIMENSIONAL MODELS OF THE UPPER OCEAN

P. P. Niiler and E. B. Kraus

10.1 Introduction

The present paper has profited from many presentations and discussions by members of the "one-dimensional working group" in Urbino. These contributors were too numerous for active participation in the writing; their names are marked by asterisks in the text.

One-dimensional models of the upper ocean can be useful because bulk temperatures or salinities tend to vary more along a vertical distance of a hundred meters than along a horizontal distance of a thousand kilometers. This holds true over many parts of the world's oceans and, where it is the case, vertical exchange processes between the air and the sea, as well as vertical mixing within the water column, are likely to affect local conditions much more rapidly and effectively than horizontal advection and horizontal mixing. It follows that it is permissible, for many purposes, to treat the upper ocean layers as being statistically homogeneous along the horizontal. This approximation is applied consistently in all models presented in this paper. It means that horizontal derivatives can be omitted in the mathematical treatment and only changes along the vertical are being considered.

It follows from the discussion in the Introduction (Chapter 1) that the prediction of the temperature is probably the most important function of upper ocean models. Temperature variations have a crucial effect on the climate, the local biological environment, and acoustic propagation. Next to temperature, we are concerned with current velocities in the upper ocean for shipping purposes, fish migration, and also for the modelling of the oceanic general circulation. However, for most of these purposes one wants to know not only the local current velocities but also their horizontal variations. The one-dimensional models are therefore generally less useful for the processing of velocity information than they are for temperature.

The temperature and velocity fields interact with each other and with the oceanic density field, which in turn is a function not only of temperature but also of salinity. In our models we have to deal therefore with the evolution of each of these properties. In the present context, this evolution can be described by a set of one-dimensional conservation equations. In particular the one-dimensional momentum equation can be reduced to the form:

$$\frac{\partial \overline{\vec{v}}}{\partial t} + f\,\vec{k} \times \overline{\vec{v}} + \frac{\partial}{\partial z}\overline{w'\vec{v}'} = 0, \qquad (10.1)$$

where $\overline{\vec{v}}$ is the averaged horizontal current velocity, f the Coriolis parameter and $\vec{k}$ a unit vector along the vertical z direction. Primes are being used here to denote turbulent deviations from the bulk velocities $\overline{w}$ and $\overline{\vec{v}}$. Their averaged product represents the vertical flux of horizontal momentum, known as the Reynolds stress. The averages can be derived from time series segments of a few hours in the oceanic cases; shorter for laboratory experiments.

Because of the assumed horizontal uniformity $\partial\bar{w}/\partial z = -\nabla_H \cdot \bar{\underset{\rightarrow}{V}} \equiv 0$.

The conservation of sensible heat or enthalpy leads to the equation:

$$\frac{\partial \bar{T}}{\partial t} + \frac{\partial}{\partial z} \overline{w'T'} = - \frac{1}{\rho c} \frac{\partial I}{\partial z} \tag{10.2}$$

($\bar{T}$=bulk temperature, ρ=density, c=specific heat, I=penetrating component of solar radiation). The fluctuating part of the temperature is denoted by T'. The last term in Eq. (10.2) is associated with a source of heat due to the absorption of solar radiation within the water column. As indicated by Ivanoff (Chapter 4), about 55 per cent of the incoming solar radiation is absorbed in the uppermost meter; the remainder penetrates deeper and is absorbed more or less exponentially with an attenuation coefficient γ which varies between about 0.03 m^{-1} in clear Mediterranean water and 0.3 m^{-1} in dirty coastal water. A value of $\gamma \sim 0.04$ m^{-1}, corresponding to a scale depth of about 25 meters or slightly less, appears to characterize much of the open tropical and sub-tropical oceans.

The equation for the salinity S is

$$\frac{\partial \bar{S}}{\partial t} + \frac{\partial}{\partial z} \overline{w'S'} = 0. \tag{10.3}$$

Finally we shall require an equation which specifies mass conservation. However, as density fluctuations appear in the following equations always multiplied by the gravitational acceleration g, it is convenient to introduce an expression for the buoyancy

$$b = -g \frac{(\rho - \rho_r)}{\rho_r} \equiv g\{\alpha(T - T_r) - \beta(S - S_r)\}.$$

The subscript r denotes constant reference values of the density, temperature and salinity; the coefficients α and β describe the logarithmic expansion of ρ as functions of T and S respectively. The last identity can be used to derive a buoyancy conservation equation from the thermal and salinity equations:

$$\frac{\partial \bar{b}}{\partial t} + \frac{\partial}{\partial z} \overline{w'b'} = - \frac{g\alpha}{\rho c} \frac{\partial I}{\partial z}, \tag{10.4}$$

with b' denoting the fluctuating part of b. Below we shall assume $\rho \doteq \rho_r$, except where the difference $\rho - \rho_r$ is needed for the specification of the buoyancy (Boussinesq approximation).

To solve the set of equations (10.1)-(10.4), one has to find explicit expressions for the turbulent fluxes. This has been attempted in a variety of ways which are listed below.

a) Deterministic solutions: Basically this approach avoids the problem by a direct determination of the fluctuating velocities from the primitive equations. This requires specifications of the initial condition on a very fine space scale and computation of their evolution with a correspondingly high time resolution. This process has been used by Deardorff (1970) but it is too expensive and time consuming to be utilized directly for routine oceanographic modelling and prediction.

b) Turbulence closure models: As discussed briefly in the following section, these models involve the so-called Reynolds flux equations which express the evolution of the averaged products in Eqs. (10.1)-(10.4) as a function of higher order moments - that is, of averaged triple products - of the fluctuating quantities. These higher moments have to be parameterized in terms of empirical coefficients and computable quantities. This parameterization introduces some uncertainties and the equations are cumbersome.

c) Eddy coefficient and mixing length hypothesis: This classical method, which is based on analogy with molecular transports, assumes that the turbulent fluxes can be expressed by the gradient of the transported quantity multiplied by an appropriate eddy diffusion coefficient:

$$-\overline{w'\underset{\rightarrow}{v}'} = K_{MV} \frac{\partial}{\partial z} \overline{\underset{\rightarrow}{v}} \quad ; -\overline{w'T'} = K_V \frac{\partial \overline{T}}{\partial z} ;$$

and so forth. Developed more than fifty years ago, the method is still widely used (see, e.g., Holland, Chapter 2), though its physical basis is precarious. The main trouble is that the eddy transport coefficients are complicated functions of space and local stability conditions, which functions have to be determined empirically. This method breaks down when gradients vanish or when coupled fluxes of two conservative quantities transport one of them against its own gradient.

d) Mixed layer models: Observations show that the top layer of the ocean is usually mixed rather thoroughly. The vertical distribution of temperature, salinity and horizontal velocity within this mixed layer is - if not uniform - at least very much smaller than the variations across the layer boundaries or variations within the thermocline below. Allowance for a uniform mixed layer permits vertical integration of Eqs. (10.1)-(10.4). This yields expressions for the turbulent transports in terms of the mean quantities and the external inputs. If $\partial I/\partial z=0$, all transports within the mixed layer are linear functions of z. The turbulence energy equation (see next section) is used to obtain an expression for the evolution of the layer depth which is needed for closure.

The following section of this paper deals briefly with the turbulence closure schemes. The whole remainder is concerned with one-dimensional mixed layer models. At this stage of the art, these provide probably the most effective tool, particularly for the prediction of sea surface temperature and upper ocean heat storage over a relatively wide range of time scales.

10.2 Turbulence closure schemes

The full governing equations for the Reynolds stress tensor can be found, for example, in a paper by Mellor (1973). Allowing for the here stipulated conditions of horizontal homogeneity and with neglect of the vertical component of the Coriolis force which cannot have any significant effect on the turbulent motion, these equations can be reduced to two separate equations: one for the mean specific turbulence kinetic energy $q^2 \equiv \overline{w'^2 + \underset{\to}{v}' . \underset{\to}{v}'}$ and a second one for the vertical transport of specific horizontal momentum $\overline{w'\underset{\to}{v}'}$.

In the present case the turbulence kinetic energy equation has the form (Kraus, 1972):

$$\frac{1}{2}\frac{\partial}{\partial t} q^2 = - \overline{w'\underset{\to}{v}'} \frac{\partial \underset{\to}{\overline{v}}}{\partial z} + \overline{w'b'} - \frac{1}{2}\frac{\partial}{\partial z}[w'(\overline{w'^2 + \underset{\to}{v}'^2}) + \rho^{-1}\overline{w'p'}] - \varepsilon. \qquad (10.5)$$

The first term on the right-hand side represents the work of the stress $\overline{w'\underset{\to}{v}'}$ on the mean shearing flow. As the kinetic energy of a shear flow is always larger than the kinetic energy of a uniform flow with the same average momentum, any reduction of the mean shear by mixing must generate an equivalent amount of eddy kinetic energy. The next term represents the rate of working of the buoyancy force. It is positive if relatively dense fluid parcels move downward while the lighter ones move upward. The following term deals with the convergence of the turbulent vertical flux which carries the energy of the turbulent velocity and turbulent pressure fluctuations. Finally, the last term represents the rate of viscous dissipation of turbulence energy. Explicit expressions for this dissipation, as found in textbooks of hydrodynamics, show that it involves also horizontal derivatives of the velocity fluctuations. However, these occur all in the form of squares and products, the averages of which do not disappear.

The corresponding one-dimensional equation for the Reynolds stress is:

$$\frac{\partial}{\partial t} \overline{w'\underset{\to}{v}'} = -\overline{w'^2} \frac{\partial \underset{\to}{\overline{v}}}{\partial z} - \frac{\partial}{\partial z} \overline{w'^2\underset{\to}{v}'} - (\overline{\underset{\to}{v}' \frac{\partial p'}{\partial z}} + \overline{w'\nabla_H p'}). \qquad (10.6)$$

The physical interpretation of the individual terms in this equation is less clear than in Eq. (10.5). The first right-hand term could be viewed as an averaged interaction between the perturbation vertical velocity and transient changes in the bulk current which are themselves produced by the vertical displacements. The second term clearly deals with the convergence of a turbulent vertical flux of Reynolds stress. The last term involves again the pressure correlations. Following an argument first made by Rotta (1951) this term can be expressed as a function of the double and triple velocity moments. It can be shown in this way that the principal effect of this term is to redistribute the energy between its components by conversion of the anisotropic disturbances associated with the shear and the buoyancy, into a more nearly homogeneous isotropic state. This has been discussed in more detail also by Lumley and Khajeh-Nouri (1974) and by Garwood *(1976).

An exact presentation of the turbulence energy and Reynolds stress equations would include also terms for a viscous transport of energy and of Reynolds stress, as well as a term involving the vertical component of the Coriolis force. These terms are exceedingly small and quite irrelevant in the present context, though they can be found in Mellor's (1973) paper. That paper also contains the equations for $\overline{b'^2}$ and $\overline{w'b'}$ which correspond to Eqs. (10.5) and (10.6).

The set of equations (10.1)-(10.6) together with those for $\overline{b'^2}$ and $\overline{w'b'}$ is still not closed, because third order moments, like $\overline{w'\underset{\to}{v}'^2}$ and $\overline{\underset{\to}{v}'w'^2}$, have been introduced as new unknowns. It would be possible, in principle, to represent the evolution of these third order moments by another set of equations which contain fourth order moments and so forth. As an alternative to such a rather futile exercise, one can try to close the system by a number of simplifying assumptions which approximate the third order moments as a function of the bulk variables, the second order moments and a variety of empirical constants. The mathematical treatment can be reduced further by assuming that the generation of the second order moments is in balance with their dissipation, that is, by omission of the time differentials and triple correlations in Eqs. (10.5) and (10.6). Mellor and Yamada (1974) have presented this whole process as a hierarchy of turbulence closure models which involve increasingly sweeping simplification. They show that it leads, at the lowest level, to the classical representation by eddy coefficients which are expressed as functions of a mixing length l and of the root mean square turbulent velocity q:

$$-\overline{w'\vec{v}'} = K_{MV}\frac{\partial\vec{\overline{v}}}{\partial z} \equiv lqS_{KM}\frac{\partial\vec{\overline{v}}}{\partial z}, \tag{10.7}$$

$$-\overline{w'b'} \cong g\alpha K_V\frac{\partial\overline{T}}{\partial z} \equiv g\alpha lqS_K\frac{\partial\overline{T}}{\partial z}. \tag{10.8}$$

The factors S_{KM} and S_K represent the influence of the static stability on the eddy coefficients. As illustrated in Fig. 10.1 they are decreasing functions of the Richardson number

$$Ri = \frac{\partial\overline{b}/\partial z}{(\partial\vec{\overline{v}}/\partial z)^2} \equiv \left(\frac{N}{\partial\vec{\overline{v}}/\partial z}\right)^2. \tag{10.9}$$

The figure shows how widely estimates by various authors differ from each other. Munk and Anderson (1948) in particular assumed lq=const. When the number Ri>0, work has to be done by the shear stress to mix a stably stratified fluid. Mellor's approach indicates that S_{KM} and S_K become zero for Ri=0.23; this means that there can be no turbulent vertical transports of momentum or heat if Ri exceeds this critical value. In conditions of instability, the factors become infinite in the Munk and Anderson model. In Mellor's model they seem to approach asymptotically some limiting large value as -Ri becomes large.

The most obscure and uncertain part of the formulation (10.7) and (10.8) is the concept of the mixing length l. At present there is no sound physical basis for a stipulation of this length as a function of the depth z. Mellor and Yamada (loc. cit.) use the interpolation formula

$$l = \frac{\kappa z}{1+\kappa z/l_\infty} \tag{10.10}$$

(κ=0.4; von Kármán's constant). In this formulation $l \to \pm l_\infty$ as $z \to \pm\infty$, with

$$l_\infty = m_l \left\{\int_0^\infty qz\,dz \Big/ \int_0^\infty q\,dz\right\},$$

where m_l is another empirical constant. Other possible formulations of l_∞ are listed by Kraus (1972), including one in terms of the Coriolis force and the gradient wind velocity by Blackadar who originally suggested the

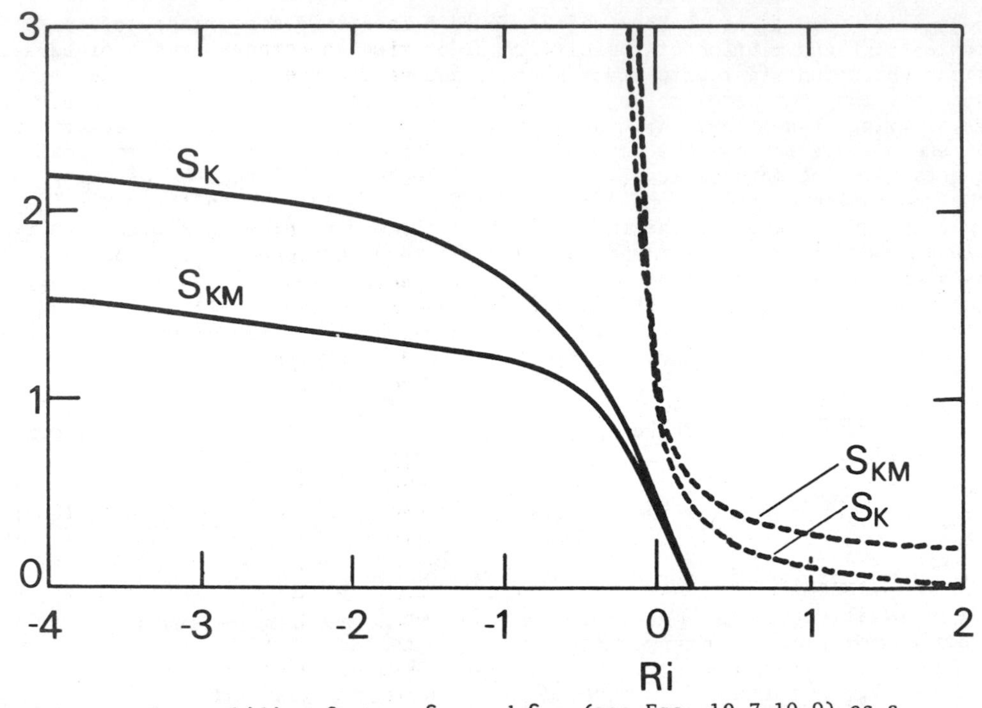

Fig. 10.1 - The stability factors S_{KM} and S_{KH} (see Eqs. 10.7-10.9) as a function of the Richardson number Ri. The full lines are based on Mellor and Durbin (1975); the dashed lines represent the corresponding parameterization by Munk and Anderson (1948).

---

expression (10.10) and a somewhat different one in terms of the Coriolis force and the friction velocity by Lettau. All these formulations are rather arbitrary until they can be justified pragmatically.

If l is stipulated, the quantity q which also enters the expression for the K's in Eqs. (10.7) and (10.8) can be obtained from the balanced turbulence energy equation (d/dt=0), if the triple and pressure correlations are neglected and the dissipation is set equal to $m_\varepsilon q^3/l$. Equation (10.5) then assumes the form

$$\overline{w'\vec{v}'} \cdot \frac{\partial \vec{\bar{v}}}{\partial z} - \overline{w'b'} = \varepsilon = m_\varepsilon l^{-1} q^3. \tag{10.11}$$

In the last term, m_ε is yet another empirically determined proportionality constant.

Equations (10.1), (10.4), (10.5), (10.7), (10.8), (10.11), together with the corresponding equation for $\overline{b'^2}$, form a closed system which can be solved to yield the evolution of a velocity and temperature profile as a function of the atmospheric inputs and of the initial conditions. Pandolfo and Jacobs (1972) and Mellor and Durbin (1975) have carried out numerical experiments on this basis. These experiments do in fact develop reasonably realistic profiles including mixed layers. However, apart from involving some rather arbitrary assumptions and at least three different empirical constants, the approach also requires a relatively complex computational scheme.

10.3 Mixed layer models

a. General considerations and assumptions

Vertically more or less uniform mixed layers can be found almost everywhere immediately below the ocean surface. We do not need a model to prove this fact. Its a priori acceptance greatly facilitates any modelling of the upper ocean. Mellor has pointed out that the resulting models are disconnected to some extent from the information which is available through general boundary layer theory, but this is compensated by the physical insight and simplicity which comes from their being tailored specifically for oceanic (and atmospheric) application.

An essential difference between the simplified equilibrium turbulence closure models and the layer models lies in the different physical importance which they assign to the transport of turbulence kinetic energy. To the extent that the equilibrium closure models omit all triple correlations, they arbitrarily set the flux of mechanical energy equal to zero. This leads to the reduced form (10.11) of the turbulence energy equation (10.5). On the other hand in the mixed layer models, turbulence energy which may have been supplied or generated indirectly by the wind near the surface is used to work against gravity at the bottom of the layer where dense water is entrained from below. The existence of a flux of turbulence energy is therefore essential to the mixed layer models. In the second order turbulence closure models, layer deepening can occur only if there is a "local" supply of energy because of a mean shearing motion at the bottom of the mixed layer.

The first mixed layer or bulk layer model developed by Keith Ball (1960) for the atmosphere above heated land did not in fact involve any mean horizontal motion. The approach was extended and modified for oceanic use by Kraus and Turner (1967). References to the following further development of mixed layer models, including consideration of horizontal shearing motions and observational evaluations, are listed in papers by Niiler (1975) and by Gill and Turner (1976).

It may be useful here to list explicitly the assumptions which have been made in most mixed layer models:

i. The mean temperature, salinity and horizontal velocity are assumed to be quasi-uniform within the layer.

ii. On the depth and time scales of the model, a quasi-discontinuous distribution can be envisaged for the same variables across the sea surface and across the lower mixed layer boundary.

iii. The rate at which the mean square turbulent velocity (velocity variance) changes locally is assumed small compared to the turbulence generating and dissipating effects.

iv. Temperature changes associated with frictional dissipation and with changes in salinity (chemical potential) are neglected.

The first two assumptions are characteristic for mixed layer models. The two others have also been made - explicitly or implicitly - in most turbulence closure or eddy coefficient schemes which have been used for oceanic modelling purposes (for an exception, see Weatherly, 1974).

Assumption (*i*), in particular allows us to integrate Eqs. (10.1)-(10.4) individually from the bottom of the mixed layer at a depth $z=-h$ to the surface at $z=0$. This permits representation of the bulk horizontal velocity $\vec{\underline{v}}$, temperature T_0 and salinity S_0 of the layer as a function of exchanges with the air above and with the interior ocean below. The system is not closed, because the time-dependent depth h has been introduced as a new variable. In the absence of a mean vertical velocity ($\bar{w}=0$), any deepening of the layer must be equal to the rate w_e with which water is entrained from the interior below. The physics of this entrainment has been discussed by Phillips (Chapter 7), including the fact that the entrainment flow can only be directed towards the more turbulent fluid region, that is, upwards in the present case. Entrainment is therefore associated with layer deepening. At times when the layer becomes shallower the entrainment must cease. Symbolically:

$$\begin{aligned} w_e &\equiv dh/dt && \text{for } dh/dt > 0; \\ w_e &\equiv 0 && \text{for } dh/dt \leqslant 0. \end{aligned} \tag{10.12}$$

To close the system, the entrainment velocity w_e is calculated from the vertical integral of the turbulence-energy equation (10.5) in its balanced form ($d/dt=0$). To evaluate the integrals of Eqs. (10.1)-(10.5) one has to know the flux boundary conditions at the surface ($z=0$) and at the bottom interface ($z=-h$).

b. Flux boundary conditions at the sea surface

The surface fluxes of sensible heat, salinity and the derived flux of buoyancy can be specified in most cases with reasonably good approximation by bulk aerodynamical formulae (Busch, Chapter 6) as functions of the existing atmospheric and mixed layer conditions:

$$\rho c\, \overline{w'T'})_{z=0} = R_0 - I_0 + H_0 + A_c Q_0 = R_0 - I_0 + \rho_a C_a u_a \{c_p[T_0 - T_a] + A_c[r_*(T_0) - r_a]\}; \tag{10.13}$$

$$\rho\, \overline{w'S'})_{z=0} = (P_0 - Q_0) S_0; \tag{10.14}$$

$$\overline{w'b'})_{z=0} = g\{\alpha \overline{w'T'})_{z=0} - \beta \overline{w'S'})_{z=0}\} \equiv B_0. \tag{10.15}$$

The subscript 0 refers to conditions at the water surface, the subscript a refers to conditions in the air at some predetermined reference level $z=a$ (usually, $a \cong 10$ m). Most of the symbols have been explained in connection with Eqs. (10.1)-(10.4); the remaining ones have the following meanings:

R_0 = surface flux of solar and terrestrial radiant energy per unit area;

I_o = surface flux of penetrating solar radiation (~45% of total solar radiation);
H_o = surface flux of sensible heat;
Q_o = surface flux of water vapor (evaporation);
A_c = latent heat of evaporation;
C_a = surface drag coefficient;
u_a = wind speed;
P_o = precipitation rate;
r = specific humidity;
r_* = saturation specific humidity.

The surface flux function B_o specifies the rate at which buoyancy is removed from the water column (or available potential energy is supplied) by surface cooling and salinity changes:

$$B_o \equiv \frac{g}{\rho} \{\frac{\alpha}{c}(R_o - I_o + H_o) - \beta P_o S_o + (\frac{\alpha A}{c}c + \beta S_o) Q_o\}. \tag{10.16}$$

All fluxes are considered positive when directed upwards from the water into the air. (This means that I_o is always negative.) Because the heat capacity and density of water are very much larger than those of air, significant gradients can be sustained only in the atmosphere above the sea surface and not in the water. It is therefore permissible, for all intents and purposes, to equate the surface temperature and salinity with the bulk mixed layer values of these quantities.

In conditions of horizontal homogeneity, that is, on a fully developed sea, the magnitude of the downward flux of momentum from the surface is equal to the wind stress τ_o. As such, it can be expressed in a variety of ways:

$$|\overline{w'\underset{\rightarrow}{v}'}| = \frac{\tau_o}{\rho} \equiv u_*^2 = \frac{\rho_a}{\rho} C_a u_a^2, \tag{10.17}$$

where u_* is known as the friction velocity. When the wind blows with a speed u_a=8m s^{-1} at a height of 10 meters above the surface, $u_* \approx 1.0$ cm s^{-1}. The corresponding value of u_* in the air would be about 28 cm s^{-1}. The direction of the stress is equal to the wind direction close over the surface.

Finally we shall need an expression for the flux of the turbulent velocity and pressure fluctuations, given by the term in square brackets in Eq. (10.5). Near the sea surface, this flux must be equal to the rate of working by the wind. This means that it has to be equal to the stress multiplied by a wind velocity. The trouble is that one cannot measure this wind velocity without ambiguity at any particular height. The wind stress works on waves at heights which are different for waves of different length. Through tangential friction it also works directly on the surface (see Phillips, Chapter 12). In the circumstances it is inconvenient to be too specific at this stage. We shall describe the rate of working by the general relation:

$$[-\tfrac{1}{2}\overline{w'(w'^2+\underset{\rightarrow}{v}'^2)} + \rho^{-1}\overline{w'p'}])_{z=0} = m_1 u_* \qquad (10.18)$$

leaving consideration of the proportionality factor m_1 for discussion below.

c. Flux boundary conditions at the bottom of the mixed layer

We shall neglect the effect of diffusion across the stable interface at the bottom of the mixed layer and assume that all mixing processes are associated with the entrainment of the lower fluid into the mixed layer. The boundary conditions for the flux of heat, salinity, and buoyancy assume then the forms:

$$\left.\begin{aligned} \overline{w'T'})_{z=-h} + w_e \Delta T &= 0 \\ \overline{w'S'})_{z=-h} + w_e \Delta S &= 0 \\ \overline{w'b'})_{z=-h} + w_e \Delta b &= 0 \end{aligned}\right\} \qquad (10.19)$$

The symbols ΔT, ΔS and Δb represent the quasi-discontinuous changes of these three quantities across the base of the mixed layer.

In the absence of entrainment, all the turbulent fluxes become zero at $z=-h$. In physical terms, this means that there is just not enough turbulence energy available in this case to overcome the stable stratification at the base of the layer and to produce any mixing with the lower water. The mixed layer becomes then effectively decoupled from the ocean interior. In the model presented here, this decoupling is assumed to be absolute.

The momentum flux boundary condition is not quite as simple. As discussed by Pollard and Millard (1970), internal waves can radiate momentum downwards through stable layers even in conditions of horizontal homogeneity. Integrated over some time, the resulting bottom drag on the mixed layer can be of the same order as the acceleration produced by the wind. It may contribute to the relatively fast attenuation of the inertial oscillation which can be produced in the mixed layer by sudden wind changes. Following Crepon*, we shall parameterize this radiation stress by the square of $\underset{\rightarrow}{v}$, multiplied by a generalized drag coefficient C, which should be a function of the particular density stratification in the fluid below the mixed layer. This parameterization is rather crude; fortunately it will be seen below that it has little effect on the features with which our model is primarily concerned. The momentum flux boundary conditions, as derived from this argument, becomes

$$\overline{w'\underset{\rightarrow}{v}'})_{z=-h} + w_e \underset{\rightarrow}{v} = C\underset{\rightarrow}{v}|\underset{\rightarrow}{v}|. \qquad (10.20)$$

Finally, the boundary condition for the flux of mechanical energy can be established similarly in the form

$$[\tfrac{1}{2}\overline{w'(w'^2+\underset{\rightarrow}{v}'^2)} + \rho^{-1}\overline{w'p'}])_{z=-h} + \tfrac{1}{2} w_e q^2 = 0. \qquad (10.21)$$

The second term represents here the rate at which turbulence energy has to be supplied by the downward flux to make the entrained (initially quiescent) water as agitated as the mixed layer water.

d. The integral relations

To obtain explicit expressions for the mixed layer momentum, temperature, salinity, and buoyancy, one integrates Eqs. (10.1)-(10.4) along the vertical. The integration is easy because the bulk variables are independent of z within the mixed layer. To deal with the effect of penetrating solar radiation we set

$$I = I_0 \exp(\gamma z),$$

as discussed in Section 10.1. The mixed layer temperature must then satisfy the equation

$$\frac{dT_0}{dt} = -\frac{w_e \Delta T}{h} - \frac{1}{\rho ch}(R_0 + H_0 + A_c Q_0 - I_0 e^{-\gamma h}). \tag{10.22}$$

The first term on the right hand side represents the temperature decrease in the layer which is caused by the entrainment of water with a temperature that is ΔT lower than T_0. The term in brackets represents the flux of heat through the sea surface minus the penetrating solar radiation which passes, through the layer into the deeper water below the level $z=-h$. In the following computations it will be assumed that $\gamma h \gg 1$, and that $\exp(-\gamma h)$ is therefore negligibly small.

The equation for the salinity has the form

$$\frac{dS_0}{dt} = -\frac{w_e \Delta S}{h} + \frac{S_0}{\rho h}(Q_0 - P_0). \tag{10.22'}$$

The corresponding integral equations for the buoyancy b and the mean momentum $\rho \underset{\rightarrow}{v}$ can be written down without difficulty.

To close the system, one needs an expression for w_e. This can be obtained from an integral of the turbulence energy equation (10.5) in its balanced form ($d/dt=0$). An integral of the vertical buoyancy flux $\overline{w'b'}$ which occurs in Eq. (10.5) can be obtained by integrating Eq. (10.4) from a depth z to the surface and again from the depth $-h$ to the surface. The time derivative db_0/dt can then be eliminated between these two integrals. After rearrangement one gets:

$$\overline{w'b'} = B_0 + \frac{z}{h}\left(B_0 + w_e \Delta b + \frac{g\alpha}{\rho c} I_0\right) + \frac{g\alpha}{\rho c} I_0 (1 - e^{\gamma z}). \tag{10.23}$$

For simplicity, we set $J_0 = \frac{g\alpha}{\rho c} I_0$ and integrate Eq. (10.23) from the bottom of the layer $z=-h$ to the surface $z=0$, obtaining

$$\int_{-h}^{0} \overline{w'b'}\, dz = \frac{1}{2} h B_0 - \frac{1}{2} w_e \Delta b h + J_0\left(\frac{h}{2} - \frac{1}{\gamma}\right). \tag{10.24}$$

The last equation represents the rate of potential energy change associated with the lifting or lowering of the center of gravity of the water column by convection. The first term on the right hand side is the contribution of the surface flux to the potential energy change. This contribution is positive when the surface is cooled ($B_o > 0$). The second term represents the reduction of potential energy caused by the lifting of the dense entrained water. The product

$$\Delta b h = c_i^2, \tag{10.25}$$

where c_i is the velocity of the long internal waves at the bottom interface $z=-h$. Finally, the last term represents the change which is due to the absorption of solar radiation in depth and to the redistribution of the resulting heat increase by convection. This redistribution increases the potential energy if the penetration scale length γ^{-1} exceeds the half depth of the mixed layer, and vice versa.

The derivation of an explicit expression for the flux and the generation of turbulence kinetic energy is not as straight forward, because the generating term $\overline{w'\vec{v}'}\cdot\partial\vec{\bar{v}}/\partial z$ is obviously zero in the interior of the layer, if the layer moves indeed uniformly like a slab. On the other hand in the transition zone at the bottom of the layer which is illustrated in Fig. 10.2, $\partial\vec{\bar{v}}/\partial z \to \infty$ as $(h'-h) \to 0$. From Eq. (10.20) we deduce that

$$\lim_{(h'-h)\to 0} \int_{-h'}^{-h} - \overline{w'\vec{v}'} \cdot \frac{\partial \vec{v}}{\partial z}\, dz \approx \frac{1}{2} w_e \vec{\bar{v}}^2 + \frac{1}{3} C|\vec{\bar{v}}|^3. \tag{10.26}$$

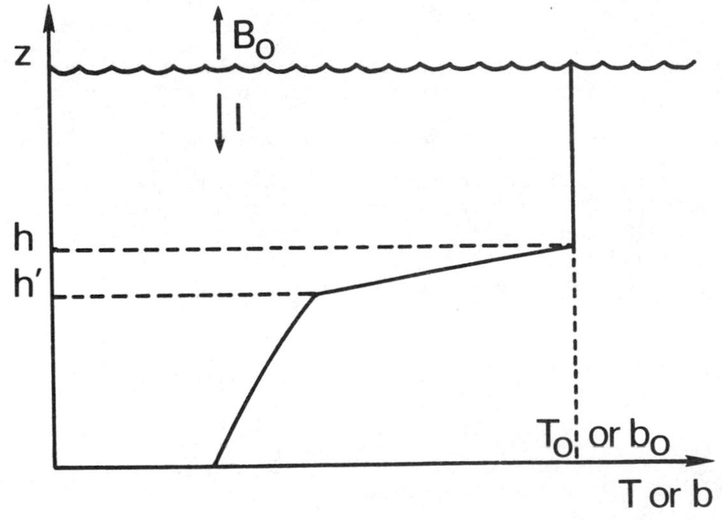

Fig. 10.2 - Schematic of mixed layer model.

Turbulence energy is also likely to be produced by the mixing of shear flows near the surface. However, in reality this must always involve running waves, and if a sea is present it cannot be modelled analytically. Fortunately, one can assume that the rate of generating of mean shearing motion near the surface will be equal to the rate of its destruction by mixing. This makes the turbulence production term proportional to the rate of working by the wind. Its effect can be accounted for by an adjustment of the proportionality constant in the relation (10.18).

On the basis of the preceding argument and with the assumption of a discontinuous interface it is now possible to write the integrated turbulence energy equation in the form

$$\underset{A}{\frac{1}{2}w_e(q^2}+\underset{B}{c_i^2}-\underset{C}{\overline{\underline{v}}^2)} = \underset{D}{m_1 u_*^3} + \underset{E}{\frac{1}{2}hB_o} + \underset{F}{\left(\frac{h}{2}-\frac{1}{\gamma}\right)J_o} - \underset{G}{C|\underline{v}|^3} - \underset{H}{\int_{-h}^{0}\varepsilon\,dz}. \quad (10.27)$$

Equation (10.27) looks cumbersome; fortunately it can be very much simplified in almost any particular case, because the individual terms tend to have unequal magnitudes. Before this is shown below, it may be useful to recapitulate once more the meaning of these terms:

A: rate of energy needed to agitate the entrained water (Eq. 10.21);

B: work per unit time needed to lift the dense entrained water and to mix it through the layer (Eqs. 10.24 and 10.25);

C: rate at which energy of the mean velocity field is reduced by mixing across the layer base (Eq. 10.26);

D: rate of working by the wind (Eq. 10.18);

E: rate of potential energy change produced by fluxes across the sea surface;

F: rate of potential energy change produced by penetrating solar radiation (Eq. 10.24);

G: rate of working of radiation stress associated with internal waves (Eq. 10.20);

H: dissipation.

e. Parameterization of the dissipation

If the most arbitrary link in the second order turbulence closure models has been the specification of a mixing length, the same might be said about the dissipation in the integral or mixed layer models. From dimensional arguments dissipation can be expected to be proportional to the third power of the mean turbulence velocity. The turbulent agitation is most intense close to the regions where turbulence energy is generated; for example, when the turbulence is due to wind stirring, its intensity tends to decrease with distance from the surface. The opposite holds when turbulence is caused by some interior shear. On the basis of such observations, it will be assumed here that the dissipation integral is composed of terms which are individually proportional to the active turbulence generating processes, that is, to the terms C and D and also to E during periods of active cooling when B_o is positive. Formally:

$$\int_{-h}^{0} \varepsilon dz = (m_1 - m)u_*^3 + (1-s)\,\frac{1}{2}\,w_e\,\overline{\underset{\rightarrow}{v}}^2 + (1-n)\,\frac{1}{2}\,h\,\frac{B_o + |B_o|}{2}. \qquad (10.28)$$

The proportionality factors have been written in the particular forms (m_1-m), $(1-s)$, and $(1-n)$ for the sake of convenience, because this simplifies the expressions to follow. They are not necessarily constant and it will be seen below that they may assume different numerical values which depend on the layer depth and the magnitude of the forcing. Actual values have to be established empirically from laboratory experiments and field observations. The last term in Eq. (10.28) should differ from zero only when turbulence is generated by surface cooling ($B_o>0$); this is the reason why it has been written in the peculiar form shown. It may be appropriate to list here some other proposed parameterization schemes. One, suggested by Niiler (1975) and in a slightly different form by Kim* (1975) has the form

$$\int_{-h}^{0} \varepsilon dz = C_\varepsilon u_*^3 \exp\left(-\frac{h}{h_o}\right) + \varepsilon_o h,$$

where C_ε is another proportionality factor, ε_o is a "background" dissipation rate, and h_o is a decay scale for dissipation as a function of depth based on field experiments by Grant, et al. (1969). This formulation relates the dissipation only to the working of the wind and not to other turbulence generating processes. It also will be seen below that even in the case of a purely mechanical energy input, it is not supported by experimental results.

Another scheme by Garwood* (1976) introduces a dissipation rate which is proportional to q^3/h. This is in keeping with other turbulence studies. He stipulates an entrainment rate of the form

$$\frac{w_e}{w_*} \propto \frac{q^2}{c_i^2},$$

where w_* is the geometric mean of the vertical component of turbulent velocity close to the layer bottom. The quantities w_* and q have to be determined separately. Readers are referred to the original paper for details of the procedure.

At this stage of the art, the formulation (10.28) is preferred. Though rather crude, it is relatively simple and it incorporates just about all that can be said with any confidence about the underlying physical processes. Introduction of Eq. (10.28) into Eq. (10.27) yields

$$\tfrac{1}{2}w_e(q^2+c_i^2-s\overline{\vec{v}}^2) = mu_*^3 + \frac{h}{4}[(1+n)B_o-(1-n)|B_o|] + (\frac{h}{2}-\frac{1}{\gamma})J_o + \tfrac{1}{3}C|\overline{\vec{v}}|^3. \qquad (10.29)$$

A B C D E F G

This expression can be simplified further, because the first and the last terms are relatively small in most circumstances. In particular, the mean square turbulent velocity q^2 in the upper ocean is typically of order 10^0 cm^2 s^{-2}. It never exceeds 10 cm^2 s^{-2}. On the other hand c_i^2 is of order 10^3 cm^2 s^{-2} even if the layer is only 10 meters deep. It becomes larger as the layer gets deeper. A layer depth of at least a few meters is assured by the slightest amount of surface cooling or wind stirring (see Sections 10.4a and 10.5a below). It follows that almost invariably $c_i^2 \gg q^2$, allowing us to neglect term (A) in all the considerations below.

In dealing with term (G) we are on weaker ground. This term is related to the rate of working of the shear stress at the layer bottom, as indicated by Eq. (10.26). Anticipating again the discussion below, one can say that most of the time in the oceans, the rate of energy input by the wind through the sea surface is likely to be very much larger than the rate of energy production by the internal shear stress. However, this may not hold true during certain stages of the development when $|\overline{\vec{v}}|$ becomes relatively large. We shall assume that when this happens term (G) remains relatively small compared to term (C); that is $w_e \gg C|\overline{\vec{v}}|$. In physical terms this implies that the energy which is generated by the working of the shear stress is either dissipated locally or used to entrain dense water from below with relatively little being radiated away. Following these considerations we shall assume that the radiational drain of energy is small and that its effect can be incorporated in the empirical factor s.

With these assumptions one can write the last equation in the form

$$w_e(c_i^2-s\overline{\vec{v}}^2) = 2mu_*^3 + \frac{h}{2}[(1+n)B_o - (1-n)|B_o|] + (h-\frac{2}{\gamma})J_o. \qquad (10.30)$$

B C D E F

The Roman letters have the same meaning as in Eq. (10.27), although the relevant terms include now the effects of dissipation. Term (E) equals hB_o when $B_o<0$; it is equal to nhB_o when $B_o>0$. It is in the form (10.30) that the turbulence energy equation will be applied below.

10.4 Mixed layer modelling of special cases and particular circumstances

The Eqs. (10.12), (10.16), (10.22), (10.22') and (10.30) form a closed system for B_o, T_o, S_o, h and w_e. Another equation for the layer momentum $\overline{\vec{v}}$ can be added. To simplify evaluation we distinguish between periods when the heating increases, and when it decreases or when cooling occurs. During the former periods, the layer tends to become shallower ($dh/dt<0$); there is no entrainment in this case ($w_e=0$) because there is not enough energy available to work against the increasing stability. The temperature and salinity will vary continuously across the layer base, though there will be a discontinuity in their gradient. Energetically, the layer is decoupled from the lower water during these periods.

During periods of decreasing heating or cooling, conditions depend crucially on the depth h. The terms (B) and (E) both tend to increase with increasing layer depth. The other terms are either independent of h or they decrease as the layer gets deeper. One can therefore expect a balance between (B) and (E) in late fall, for example, when the layer is relatively deep. Another controlling factor is the relative magnitude of the wind stirring and the thermal forcing, as expressed by the ratio

$$\frac{2u_*^3}{B_0 h} \equiv \frac{L^*}{h}, \qquad (10.31)$$

where L^* can be interpreted as a generalized Monin-Obukhov length. When L^*/h is large, the turbulence is dominated by wind stirring; thermal convection becomes the dominant factor when the ratio is small and positive.

We shall now consider some of the various possibilities:

a. Increasing stability - no wind

On a calm morning or during windless spring days when the sea is being heated by radiation ($R_0<0$), the layer depth tends to decrease ($dh/dt\leq 0$). There can be no entrainment in this case ($w_e=0$), and all the terms in Eq. (10.30) except (E) and (F) are therefore zero. It follows that the depth h and the flux function B_0 must adjust themselves in a way which make the sum of these two terms also equal to zero. In the absence of wind, the evaporation and the turbulent flux of sensible heat from the sea surface are likely to be small as well. The value of B_0 in Eq. (10.16) is then determined approximately by radiation alone.

Two cases have to be distinguished depending on the sign of the surface infrared radiation balance (R_0-I_0), which is also the sign of B_0. If (R_0-I_0), and therefore B_0, is negative, the surface is being heated by infrared radiation. Without any heat sink within the water, the enthalpy flux and the buoyancy flux must be directed downward ($\overline{w'b'}<0$). The balanced turbulence energy equation (10.5) can then not be satisfied, because there is no energy source available to transport buoyancy downward. In other words there can be no mixed layer present: $h=0$ if $B_0<0$ and $u_*=0$.

If R_0-I_0 and B_0 are positive, one gets from the sum of the terms (E) and (F) after rearrangement

$$h = \frac{2}{\gamma}\frac{J_0}{nB_0+J_0} \approx \frac{2}{\gamma}\frac{I_0}{n(R_0-I_0)+I_0}. \qquad (10.32)$$

If there is no dissipation (n=1), one sets

$$h = \frac{2}{\gamma}\frac{I_o}{R_o} .$$

On the other hand, if all the convectively generated energy is dissipated locally (n=0), the layer depth tends to become equal to double the penetration scale length

$$h = \frac{2}{\gamma} .$$

To compute the temperature rise, one eliminates h between Eqs. (10.22) and (10.32). With the radiation flux through the base of the layer considered very much smaller than the radiation balance at the surface, one gets, with w_e=0:

$$\frac{dT_o}{dt} = -\frac{R_o\gamma}{\rho c}\frac{n(R_o-I_o)+I_o}{2I_o} . \qquad (10.33)$$

The case represented by the Eqs. (10.32) and (10.33) was first modelled by Kraus and Rooth (1961) who also list empirical expressions for R_o as a partial function of T_o. Although Eq. (10.32) is a balance equation, h is not necessarily constant, but changes with R_o and therefore with T_o. The temperature change dT_o/dt is always positive during the adjustment process (active heating).

b. Decreasing stability - no wind

During night I_o=0; the convective energy generation minus dissipation which is represented by term (E) in Eq. (10.30) is balanced by the entrainment of dense water as represented by term (B). It follows that:

$$w_e = \frac{nhB_o}{c_i^2} = \frac{nB_o}{\Delta b} . \qquad (10.30')$$

With active entrainment, w_e=dh/dt>0 (see Eq. 10.12). The form of B_o is specified by Eq. (10.16). In the circumstances envisaged here, B_o is dominated by infrared cooling ($B_o \propto R_o$; $\Delta b \propto \Delta T$). The last equation is therefore equivalent to:

$$\frac{dh}{dt} = \frac{n}{\rho c}\frac{R_o}{h} . \qquad (10.34)$$

Introduction of this expression into Eq. (10.22) yields:

$$\frac{dT_o}{dt} = \frac{1+n}{2\rho c}\frac{R_o}{h} . \qquad (10.35)$$

It is easy to add the appropriate terms for the sensible heat loss and evaporation to R_o on the right hand side of the Eqs. (10.34) and (10.35) if these processes are significant. However, it should be noted, that even in the present simple case the pair of Eqs. (10.34) and (10.35) generally can only be integrated numerically. In particular, R_0 is a nonlinear function of T_0 and also depends on the atmospheric conditions; ΔT depends not only on T_0 but implicitly also on the layer depth h and on the temperature gradient below.

The argument leading to the expressions (10.34) and (10.35) was first developed by Ball (1960) in a model of atmospheric temperature changes over heated land. Instead of cooling at the upper surface of the fluid, the forcing is brought about in his case by the convective heating from the isolated land surface. Ball assumed that the resulting free convection is carried by large eddies which are not affected at all by dissipation. With our present terminology this would imply $n=1$.

The opposite extreme would be to assume that all the convectively generated energy is dissipated within the layer. This is tantamount to setting $n=0$. There is then no energy available for entrainment and the rate of temperature change dT_0/dt is only half as fast as for $n=1$. However, as shown, for example, by Lilly (1968), with somewhat different nomenclature for the atmospheric case, the final mixed layer temperature which is established may not differ very much in the two cases.

c. Increasing stability - with wind

We shall assume that the wind stirring is insufficient to produce entrainment ($w_e=0$) in the presence of strong radiational heating. The layer depth must then become shallower until h has decreased to a value which keeps the right-hand side of Eq. (10.30) equal to zero.

For the case of surface heating ($B_0<0$), one gets from Eq. (10.30)

$$h = \frac{-2mu_*^3+2J_0/\gamma}{B_0+J_0}. \tag{10.36}$$

During night or in heavy overcast conditions $J_0=0$. Equation (10.30) then reduces - with consideration of the expression (10.31) - to

$$h = -\frac{2mu_*^3}{B_0} = -mL^*. \tag{10.37}$$

This means that the layer should become stabilized when it has a depth which corresponds to the Monin-Obukhov length. Kitaigorodskii (1960) suggested that this should be the fundamental scale depth of the ocean mixed layer - one in which just sufficient mechanical energy is available from the wind to mix the heated surface water uniformly down to a depth $h \propto L^*$.

The layer temperature T_0 and surface salinity S_0 can be evaluated by introducing Eq. (10.36) or (10.37) into (10.22) and (10.22') with $w_e=0$ and with an explicit expression for B_0 as specified in Eq. (10.16).

When the surface is cooled ($B_o>0$) in the presence of wind, the layer tends to deepen ($w_e \neq 0$). This case will be discussed in Section 10.4e below.

d. The effect of inertial currents in the mixed layer

Any wind change generates inertial currents in the upper ocean. Their presence tends to produce a sharp velocity gradient at the layer bottom interface. When this happens, the term (C) in Eq. (10.30) must be considered.

The contribution of shear stresses at the bottom of the layer, to the work of layer deepening was investigated first by Pollard, Rhines and Thompson (1973). They argued that these stresses feed energy into finite-amplitude perturbations, which break up the interface when the ratio of the hydrostatic stability to the shear instability - that is the local Richardson number - drops below some critical value. The break-up speeds up the entrainment process, causing an adjustment of the layer depth which prevents any further increase of the velocity shear beyond its critical value. While the deepening proceeds, the work of the shear stress is used mainly to lift the dense water from below. Pollard, et al. equated the critical local Richardson number with the bulk Richardson number

$$Ri_* = \frac{\Delta b\ h}{\vec{\overline{v}}^2} = \left(\frac{c_i}{\vec{\overline{v}}}\right)^2. \qquad (10.38)$$

Using this expression to eliminate c_i from Eq. (10.30) one gets

$$w_e(Ri_*-s)\vec{\overline{v}}^2 = 2mu_*^3 + nhB_o. \qquad (10.30'')$$

For a given rate of surface forcing, a small value of the difference Ri_*-s must be associated with large values of w_e and therefore with rapid layer deepening.

For simplicity's sake, the solar heating term (F) has been omitted in Eq. (10.30''). It represents a process which is unlikely to be important during storm driven layer deepening. Retention of this term, when that is desirable, does not involve any additional analytical difficulties, but it does make the resulting expressions longer and more cumbersome.

To obtain an explicit expression for $\vec{\overline{v}}^2$ one integrates Eq. (10.1) over the layer depth with the boundary conditions (10.17), (10.20) and over time with the initial condition $\vec{\overline{v}}=0$ for t=0. The result in symbolical form is:

$$|\vec{\overline{v}}| = \frac{u_*^2}{fh}(e^{-ift}-1). \qquad (10.39)$$

The radiation term $C\vec{\overline{v}}^2$ which occurs in the boundary condition (10.20) has been neglected in this derivation. The squared modulus of Eq. (10.39) is

$$\vec{\overline{v}}^2 = \frac{2u_*^4}{h^2f^2}(1-\cos ft) = \frac{\Delta b\ h}{Ri_*}. \qquad (10.40)$$

The last equality follows directly from (10.38). Rearrangement yields

$$h = [Ri_* \frac{2u_*^4}{\Delta b f^2} (1-\cos ft)]^{1/3}, \tag{10.41}$$

In their original paper, Pollard, et al. assumed arbitrarily that

$$Ri_* = 1, \tag{10.38'}$$

which is tantamount to setting $\overline{\vec{v}}^2 = c_i^2$. This restriction was abandoned in later papers.

To evaluate Eq. (10.41) explicitly one has to use a second independent relation which connects the two dependent variables h and Δb. Regardless of this complication, the expression (10.41) would suggest that the layer reaches a maximum limiting depth h_f when $t=\pi/f$, that is, one half of a pendulum day after the onset of the inertial oscillation. Inertial motions can only play an important role in the deepening process if the initial depth $h(t=0)<h_f$.

e. The erosion of deep mixed layers

As the layer deepens further, c_i^2 tends to increase with h while $\overline{\vec{v}}^2$ decreases in inverse proportion to h^2. The contribution of the velocity shear to the layer deepening tends therefore to become small as h increases beyond h_f. Omission of $\overline{\vec{v}}^2$ at this stage and division by $c_i^2 = h\Delta b$, changes Eq. (10.30'') into

$$\frac{dh}{dt} = w_e = \frac{2mu_*^3}{h\Delta b} + \frac{nB_o}{\Delta b}. \tag{10.30'''}$$

This was the equation used by Kraus and Turner (1967) in their model of the deepening layer.

Consider first the case when wind stirring predominates. In terms of Eq. (10.31) one has then $h<<L^*$ which allows us to neglect the last term in Eq. (10.30'''). We divide by u_* and obtain then

$$\frac{w_e}{u_*} = 2m \frac{u_*^2}{h\Delta b} = 2m\ Ri^{-1} \tag{10.42}$$

This equation is equivalent to Eq. (7.6) in Phillips' treatment of entrainment, with Ri the bulk Richardson number as defined there. From the discussion there - to be recapitulated below - it follows further that m is not constant, but becomes smaller at high Richardson numbers. With m known empirically from the laboratory experiments, one can again compute the changing layer depth and temperature by integration of Eqs. (10.22) and (10.42).

Continued deepening must ultimately cause the first term on the right-hand side of Eq. (10.30''') to become small compared to the last one. In other words, as h continues to grow it must become larger than L^*. Wind stirring then becomes irrelevant for further deepening, and the entrainment equation assumes again the form (10.30') which as established in Section 10.4b for a shallow layer with no wind above:

$$\lim_{h \to \infty} (w_e) \to n \frac{B_o}{\Delta b}.$$

If one sets $n=1$, the last expression approaches once more Ball's (1960) original formulation. Gill and Turner (1976) argue convincingly that this original formulation cannot be representative at all times. They show that neglect of dissipation during layer deepening - in our terms $n=1$ - would be associated with an ever increasing potential energy. They propose, therefore, that at this stage there is a balance between convective energy generation and dissipation. Such a balance would imply that $n=0$, and as a discontinuity could not be maintained in these circumstances, one also has $\Delta b=0$. Below it will be argued that $n=1$ and $n=0$ are extremes which are approached at different stages of the development with $n \to 0$ as $h \to \infty$.

10.5 The proportionality factors - comparison with laboratory and field experiments

The preceding analysis involved three proportionality factors m, n and s, which have to be established empirically. Most of the experimental work so far has concentrated on the determination of the factor m. Laboratory experiments by Kato and Phillips (1969) suggested a value $m=1.25$. Later experiments by Kantha* (1975) are described in Phillips' Chapter 7. They do show that m is not constant. To compare the results with the present analysis, one must consider that B_0 was zero in Kantha's experiments. Equation (10.42) is relevant in these circumstances. When this equation is compared with the plot of Fig. 7.2, one finds that a line with a constant slope of $2m=2.5$ coincides with the axis of the shaded area which represents Kato's and Phillips' original experiments. The later results indicate that m is rather larger than originally estimated and that it is a decreasing function of the ratio $c_i^2/u_*^2=h\Delta b/u_*^2$.

$$m \equiv m \left(\frac{h\Delta b}{u_*^2}\right). \qquad (10.43)$$

When this ratio is of order one hundred, m is about 3.3. The experimentally determined values of m become smaller as the ratio $h\Delta b/u_*^2$ increases.

For the interpretation of field observations, it is more convenient to express the working of the wind not in terms of u_* but in terms of the actually measureable wind velocity u_a at the level $z=a$ above the surface. Denman and Miyake (1973), for example, use the expression

$$mu_*^3 = m_a(\rho_a C_a/\rho)^{1/2}\, u_a^3.$$

They calculated that a value of $m_a=0.0015$ corresponded to a value of $m=1.25$. A slightly smaller value of $m_a=0.0012$ gave a reasonably good fit for their analysis of records from ocean weather station Papa (50°N, 145°W). The same value also fitted the results of Denman's (1973) idealized wind mixing studies. On the other hand, Turner (1969) found that a considerably larger value of $m_a \approx 0.01$, corresponding to $m \approx 8$ would be necessary to account for rapid storm-induced deepening of the mixed layer which had been observed by Stommel, et al. (1969) in the Sargasso Sea. In a similar way a rapid increase in the mixed layer depth which had been reported by Halpern (1974b) to have occurred in August off the west coast of N. America, could be interpreted as due directly to wind mixing only if one were to stipulate a value of the proportionality factor m which is much in excess of the laboratory experiments.

The contradiction between the laboratory experiments and some of the oceanic observations - particularly those characterized by rapid deepening - may be caused by the interface destabilization which is produced by the working of the shear stress at the bottom of the mixed layer. This can be associated in turn with transient shears produced by internal waves as suggested in Kantha's note below, or with inertial currents in the mixed layer. If allowance is made for these additional processes, which can be important if $h<h_f$ initially, one is faced with the need to evalue the factor s. Unfortunately we have hardly any experimental data for this purpose.

It is encouraging that Pollard and Millard (1970), Halpern (1974) and Begis and Crepon* in the Mediterranean all found that during the first few inertial oscillations after a storm passage, the mixed layer does in fact tend to move like a slab with little vertical velocity shear. This means that all the shearing motion is concentrated at the bottom of the layer and Eq. (10.40) is applicable. The question is how much of the shear-produced energy release remains available for entrainment?

Pollard and Wyatt* (1976) based a tentative answer on the assumption

$$m \approx s \approx Ri = \Delta b \, h/\overline{v}_i^2 \, .$$

In the absence of surface cooling $B_0 \approx 0$, Eq. (10.39) assumes then the form

$$\frac{w_e}{u_*} = \frac{2s}{1-s}\left(\frac{u_*}{\overline{\vec{v}}}\right)^2 . \qquad (10.44)$$

Equation (10.44) allows determination of s as a function of time during periods when the layer deepening is in fact dominated by the production of turbulence kinetic energy from shearing motion at the lower layer boundary.

To carry out their assessment, Pollard and Wyatt used measurements of dh/dt, u_* and $\vec{v}$ which had been obtained during the 1972 JASIN experiment in the North Atlantic. The results suggest $s<1$. This is in agreement also with the results of a study by Shonting and Goodman* in the Mediterranean. As of present, the best series of observations were obtained probably by Price*, Mooers, and Van Leer (1976) in the Gulf of Mexico. From silumtaneous measurements of the temperature structure and of the inertial current velocity, Price found that $s \approx 0.7$ gave the best fit. As mentioned above, the factor s is affected not only by the dissipation but also by the drain of energy caused by internal wave radiation downward from the mixed layer base. This being the case, one would expect s to be a function not only of $\vec{v}^2$ but also of the stability in the water immediately below the depth $z=-h$.

As regards the factor n, several authors - starting with Ball - assumed

$n=1$. This assumption was based on the rationale that the convectively produced eddies are too large in size to be much affected by dissipation. On the other hand, Gill and Turner found that $n=0$ gave a good fit to the data presented here as Fig. 10.4 (below). Deardorff, Willis, and Lilly (1969) carried out laboratory tank experiments which yielded $n=0.015$. Farmer (1975), observing the deepening of a mixed layer under the ice of a frozen lake, found values of $0.003 \leq n \leq 0.113$ with a mean of $n=0.036$ for his series of twelve observations.

Tentatively, we suggest that the factor n exhibits a behavior which is similar to that of m, and that it is a decreasing function of the ratio between the relevant kinetic energy consuming and energy producing terms. It therefore should go towards zero as h becomes large. This suggestion is speculative. If it is thought acceptable, one has still to distinguish between forced and free convection.

The case of forced convection is represented by Eq. (10.30'''). In this case w_e can be scaled by u_*, and n is then presumably a function of the form

$$n \approx n\left(\frac{B_0}{u_* \Delta b}\right). \tag{10.45}$$

In the case of free convection ($u_*=0$) which applies to the quoted observational studies, one has to scale w_e by a convective velocity which is proportional to the cube root of (hB_0). If n is to be a function of the work terms, it must have the form

$$n \approx n\ (B_0^{2/3} h^{-1} \Delta b^{-1}). \tag{10.46}$$

From the preceding discussion one would expect n to be a decreasing function of its argument, that is $n \to 0$ as $h \to \infty$, in analogy with the behavior of the factor m.

In view of the uncertainties about the factors m, n and s, it is rather fortunate that the one-dimensional mixed layer models are not very sensitive to the actual value of these quantities. Particularly in those cases where a new equilibrium is being developed by the model, as a result of changes in the forcing function, it will be the rate of approach to this equilibrium rather than its final character which can be affected by the actual numerical value of the proportionality factors.

It is obvious that there is scope for further research on the functional form of the factors m, n and s; or for the matter on the whole problem of representing dissipation as a parametric function of the bulk variables. An elucidation of these problems is most likely to come from further laboratory experiments and from oceanic observations carried out over a wide range of different circumstances. Work on the downward mixing of fresh water lenses produced by tropical showers as illustrated by Fig. 10.3 and the studies of mixed layer evolutions below ice covers as reported by McPhee* (1975) or Farmer (loc.cit.) are of particular interest in this context.

10.6 Explicit simulation of time-dependent developments

The actual evolution of mixed layers and surface temperatures with time depends on the vagaries of the wind, air temperature, radiation and so forth.

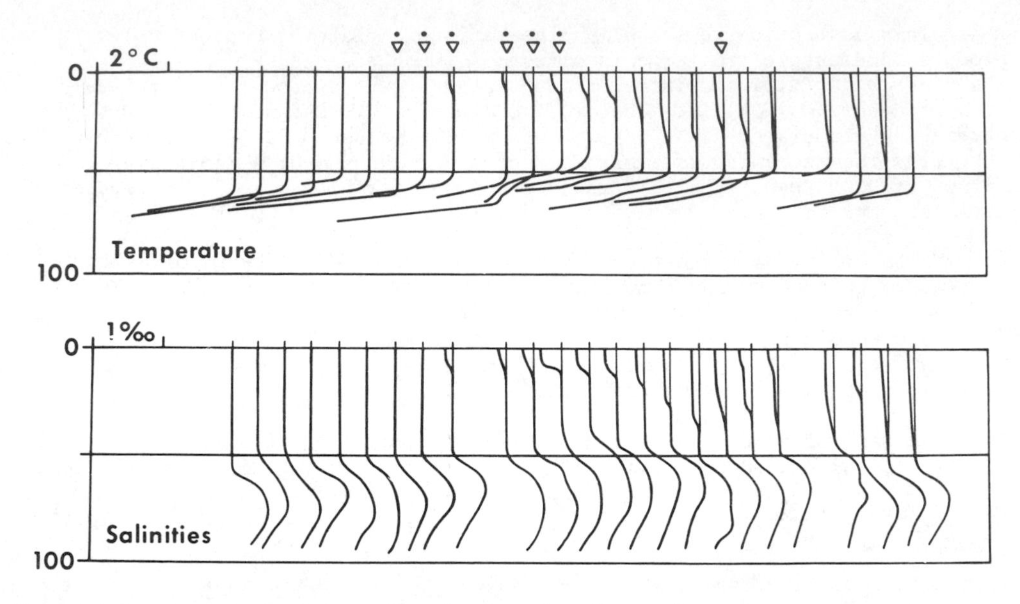

Fig. 10.3 - Downward mixing of a fresh water lens after showers, based on hourly sounding at about 5°N, 21°W in July 1972. Showers are marked by the conventional international symbol. Note the characteristic salinity maximum below the mixed layer base. The figure has been kindly supplied by F. Ostapoff. (See also Ostapoff, Tarbeyev and Worthem, 1973).

In other words, $u_* \equiv u_*(t)$ and the same applies to other forcing processes. The rate of layer deepening depends in addition on the stratification in the water below. Two special cases which have drawn particular attention are the deepening of a mixed layer following the sudden onset of a storm and the simulation of diurnal or annual cycles.

a. The deepening of a wind-stirred layer

A study of layer deepening as a function of time shows how the different processes and equilibria, which were discussed in Section 10.4, necessarily dominate the development. An orderly time sequence of separate physical regimes becomes apparent. This separation allows us in principle to evaluate the free model parameters (m, n, s) independently from different time segments of a single record.

Following Niiler (1975) and de Szoeke* and Rhines (1976) we shall consider a highly simplified model, which is characterized initially by a fluid at rest with a surface buoyancy b_0 and buoyancy gradient

$$\frac{\partial b}{\partial z} = N^2 = \text{constant.} \qquad (10.47)$$

At a time $t=0$ a constant stress ρu_*^2 and a constant cooling rate B_0 are applied to the surface. The penetration of solar radiation is neglected ($\gamma = \infty$).

The assumptions of a constant N and constant forcing after the initial impulse represent a highly idealized state. Real oceanic conditions are never quite as simple. For example, B_o cannot remain constant, in general, when the temperature of the deepening layer changes. In spite of its artificiality - or perhaps because of it - the simplistic model with its constant forcing and uniform initial lapse rate reveals rather clearly the succession of different regimes.

Following the beginning of the stirring process at $t=0$, a mixed layer with constant buoyancy $\bar{b}$ is established. This leads to the establishment of a density discontinuity at the bottom of the layer

$$\Delta b = \bar{b} - b_h = \bar{b} - b_o + hN^2. \qquad (10.48)$$

The value of b_h - the original buoyancy at the level h - is obtained from the integral of Eq. (10.47). The mixed layer buoyancy must be equal to the mean of the buoyancy which existed in the layer between $z=0$ and $z=-h$ before mixing, minus the time integral of the surface buoyancy flux B_o. With B_o assumed constant, one gets:

$$h\bar{b} = \frac{h}{2}(b_o+b_h) - B_o t.$$

Elimination of $\bar{b}$ between the last two equations yields:

$$c_i^2 = \Delta bh = \frac{1}{2} N^2h^2 - B_o t. \qquad (10.49)$$

We introduce Eqs. (10.40) and (10.49) into the turbulence energy equation (10.29) with $I_o=0$, $B_o \geq 0$ and with $C|\underset{\to}{v}| << sw_e$. This equation then assumes the form

$$\frac{1}{2}\frac{dh}{dt}[q^2 + \frac{1}{2}h^2N^2 - B_o t - 2su_*^4 f^{-2}h^{-2}(1-\cos ft)] = mu_*^3 + \frac{1}{2}B_o h. \qquad (10.29')$$

A B B' C D E

For a scale analysis, the magnitude of the external parameters has to be specified. We shall use typical mid-latitude values of $u_*=1.5$ cm s^{-1} (corresponding to a fresh breeze with a velocity of about 12 m s^{-1}), $B_o=2\times10^{-3}$ cm^2 s^{-3} (corresponding to a surface heat loss of 0.01 cal cm^{-2} s^{-1}), $N^2=2\times10^{-4}$ s^{-2} and $f=10^{-4}$ s^{-1}.

Immediately after the onset of the stress, when h and t are both small, term (D) can be balanced only by term (A). At this stage the mixed layer is of the same order or smaller than the depth of the constant stress layer and therefore $q^2=u_*^2$. Integration of Eq. (10.29'), with only terms (A) and (D) different from zero, yields

$$h = 2mu_* t. \qquad (10.50)$$

The depth grows linearly with time at this state. This regime must continue until term (B) becomes comparable to (A). This happens when the layer depth has reached a value h_1 at which $u_*^2=c_i^2$. To compute h_1, we introduce the value of c_i^2 from Eq. (10.49) into this equality and substitute for t from Eq. (10.50).

This yields

$$h_1 = \frac{u_*}{N}\left[\sqrt{2} + \frac{B_o}{2mNu_*^2} + \sqrt{2}\left(\frac{B_o}{2mNu_*^2}\right)^2 + - - -\right] \approx \frac{u_*}{N}\sqrt{2}. \qquad (10.51)$$

The last approximation is a consequence of the chosen numerical values of B_o, N and u_*, with $m>1$. With these numerical values, the depth h_1 is only about 1.5 meters, which shows that the mixed layer depth is indeed comparable to the depth of the quasi-constant stress layer during the validity of this regime. The layer reaches the depth h_1 within a time

$$t_1 = \frac{h_1}{2mu_*} = \frac{1}{\sqrt{2}mN}; \qquad (10.52)$$

assuming $m=1.25$ at this state, one finds that t_1 is only about 40 seconds. In laboratory experiments where N is usually larger than in nature the development proceeds even faster. For $t>t_1$ the first term in Eq. (10.29') becomes neglibibly small and this justifies the omission of this term in the discussions presented in Section 10.4.

At the time t_1 the term (B) in Eq. (10.29') is large compared to (B'). By expanding cos(ft) it can be shown readily that it is also larger than (C) which becomes important only after a significant fraction of one inertial period has passed. Until that happens there must be a balance between terms (B) and (D), which makes the ratio of dh/dt to u_* proportional to the inverse Richardson number $(u_*/c_i)^2$. Integration of Eq. (10.29') with all terms except (B) and (D) negligible yields:

$$h = \frac{u_*}{N}(12mNt)^{1/3}. \qquad (10.53)$$

The layer deepens now at a rate which is proportional to the cube root of t.

This cube root regime continues to prevail until either (C) or (E) becomes relatively large. With the chosen numerical values of f and B_o the former will occur first. Terms (B) and (C) become about equal when the layer has expanded to a depth

$$h_2 \approx \frac{u_*}{N}(6ms^{-1/2}). \qquad (10.54)$$

This occurs at a time

$$t_2 \approx \frac{1}{N}(18m^2\ s^{-3/2}). \qquad (10.55)$$

It is interesting to note that the transition time t_2 is independent of u_* or f.

Gradually the mean flow, or inertial motion, accelerates to make term (C) comparable to term (B). The sum in brackets then becomes very small. To achieve balance with the energy input from the wind, dh/dt has to become very large, at least during some interval in this regime. In other words, we should expect a very rapid deepening of the mixed layer when the inertial current velocity is of the same order as the internal wave velocity c_i. Exceptionally rapid deepening in such circumstances has indeed been observed by Price*, et al. (1976) in the Gulf of Mexico.

The term (C) reaches its largest value after one-half pendulum day, that is, at a time

$$t_3 = \pi/f. \tag{10.56}$$

If we accept the arguments which were put forth by Pollard, et al. (1973), the corresponding mixed layer depth can be computed by substituting Eqs. (10.49) and (10.56) into Eq. (10.41). This yields:

$$h_3 = \frac{u_*}{N}(2N/f)^{1/2} s^{1/4}. \tag{10.57}$$

For $t>t_3$ the value of the term (C) becomes smaller again and the regime corresponds then to the condition discussed in Section 10.4e. There will be a renewed balance between terms (B) and (D) and continuing deepening with the cube root of time until the layer depth exceeds the value of

$$h_4 = \frac{m}{n} L* = \frac{2mu_*^2}{nB_o}. \tag{10.58}$$

Further deepening beyond h_4 would be described by an equation of the form

$$\frac{dh}{dt}(N^2h^2 - B_o t) = 2nB_o h. \tag{10.59}$$

At this depth, however, the assumption of a constant N becomes very unrealistic indeed.

The successive dominance of different regimes is summarized in Table 10.1.

TABLE 10.1

Range of transitional regimes in a deepening wind-stirred layer (the computation of particular transition depths and times is based on the following specified values of the external and internal parameters: $u_*=1.5$ cm s^{-1}, $B_o=2\times10^{-3}$ cm^2 s^{-3}, $N=1.4\times10^{-2}$ s^{-1}, $f=10^{-4}$ s^{-1}, m=1.25, s=0.7, n=0.7, C=0.)

Transition Time	Transition Depth	Regime
0	0	
		{linear growth of h
$N^{-1}(\sqrt{2}m)^{-1}$ (40 sec)	$\frac{u_*}{N}\sqrt{2}$ (1.5 cm)	
		{$h \propto t^{1/3}$
$N^{-1}(18m^2s^{-3/2})$ (1 hour)	$\frac{u_*}{N}(6ms^{-1/2})$ (10 m)	
		{inertial deepening
$f^{-1}\ \pi$ (9 hours)	$\frac{u_*}{N}(2N/f)^{1/2}s^{1/4}$ (17 m)	
		{$h \propto t^{1/3}$
$N^{-1}\ \frac{2}{3}\ m^2\ (\frac{u_*^2N}{nB_o})^3$ (10 days)	$\frac{2mu_*^3}{nB_o}$	
		{deep convective erosion

It should be noted that different values of the forcing parameters and initial conditions would lead to different transition times and depths. In the present model, the choice of a relatively small value for B_0 allowed us to ignore the term (B') in Eq. (10.29') through most of the development. This will not always be the case. The internal wave radiation, which has been neglected in our computation (C=0), must have some influence particularly on the development of the inertial regime. Values of C could be derived possibly from observations of the phase and amplitude of the inertial oscillations. They cannot be derived directly from the present model. Lastly, the chosen value of m is that based on the old Kato-Phillips experiments. The choice of a larger, variable m would have stretched out the validity of the second and fourth regimes in both time and depth.

b. Cyclical changes

In studies of the diurnal and seasonal mixed layer evolution, the forcing functions B_0 and u_* cannot be constant; in fact, the former in particular has necessarily a cyclical component. It is also inappropriate now to stipulate, a priori, the density structure of the water column below the mixed layer. Instead, modellers have assumed usually that this structure was established during the spring heating season, when the mixed layer reached down to the relevant depth. As the heating increases, the layer becomes shallower as indicated by the Eqs. (10.32) or (10.36) leaving a stably stratified fluid below. During this process, the layer is decoupled from the lower water, which remains more or less undisturbed, at the temperature which it had when it was last directly in contact with the surface. The corresponding density structure is conserved until it is again swallowed up by the deepening mixed layer in the cooling season.

In the real ocean, conditions at depths below the mixed layer can obviously not be exactly conserved. Diffusion tends to smooth out gradient changes. More important even is the fact that the seasonal development is not a smooth process, but one which is punctuated by storms which can mix surface water down to considerable depths within a relatively short time. However, this process can be included in models and if it is allowed for, the stipulated conservation of conditions at depths below the mixed layer is in fact reproduced, at least qualitatively, over many parts of the ocean. In general, the analysis has to be carried out numerically unless one stipulates a very special form of the surface heating B_0 as a function of time. Obviously, if the ocean below the layer "remembers" how its temperature structure was established, the computer has to do the same. In other words, numbers for the temperature distribution which had been established by the receding mixed layer during the heating season, have to be stored in memory until the deepening layer reaches again down to the same levels. The process has been simulated in laboratory experiments (Turner and Kraus, 1967).

It is characteristic for the cyclical changes that the evolution of the surface temperature is not in phase with the layer depth. This fact is well simulated by the mixed layer models. For example, Eq. (10.36) shows that during surface heating ($B_0<0$) the layer depth reaches its smallest value when the heating $|B_0|$ reaches its maximum - that is at the time of the summer solstice for the seasonal case, or at noon for the diurnal evolution. At that time $w_e=dh/dt=0$, but from Eq. (10.22) it can be seen that then $dT_0/dt\neq0$;

in fact, the surface temperature rise will tend to be largest at that time.

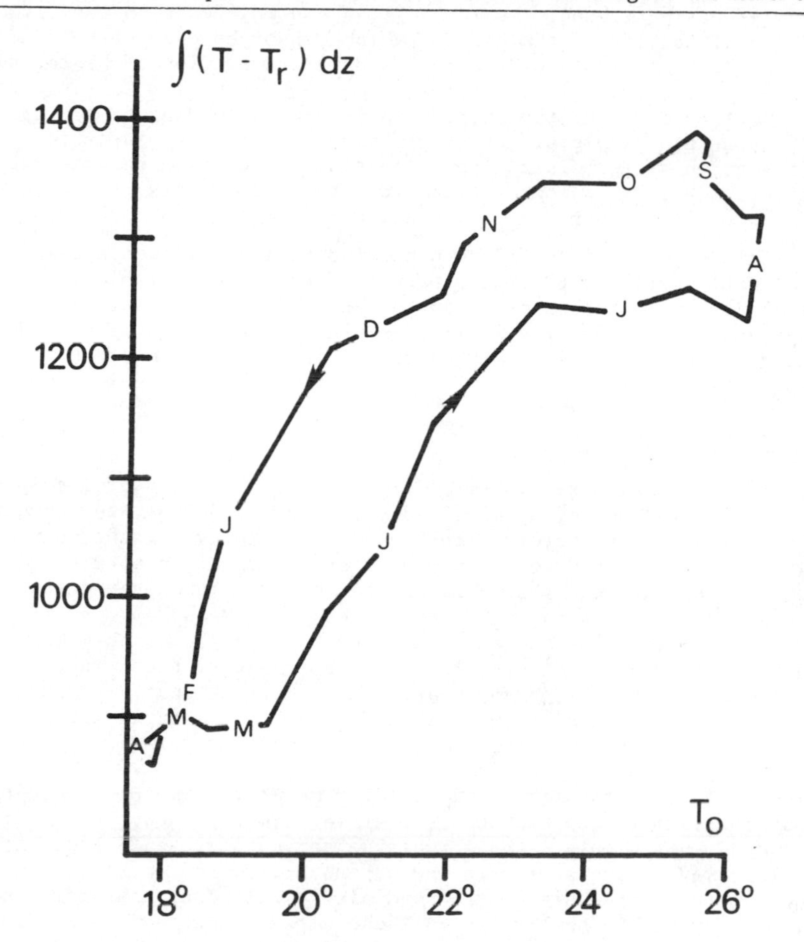

Fig. 10.4 - Heat content as a function of surface temperature at Ocean Weather Station Echo (35°N, 48°W). The integral extends from a fixed depth of 250 m to the surface. The reference temperature T_r represents the mean of the temperature at 250 m and 275 m depth. (After Gill and Turner, 1976).

This out-of-phase relationship of T_0 and h has been noted in many series of lake and ocean observations (see Kraus (1972) for diagrams and references). Because of it, any plot of h against T_0 or B_0 as a function of time will exhibit a hysteresis loop. Instead of plotting h against T_0 directly, one can plot the total heat content (or the corresponding mass deficit) of a water column in the upper ocean against the surface temperature. Such a plot is shown in Fig. 10.4 which is drawn from Gill and Turner (1976) and is based on data from ocean station Echo (35°N, 48°W). It can be seen, for example, that the surface temperature is much higher in August than in November, but the mixed layer is shallower in summer and the heat content of the water column in the two months is about the same. The existence of this hysteresis loop is, of course, simply a manifestation of the ocean's capacity to store heat beyond the heating season. As such, it has an influence on the climate, and indeed the livability of the planet, which cannot be overestimated. (See also Holland, Figs. 2.6 and 2.15 for computer simulation of hysterisis loops.)

As mentioned above, the original models with forcing by a cyclical B_0 did not produce a completely cyclical response. Without suitable parametrization of the dissipation, the convective layer becomes deeper and the potential energy increases from cycle to cycle. This is due to the continuing positive energy input by the wind ($u_*^3>0$). To model real cyclical developments one has to let m and n both go to zero as the layer becomes deep. All the input energy is then dissipated and none is available to increase the potential energy of the water column excessively. Alternatively one can allow for a general slow upwelling $\bar{w}$. Such a widespread upwelling is indeed needed over much of the oceans, to compensate for the production of bottom water by sinking in high latitudes.

10.7 Concluding remarks

One-dimensional mixed layer models may well become useful components in more comprehensive ocean, climate and biological models. They have the advantage of being relatively simple and of providing a rather direct physical insight. They can yield reasonably realistic simulation of diurnal and of seasonal temperature changes in lakes and in suitable chosen ocean areas where horizontal advection is insignificant. The realism is improved if one chooses some functional form for the factors m and n which causes them to go towards zero as the layer becomes very deep. A zero value of these factors implies a balance between turbulence generation and dissipation. It does not imply an invariant layer depth, though it must cause the disappearance of the discontinuity at the layer base.

The ability of the simple mixed layer model to simulate rapid changes after storms is more questionable. They seem to work sometimes, but at other times they do not simulate the actually observed oceanic development very well. This may be due partly to the dependence of the development on the exact character of the dissipation and also to the fact that horizontal variations may not be negligible in these circumstances.

We wish to thank once more all the contributors to the Urbino working group on one-dimensional modelling. In preparing this report, we had the advantage of support from the National Science Foundation under Grant No. NSF GA 33550X and from the Office of Naval Research, Grant No. N0014-75-C-0299, NR083-315.

Chapter 10a

NOTE ON THE ROLE OF INTERNAL WAVES IN THERMOCLINE EROSION

L. H. Kantha

In the preceding chapter by Niiler and Kraus, (referred to below as N&K), the influence of internal waves on entrainment processes has been essentially ignored. There is reason to suspect, however, that their influence could be quite substantial, at least under some conditions. It is the purpose of this note to discuss the possible effect of internal waves on the entrainment rate and on the associated energy flux.

Internal waves can be generated locally by turbulence or externally by a distant storm. Waves always radiate energy away from the area where they are being generated. They can cause thus a local energy loss from the mixed layer in the area below a storm. This tends to reduce entrainment, simply because less energy is available for the lifting of the dense lower fluid into the mixed layer. On the other hand, waves which propagate into the region can make additional energy available, which might increase the rate of layer deepening.

The energy radiated away by locally generated waves was parameterized in the Eqs. (10.26) and (10.27) of N&K in the form $E_i/\rho \equiv c|\vec{\bar{v}}|^3$ and lumped together with the shear production of turbulence kinetic energy at the bottom of the mixed layer. The term was subsequently dropped. However, in any case, this parameterization is rather arbitrary. The energy supplied to the waves must depend upon the amplitude and the length scale of the interface perturbations and upon the stratification in the fluid below. If the region below the interface is not stratified, internal waves propagating into the interior cannot be generated. It follows that the energy loss might be expressed more appropriately in the form:

$$\frac{E_i}{\rho} \equiv \frac{E_i}{\rho}(q_z, l_m, N, \Delta b),$$

with q_z the root mean square vertical component of the turbulent velocity, l_m the integral length scale of turbulence within the mixed layer; N is the bouyancy frequency in the fluid below and Δb has the same meaning as in N&K. Using dimensional analysis:

$$\frac{E_i}{\rho q_z^3} = g\left(\frac{\Delta b h}{q_z^2}, \frac{N q_z}{\Delta b}, \frac{l_m}{h}\right). \qquad (10a.1)$$

For brevity's sake we write $Nq_z/\Delta b \equiv \delta$. With the interface treated as a sharp discontinuity $\Delta bh = c_i^2$; and with the integral scale length proportional to the layer depth in fully developed mixed layer turbulence $l_m \propto h$, this allows Eq. (10a.1) to be rewritten in the form:

$$\frac{E_i}{\rho q_z^3} = g_1\left(\frac{c_i}{q_z}, \delta\right). \qquad (10a.2)$$

The actual form for the function g_1 cannot, however, be deduced by dimensional analysis and recourse must be made to detailed examination of the mechanics of internal wave generation by turbulent eddies.

Let A be the typical amplitude and λ the typical horizontal wave length of the internal waves generated at the edge of the mixed layer. λ is also the horizontal scale of the eddies. Following Thorpe (1973), the vertical energy flux per unit horizontal area to these waves can be written as:

$$E_i = \rho A^2 N^2 \vec{c}_g \cdot \vec{k},$$

where $\vec{c}_g \cdot \vec{k}$, the vertical component of the group velocity of these waves, is a maximum when the ratio of the horizontal to vertical wave number is $\sqrt{2}$. The maximum energy flux to these waves is, therefore:

$$E_i(\max) = \rho A^2 N^3 \lambda / 3\sqrt{3}\,\pi. \tag{10a.3}$$

The amplitude of the internal waves excited should be deduced from the response of the interface-stratified fluid system to excitation by turbulent eddies whose frequency wave number spectrum is known. This is a formidable task certainly not justified for use in simple slab models. Instead, we will use scaling arguments to arrive at a simple, yet plausible parameterization.

The horizontal wave length of the waves is the integral length scale of the eddies which is proportional to the depth of the mixed layer:

$$\lambda \sim h. \tag{10a.4}$$

What is the typical amplitude A of the excited waves? It should scale with the amplitude of the indentations of the interface resulting from the impinging eddies. Let us first consider the case when the buoyancy jump across the interface is zero ($\Delta b=0$). Townsend (1966) has shown that an eddy with a characteristic vertical velocity q_z impinging on a stratified fluid would be stopped in a distance of the order q_z/N if it does not mix with the surrounding fluid. This is equivalent to saying that the potential energy $N^2x^2/2$ due to the deflection x of the interface between the buoyant and non-buoyant fluids is of the order of the vertical kinetic energy of the eddy. We can therefore take the typical amplitude of internal waves generated to be proportional to q_z/N.

The validity of this expression is limited to the case of $\Delta b=0$. Now consider the case of a finite interfacial buoyancy jump $\Delta b \neq 0$, but no stratification in the fluid below ($N=0$) that is, a two layer system. Linden (1973) has shown both theoretically (using energy arguments) and experimentally that the maximum extent of the interface deflection produced by a vortex ring projected against the interface with velocity u is proportional to $u^2/\Delta b$. This is again equivalent to the statement that the potential energy due to the deflection of the interface which is proportional to $\Delta b\, x$ is of the order of the kinetic energy of the impinging vortex ring. If we assume that the turbulent eddies behave similarly, the typical amplitude of the internal waves generated should scale as $q_z^2/\Delta b$.

We have two expressions for the typical amplitude of the internal waves generated valid in different limits:

$$A \sim q_z/N \qquad (N \to 0,\ \Delta b = 0),$$

$$A \sim q_z^2/\Delta b \qquad (N = 0,\ \Delta b \to 0). \tag{10a.5}$$

In the general case when neither N nor Δb vanish, the combined potential energy should match the kinetic energy $q_z^2/2$ of the vertically moving parcel:

$$C\Delta bx + N^2x^2/2 \sim q_z^2/2,$$

where C is an empirical constant approximately equal to 1. Assuming the internal wave amplitude A to be proportional to the interface deflection x, one gets

$$A \sim \frac{\{(C\Delta b)^2 + N^2q_z^2\}^{1/2} - C\Delta b}{N^2}. \tag{10a.6}$$

Combining Eqs. (10a.3), (10a.4), and (10a.6),

$$\frac{E_i}{\rho} = C_0 \frac{2(C\Delta b)^2 + N^2q_z^2 - 2C\Delta b\{(C\Delta b)^2 + N^2q_z^2\}^{1/2}}{N} .h .$$

Expressed in non-dimensional form,

$$\frac{E_i}{\rho q_z^3} = C_0\left(\frac{c_i}{q_z}\right)^2 \frac{2C[C-(C^2+\delta^2)^{1/2}] + \delta^2}{\delta}. \tag{10a.7}$$

Equation (10a.7) is the explicit form of the Eq. (10a.2). It can be said to provide a different form for the term (G) in Eq. (10.29) of N&K.

The expression in Eq. (10a.7) has the right behavior in the limits N=0 and $\Delta b \to \infty$, that is, when δ=0. Under these conditions, energy loss to internal waves should vanish, because for N=0 internal waves cannot be generated in the non-stratified fluid below, and for $\Delta b = \infty$ the impinging eddy cannot produce any finite interface deflection. E_i=0 in both these limits.

In the derivation of Eq. (10a.7), we have implicitly assumed that the frequency associated with the eddies generating the dominant internal waves is less than the buoyancy frequency N and most of the vertical turbulent kinetic energy is associated with these eddies. A situation where the dominant frequency in the turbulence spectrum is greater than N could occur. Strictly speaking, q_z^2 should be the kinetic energy associated with the region of the turbulence spectrum with frequencies below N and the scale of the waves, the turbulence length scale associated with this region of the spectrum. An analysis of this general case will not be undertaken here.

We have also assumed that the region below is unbounded and has uniform stratification. These conditions are however unduly restrictive. Following Townsend (1966), it can be shown that the internal wave amplitude falls off rapidly due to viscous dissipation at distances z from the interface beyond a critical distance which is of the order of the product of the vertical propagation velocity and the viscous decay time. It can be shown that this critical distance z_c is:

$$z_c = C_1 \lambda^3 N/\nu.$$

A rapid attenuation of bodily waves generated at the edge of the mixed layer beyond a certain distance has indeed been observed in laboratory experiments (Linden, 1975). The value of C_1 in Linden's experiments seems to be about 1.5×10^{-3}. Using this value for the constant and typical values of 2 m for λ and 10^{-4} s^{-2} for N^2, z_c turns out to be about 86 m, roughly of the same magnitude as the mixed layer depth. It is safe to conclude that this critical distance could be of the same order as the mixed layer depth in the ocean. As long as the bottom boundary is at a distance z away from the interface large compared to this distance, reflection of internal waves from the bottom can be ignored and energy flux E_i regarded as a total loss. This is invariably the case in the deep ocean. Further, Eq. (10a.7) can be applied to the general case of non-uniform stratification below as long as the scale of variation of N is much greater than this critical distance.

Using Eq. (10a.4), z_c can be written as:

$$z_c/h \sim \frac{Nh}{q_z} Re, \tag{10a.8}$$

where $Re = h\,q_z/\nu$ is the Reynolds number.

Usually, in the ocean $\delta \ll 1$. The expression (10a.7) can therefore be simplified considerably by expansion in terms of δ:

$$\frac{E_i}{\rho q_z^3} = C_2 \left(\frac{c_i}{q_z}\right)^2 \delta^3, \tag{10a.9}$$

where C_2 is another constant. Linden's experiments suggest a value of roughly 300 to 500 for this constant, although the precise value is rather uncertain. For typical oceanic conditions, say $N^2 = 10^{-4}$ s^{-2}, h=50 m and $\Delta\rho/\rho = 0.002$, let us compute the magnitude of E_i/ρ using the lower value of 300 for the constant C_2. For a wind of about 8 m s^{-1}, the friction velocity u_* is about 1 cm s^{-1} and q_z can be taken as 1.1 cm s^{-1} using neutral flow data. For these values E_i/ρ is about 0.57 cm^3 s^{-3}. Taking a value of 3 for the constant m in N&K Eq. (10.29), E_i turns out to be almost 20% of the rate of working by the wind! The reader should however be reminded that E_i/ρ depends rather sensitively on the value of N immediately below the interface and could be substantially more or less than this value depending on the value of N. Also, Linden's results are for flow without mean shear, whereas oceanic conditions invariably involve mean shear.

There is no doubt that under appropriate conditions, energy radiated to internal waves could be substantial. Linden (1975) has shown that internal waves reduced the entrainment rate by as much as 50% in his experiments on entrainment at a density interface with turbulence produced by mechanical stirring. The increased entrainment rates observed in the two-layer experiments of Kantha (1975) as compared to the original Kato-Phillips experiments on entrainment due to an applied shear stress (Kato and Phillips, 1969) are partly due to the absence of internal waves in Kantha's experiments. The large amount of scatter observed in field experiments (Farmer, 1975) and some laboratory experiments (e.g., Deardorff, et al., 1969) on penetrative convection in stratified fluids could very well be due to the unaccounted effects of internal waves. Although there is thus some evidence to suggest that internal waves could substantially reduce the entrainment rates, there is at present not much reliable data for a better quantitative assessment of this reduction or verification of Eq. (10a.7) or (10a.9). If this note attracts attention to the difficult task of careful determination of the reduction in entrainment due to internal waves, either from field observations or laboratory experiments, it will have served its purpose. Similar studies are desirable for the opposite case where the entrainment rate could be enhanced by interfacial waves that have been generated elsewhere.

Acknowledgement: The author wishes to thank Professor O. M. Phillips for his help in completing this note. This work was supported by Office of Naval Research (Physical Oceanography) under contract No. N00014-75-C-0700.

Chapter 11

UPWELLING IN THE OCEAN: TWO-AND THREE-DIMENSIONAL MODELS OF UPPER OCEAN DYNAMICS AND VARIABILITY

James J. O'Brien, R. Michael Clancy, Allan J. Clarke, Michel Crepon, Russell Elsberry, Tor Gammelsrød, Malcolm MacVean, Lars Petter Röed and J. Dana Thompson

11.1 Introduction

The NATO conference at Urbino was directed toward understanding the state of our knowledge of modelling the upper ocean. Niiler and Kraus (Chapter 10) have reviewed the one-dimensional models. We shall review several specific examples of two and three-dimensional models. The strong influence of horizontal boundaries, bottom topography, and spatial and temporal variability of wind stress on the horizontal momentum distribution and vertical motion patterns are illustrated by specific examples of coastal as well as open ocean upwelling, baroclinic shelf-waves, ice edge upwelling, and ocean front formation. These are only a few examples of interesting upper ocean baroclinic circulation problems. The important storm surge problem, a barotropic phenomenon, is specifically not addressed. Large-scale ocean modelling is not reviewed. The important mesoscale eddy problem is discussed by Bretherton (1975) and Holland (Chapter 2). Holland also discusses parameterization of vertical and horizontal mixing in the upper ocean in global scale models. Almost all of the research described in this paper has appeared in the literature or will appear soon. Our intention is to discuss many interesting problems and lead the reader to the latest literature for the details.

Because of the economic importance of fisheries in upwelling regions, several large oceanographic field programs have been conducted off Oregon and Northwest Africa and are planned for Peru. These experiments provide the basis for verifying analytical and numerical models of coastal upwellings. Section 11.2 discusses the coastal upwelling problem from a two-dimensional viewpoint. Because the sea breeze circulation depends on the land-sea temperature contrast for its existence, coastal upwelling must have a strong effect on it. The pioneering work of coupling a mesoscale ocean circulation to a mesoscale sea breeze circulation is outlined in Section 11.3.

Coastal upwelling is rarely two-dimensional. The physics of the three-dimensional circulation can be cast in terms of trapped baroclinic shelf waves (Section 11.4). The presence of coastline irregularities such as capes and bays and the influence of bottom topographic features such as canyons and ridges create three-dimensional circulations which are not completely understood or observed (Section 11.5). At the equator, the vanishing of the Coriolis acceleration creates a one-sided divergence which forces strong upwelling. We have chosen not to review equatorial models in this paper. The reader is referred to the recent reviews by Gill (1975), Philander (1973) and Moore and Philander (1976). The basic time dependent motions at the equator were first discussed by Yoshida (1959). O'Brien and Hurlburt (1974) simulated the Wyrtki (1973) Indian Ocean jet using a two-layer non-linear model. Hurlburt, Kindle, and O'Brien (1976) have studied the mechanisms leading to the El Niño disturbances. These papers and the references therein should suffice to show our state of knowledge for modelling equatorial circulations.

11.2 Two-dimensional numerical models of coastal upwelling

Ideally, the theoretical study of coastal upwelling should culminate in a continuously stratified, non-linear, time dependent analytical model. Historically, most analytical models have been steady, linear, and two-dimensional, with great mathematical difficulties encountered in modelling the coastal corner region. While recent analytical contributions of Allen (1973), Pietrafesa (1973), Pedlosky (1974), and others have overcome some of the mathematical impediments, analytical models of coastal upwelling remain restricted in their parameter range and realism. While they are valuable for elucidating physical processes in upwelling systems, they do not provide the capability of simulating non-linear upwelling circulations driven by highly time dependent forcing. It is these circulations which are of greatest interest to the ecosystem modeller.

Numerical models provide an alternative approach to understanding coastal upwelling. While they too may suffer from lack of realism, a careful formulation can lead to a useful tool for understanding upwelling dynamics, while avoiding the mathematical limitations of the analytical models. In this section two-dimensional numerical models of coastal upwelling will be discussed. Since virtually all published literature on numerical models of coastal upwelling has involved vertically integrated hydrostatic layers (Lagrangian vertical coordinate) in contrast to vertical grid point (Eulerian vertical coordinate) models, the subsequent discussion will be strongly biased toward the former formulation. Recent work by Hamilton and Rattray (1975) does show promise for studying some important details of the upwelling circulation using the vertical grid point approach.

Consider a stably stratified, rotating, incompressible fluid on a continental shelf slope cross-section near a north-south coastline. Assume the fluid is represented by two hydrostatic layers having a density jump at the fluid interface. This is the simplest model to retain the barotropic and first baroclinic modes. Also assume that the interface may be permeable to mass, heat, and momentum, such that turbulent entrainment across the interface is permitted. On retaining the first term in the perturbation expansion for the advective terms, the vertically integrated momentum equations are (Thompson, 1974)

$$\frac{\partial \vec{v}_1}{\partial t} + (\vec{v}_1 \cdot \nabla_H)\vec{v}_1 + f\vec{k}\times\vec{v}_1 = -g\nabla_H(h_1+h_2+h_b) + \frac{\vec{\tau}_o - \vec{\tau}_I}{\rho_r h_1}$$

$$+ \frac{[\nabla_H \cdot (\rho_1 h_1 K_{MH} \nabla_H)]\vec{v}_1}{\rho_r h_1} - \frac{g h_1}{2\rho_r} \nabla_H \rho_1 + \vec{S}_1 ; \qquad (11.1)$$

$$\frac{\partial \underset{\to}{v}_2}{\partial t} + (\underset{\to}{v}_2 \cdot \nabla_H)\underset{\to}{v}_2 + f\underset{\to}{k}\times\underset{\to}{v}_2 = -g\nabla_H(h_1+h_2+h_b) + g'\nabla_H h_1 + \frac{\underset{\to}{\tau}_I - \underset{\to}{\tau}_b}{\rho_r h_2}$$

$$+ \frac{[\nabla_H \cdot (\rho_2 h_2 K_{MH} \nabla_H)]\underset{\to}{v}_2}{\rho_r h_2} - \frac{g h_1}{2\rho_r}\nabla_H \rho_1 - \frac{g h_2}{2\rho_r}\nabla_H \rho_2 + \underset{\to}{S}_2; \qquad (11.2)$$

where

$$g' = g\frac{(\rho_2-\rho_1)}{\rho_2}; \qquad \underset{\to}{\tau}_I = \rho_r c_I |\underset{\to}{v}_1 - \underset{\to}{v}_2|(\underset{\to}{v}_1 - \underset{\to}{v}_2);$$

$$\underset{\to}{S}_1 = \rho_2 w_{e_1}\frac{(\underset{\to}{v}_2-\underset{\to}{v}_1)}{\rho_1 h_1}; \qquad \underset{\to}{\tau}_b = \rho_r c_b |\underset{\to}{v}_2|\underset{\to}{v}_2;$$

$$\underset{\to}{S}_2 = -\rho_1 w_{e_2}\frac{(\underset{\to}{v}_2-\underset{\to}{v}_1)}{\rho_2 h_2}; \qquad (11.3)$$

$$\Delta\rho = \left(\frac{\rho_2-\rho_1}{\rho_1}\right);$$

$$f = f_o + \beta(y-y_o).$$

The subscripts 1 and 2 refer to upper and lower layers, respectively. The usual notation and right-handed coordinate system is used. The stresses are at the surface (O), interface (I) and bottom (b). The height at the bottom topography above a reference level is given by h_b. w_{e_1} and w_{e_2} represent turbulent entrainment rates, where w_{e_1} represents entrainment of lower layer fluid upward into the turbulent upper layer. These entrainment rates have the dimensions of velocity and will be discussed further below. Explanation of the interfacial and bottom stress formulations can be found in Thompson and O'Brien (1973).

Density and layer thicknesses are predicted by

$$\frac{\partial}{\partial t}\rho_1 + \underset{\to}{v}_1 \cdot \nabla_H \rho_1 = \frac{w_{e_1}\Delta\rho}{h_1} - \frac{\alpha H_o}{\rho_1 c h_1} + \frac{1}{h_1}\nabla_H \cdot (K_H h_1 \nabla_H \rho_1); \qquad (11.4)$$

$$\frac{\partial}{\partial t}h_1 + \nabla_H \cdot (h_1 \underset{\to}{v}_1) = w_{e_1} - w_{e_2} + \frac{\alpha H_o}{\rho_1^2 c}; \qquad (11.5)$$

$$\frac{\partial}{\partial t} \rho_2 + \underset{\to}{V}_2 \cdot \nabla_H \rho_2 = - \frac{w_{e_2} \Delta\rho}{h_2} + \frac{1}{h_2} \nabla_H \cdot K_H h_2 \nabla_H \rho_2 \; ; \qquad (11.6)$$

$$\frac{\partial}{\partial t} h_2 + \nabla_H \cdot (h_2 \underset{\to}{V}_2) = w_{e_2} - w_{e_1} \; ; \qquad (11.7)$$

where

$$\rho_1 = \rho_r - \alpha T_1;$$

$$\rho_2 = \rho_r - \alpha T_2.$$

The effective thermal expansion coefficient for sea water is represented by α and ρ_r is a reference density. The horizontal eddy diffusion coefficients are K_H for heat and K_{MH} for momentum. Diabatic effects (solar isolation, back radiation, and air-sea exchange of sensible and latent heat) are represented by H_o (defined positive upwards).

Except for Yoshida's (1967) analytical work and Thompson's (1974) numerical work, all researchers have prohibited mass and heat transfer across the fluid interface or the sea surface. Then $w_{e_1}=w_{e_2}=0$ and thermodynamic effects are ignored, hence the horizontally averaged density of each layer remains constant in time. In that case, the last two terms of Eqs. (11.1) and (11.7), the last three terms of Eqs. (11.2) and (11.5) and Eqs. (11.4) and (11.6) vanish. Such models are termed hydrodynamic models, in contrast to hydrodynamic-thermodynamic models where the density field varies spatially and temporally. Implicit in the hydrodynamic models is the assumption that vertical mixing processes have time scales long compared to the upwelling time scale of several days.

It is unlikely that any coastal upwelling regime is consistently two-dimensional on time scales of a season or longer. However, some may be two-dimensional to first order in some regions on the "event" or weekly time scale. It is certainly reasonable to begin the study of upwelling dynamics under the assumption that longshore length scales are much greater than offshore scales and longshore variations in coastline, wind stress, and bottom topography are negligible. Then the velocity field is independent of the y coordinate for a north-south coastline.

The two-dimensional, two-layer, hydrodynamic model of coastal upwelling, involving circulation in a vertical plane, has been studied analytically by Yoshida (1967) and numerically by O'Brien and Hurlburt (1972), Hurlburt and Thompson (1973), Thompson and O'Brien (1973), and McNider and O'Brien (1973). A hydrodynamic-thermodynamic version of this model has been developed by Thompson (1974). Three-dimensional (x-y two-layer) models have been studied analytically by Yoshida (loc. cit.), Allen (1975), and by Gill and Clarke (1974). Numerical investigations were carried out by Hurlburt (1974), Suginohara (1974), and Peffley and O'Brien (1976). The remainder of this section will focus on the formulation, progress, and limitations of the two-dimensional numerical models of coastal upwelling.

It should be recognized that coastal upwelling occurs in a narrow boundary region, where the appropriate horizontal offshore scale is the baroclinic radius of deformation (Yoshida, 1955). In a linear two-layer model, one can show that a wind stress parallel to the shore generates a strong upwelling by the process of internal geostrophic adjustment. If the wind stress is uniform in space and a step function in time, the asymptotic expansion of the elevation at the interface is of the form

$$h_1(x,t) = \frac{\tau_o^y}{\rho_r c_i f}\left\{ft \exp\left[-\frac{xf}{c_i}\right] + 0\left[\frac{\cos(t-\pi/4)}{(ft)^{1/2}}\right]\right\},$$

where

τ_o^y is the wind stress parallel to the shore;

c_i the velocity of long internal gravity waves;

x the distance from the shore.

The elevation is maximum at the shore. This geostrophic adjustment process provides a radiation of energy from the shore in the form of the long internal gravity waves (Crepon, 1972; O'Brien and Hurlburt, 1972; McNider and O'Brien, 1973; Crepon, 1974). The wave length λ at the front of the radiational disturbance is of the order of the internal radius of deformation. Thus, in order to include this fundamental mechanism in the numerical models, it is necessary that the spatial size of the grid should be of the order of $\lambda/4$ or less, or 0(1 km). In mid-latitudes λ is 0(10 km). Since the problem is highly time-dependent, small time steps 0(1 hr) are necessary.

The first numerical model of coastal upwelling was constructed by O'Brien and Hurlburt (1972). Their non-linear, time-dependent f-plane model allowed the efficient retention of the free surface in the problem (instead of a "rigid lid") in order to account for the free surface gravity waves important to geostrophic adjustment. Their principal contribution was the prediction of an equatorward surface jet occurring within a baroclinic radius of the coast during active upwelling. This jet was originally explained by Charney (1955) in terms of conservation of potential vorticity. For a system initially at rest driven by a uniform equatorward wind (parallel to a coast to the east) the time integrated inviscid potential vorticity equations become

$$\frac{\partial v_1}{\partial x} = f\,\frac{(h_1-h_1^0)}{h_1^0}\,;$$

$$\frac{\partial v_2}{\partial x} = f\,\frac{(h_2-h_2^0)}{h_2^0}\,; \qquad (11.8)$$

where h_1^0 and h_2^0 are the undisturbed layer thicknesses. Clearly, as upwelling proceeds, h_1 decreases, h_2 increases, v_1 is a minimum near the coast and v_2 is a maximum. The existence of a frictional boundary layer bringing the longshore flow to zero at the coast forces the equatorward surface jet minimum to occur offshore. The frictional boundary width, $0[\sqrt{K_{MH} v_1/fu_1}]$ indicates that if $K_{MH}=0[10^6\ cm^2s^{-1}]$, the potential vorticity argument is invalid and the equatorward surface jet may be engulfed by the friction layer. All numerical modellers of coastal upwelling have chosen to set K_{MH} small to minimize the importance of the sidewall frictional boundary.

Although O'Brien and Hurlburt successfully predicted the observed equatorward surface jet, they failed to produce the oft observed poleward undercurrent. Instead, a large barotropic mode developed in their model. Previously Yoshida (1967) and Garvine (1971) had postulated the existence of a small north-south barotropic pressure gradient as a necessary component of upwelling dynamics. This pressure gradient acts to inhibit the development of the barotropic mode and geostrophically balances the onshore return flow feeding the upwelling. Observationally, this pressure gradient is exceedingly difficult to measure, since a sea surface slope of only a few centimeters per 1000 km is required. Theoretically, the explanations for the existence of the sea surface slope remain a matter of much controversy.

Extending the two-dimensional hydrodynamic model to the β-plane, Hurlburt and Thompson (1973) suggested one possible mechanism for introducing the barotropic north-south pressure gradient. Using a technique originally employed by Veronis and Stommel (1956), they were able to neglect longshore deviations of the velocity field while retaining the longshore barotropic and baroclinic pressure gradients. They found that a Sverdrup interior ocean matched to the coastal upwelling boundary layer required a north-south sea surface slope of just the right magnitude to balance the onshore flow outside the upwelling zone. The simplest possible dynamical argument for this result consists of assuming the existence of a Sverdrup interior ocean and geostrophic meridional flow,

$$\beta v D = \frac{1}{\rho_r} \vec{k} \cdot \text{curl}\ \vec{\tau}_0 ; \tag{11.9}$$

$$f v = g \frac{\partial D}{\partial x} ; \tag{11.10}$$

where D is the barotropic fluid depth. For constant wind stress curl, the meridional flow is constant at all latitudes. However, if the meridional flow is geostrophic, variations in latitude require variations in the east-west barotropic pressure gradient. Clearly for negative wind stress curl the east-west pressure gradient must become relatively more negative at higher latitudes, implying a sea surface sloping downward to the north in the eastern ocean relative to that in the ocean interior. The time scale for the development of the north-south pressure gradient at the eastern boundary is several inertial periods while the time scale for development of the Sverdrup balance at a given point in the eastern ocean is the time required for a barotropic Rossby wave to propagate from the eastern boundary to the point in question.

Figure 11.1 (from Hurlburt and Thompson, 1973) compares the f-plane and β-plane solutions after 15 days of integration for the case of a uniform depth ocean driven from rest by a simple northerly wind stress. The details of this mid-latitude case are not required for the comparison shown here. The principal point of the figure is that while β exerts a secondary influence on the vertical mass transport in the upwelling zone it has a profound effect on the longshore flow. The importance of the ocean interior to the upwelling dynamics is clear from this comparison and underscores the need for adequately including the interior dynamics in upwelling models (Garvine, 1974).

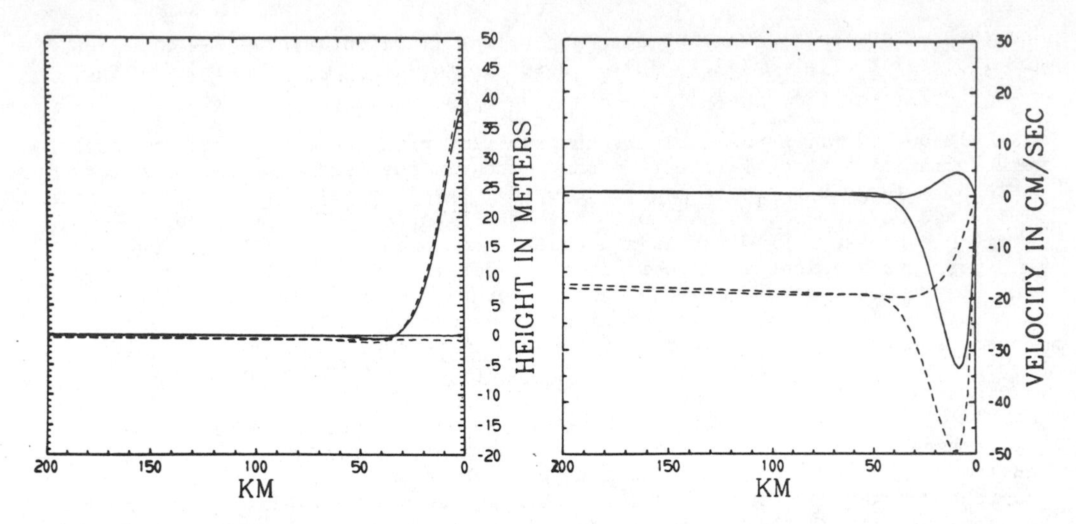

Fig. 11.1 - The f-plane versus the β-plane solutions at day 15: pycnocline and free surface height anomalies (left) and longshore velocity profiles (right) versus distance offshore for $\beta=2\times10^{-13}$ $cm^{-1}s^{-1}$ (solid) and $\beta=0$ (dashed). The upper layer equatorward surface jet appears in both cases. Note the absence of the barotropic mode in the β-plane model.

For the flat bottom case Hurlburt and Thompson demonstrate that the local integrated longshore mass transport in the upwelling zone is zero, implying the same width scale for the undercurrent as for the surface jet. Further, they show that the offshore transport in the surface layer is

$$\rho_r u_1 h_1 = \frac{\tau_0^y}{f} - \frac{\rho_r g h_1}{f} \frac{\partial}{\partial y} (h_1+h_2); \qquad (11.11)$$

i.e., the Ekman transport reduced by a geostrophic component. The return flow is shown to be geostrophic. When significant bottom topographic influences are introduced, the barotropic mode does not vanish and the dynamical balances become more complicated and non-linear. In that case the undercurrent does not generally have the same width scale as the surface jet.

Recent work (unpublished research by Hurlburt and Thompson) has verified the two-dimensional results of the β-plane model by comparison with a closed basin three-dimensional layered model. For a straight north-south coast the results of the two-dimensional model are virtually identical to those of the three-dimensional closed basin model at the same latitude. Only after three or four weeks do the model solutions diverge, consistent with the disturbance of the upwelling by northward propagating Kelvin waves generated near the southern boundary in the three-dimensional model.

Purely hydrodynamic upwelling models are valid only within the time scale in which vertical mixing processes are unimportant compared to vertical advection. Observations suggest that vertical mixing may become significant within several days of the onset of a coastal upwelling event. Turbulent motions produced by boundary stresses or shear instabilities can produce vertical entrainment of heat, mass, and momentum. Based on the laboratory work of Rouse and Dodu (1955), Kato and Phillips (1969) and others, as discussed by Phillips in

Chapter 7, this entrainment w_e of nonturbulent fluid into turbulent fluid may be expressed as a function of friction velocity and Richardson number, $w_e=u_* f(Ri)$. For the range of Richardson numbers found in the coastal upwelling regime off Oregon, for example, the entrainment rate, from the Kato and Phillips study, has the form $w_e=mu_*(Ri)^{-1}$. For the layered model represented in Eqs. (11.1)-(11.7), the entrainment of lower layer fluid into the upper layer due to wind stirring is expressed as

$$w_{e_1} = \frac{\rho_r m u_*^3}{g\Delta\rho h_1}. \qquad (11.12)$$

The constant m is proportional to the fraction of wind energy available for turbulent mixing (Denman, 1973). The friction velocity is $(|\underset{\rightarrow}{\tau}_0|/\rho_r)^{1/2}$. Entrainment of warm, upper layer fluid into the lower layer due to bottom generated turbulence, w_{e_2}, is identical to w_{e_1}, with h_1 replaced by h_2 and u_* represents the bottom friction velocity.

Admittedly, this parameterization of turbulent mixing is simplistic. More detailed formulations are discussed by Niiler and Kraus (Chapter 10). It has the distinction, however, of being related to the bulk dynamic stability of the water column, and is self determined during the numerical integrations.

A further extension of the earlier hydrodynamical models involves an attempt to reconstruct the Ekman-like boundary layers smeared out by the original vertical averaging process. In this reconstruction, departures from the vertically averaged velocity field within each layer are calculated under the assumption of hydrostatic balance, small Rossby number, and quasi-equilibrium conditions. Mathematically, the problem is reduced to solving the equation

$$\underset{\rightarrow}{k} \times \frac{\partial \underset{\rightarrow}{v}'_j}{\partial z} = \frac{g}{\rho_r} \nabla_H \rho_j + K_{MV} \frac{\partial^3 \underset{\rightarrow}{v}'_j}{\partial z^3}, \qquad (j=1,2) \qquad (11.13)$$

where K_{MV} is the vertical diffusion coefficient for momentum. The primed quantities are departures from the vertical average within a layer and the subscript refers to the upper or lower layer. We require the boundary conditions

$$K_{MV} \frac{\partial \underset{\rightarrow}{v}'_1}{\partial z} = \underset{\rightarrow}{\tau}_0 \quad \text{at the sea surface;}$$

$$\overline{\underset{\rightarrow}{v}}_1 + \underset{\rightarrow}{v}'_1 = \overline{\underset{\rightarrow}{v}}_2 + \underset{\rightarrow}{v}'_2 \quad \text{at the interface;}$$

$$\frac{\partial \underset{\rightarrow}{v}'_1}{\partial z} = \frac{\partial \underset{\rightarrow}{v}'_2}{\partial z} \quad \text{at the interface;} \qquad (11.14)$$

$$\overline{\underset{\rightarrow}{v}}_2 + \underset{\rightarrow}{v}'_2 = 0 \quad \text{at the bottom;}$$

and the integral constraints

$$\int_{\substack{\text{upper}\\\text{layer}}} \underset{\to}{v}_1' \, dz = \int_{\substack{\text{lower}\\\text{layer}}} \underset{\to}{v}_2' \, dz = 0. \tag{11.15}$$

Note that the perturbation velocities can only be calculated if the averaged flow and density fields are known. In practice these values are obtained during the course of the integration.

The perturbation velocity approach is not generally valid for highly nonlinear, time-dependent conditions. It does serve as a first order representation of the flow field, and is useful for flow visualization. Note in the above problem that vertical eddy viscosity is assumed constant. This restriction may be relaxed to make K_{MH} a function of dynamic stability. Also, a preferred strategy is to allow the mean fields and perturbation fields to interact during the integrations. Unfortunately, using this technique increases computer time drastically.

A third extension of the previous hydrodynamical model is to include the diabatic processes of solar heating and infrared back radiation. For event time scales in mid-latitude coastal upwelling, warming of the upper layer is not significant. At lower latitudes solar insolation does influence sea surface temperature and stratification on time scales as short as one week.

Briefly then, the layered hydrodynamic models have recently been extended in three limited ways: (*i*) by including transfer of heat, mass and momentum across the interface due to turbulent entrainment, (*ii*) by including perturbation corrections to the vertically averaged flow, and (*iii*) by including the diabatic effects of short and long wave radiation. A few results from a model with those extensions is discussed in the following paragraphs.

An obvious limitation of the hydrodynamical models has been the necessity for termination of the numerical integrations when the internal interface reached the sea surface. With the hydrodynamical-thermodynamical model this problem does not arise, since the turbulent mixing tends to deepen the interface, acting against vertical advection. Not only does this allow the hydrodynamical-thermodynamical model to be integrated much longer in time, it also enables the model to reproduce the rapid drops in sea surface temperatures and the strong sea surface temperature gradients often observed in upwelling zones.

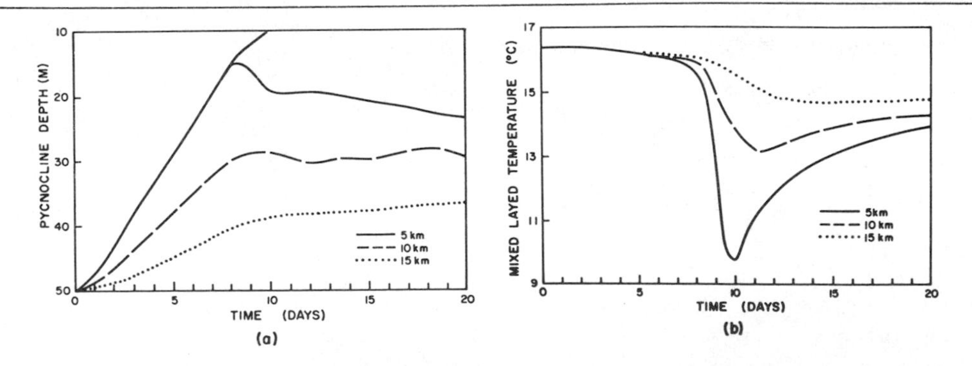

Fig. 11.2 - Time series of upper layer (a) depth and (b) temperature at 5, 10, and 15 km offshore for the event case. The solid line extending upward in (a) represents the solution when no vertical mixing across the interface is included.

To demonstrate this capability, consider Fig. 11.2. The numerical model was started from rest, driven by a northerly wind stress whose magnitude increased linearly to 1 dyne cm^{-2} during the first day, remained constant to day 10, then linearly decreased to zero on the eleventh day. This temporal variation is designed to simulate an upwelling event. The offshore wind distribution was chosen to simulate the large-scale wind stress curl in the ocean interior. Basin width was 3100 km, with grid resolution of 1 km in the nearest 100 km of the coast.

Figure 11.2 shows time series of upper layer thickness and upper layer temperature versus time. Note that during the first week vertical mixing is unimportant. Near day 8 mixing tends to deepen the interface as compared to the no mixing case, which allows the interface to continue rising toward the sea surface. The upper layer temperature also is influenced by vertical mixing. Two non-linear processes act to produce the rapid temperature drop near day 8. First the amount of fluid in the upper layer water column decreases during upwelling. Also, as the layer thins, the "pycnocline" moves closer to the source of the turbulent energy at the sea surface, intensifying the entrainment rate. Figure 11.26 is similar to the sea surface temperature time series presented by Holladay and O'Brien (1975) from data obtained during CUE-II off Oregon.

A second aspect of the above described numerical model is its ability to reconstruct the boundary layer flow. Figure 11.3 (Thompson, 1974) presents the velocity fields at day 10 of the integrations for the case just described with a "continental shelf." Of note in the figure is the two-cell circulation over the shelf, the offshore motion at the base of the pycnocline, the secondary upwelling at the shelf break, and the poleward undercurrent and equatorward surface jet within 20 km of the coast. Each of these features has a counterpart in observational data (Mooers, Collins, and Smith, 1976). In general, the cellular circulations are very sensitive to wind stress, stratification, and the shape of the bottom.

In summary, the most successful two-dimensional numerical models of coastal upwelling to date have retained a Lagrangian vertical coordinate, with two or more layers. The time scales of validity of the hydrodynamical layered models have been extended by including effects of vertical turbulent mixing and diabatic features. A host of features consistently observed in mid-latitude coastal upwelling have been reproduced by these models. Presently there is a need for greater vertical resolution in the numerical models, as well as a more sophisticated parameterization of vertical turbulence.

The modelling efforts discussed so far have neglected mesoscale variability in the wind field and have allowed no air-sea feedback coupling. In the real world these effects are expected to be important since almost all coastal upwelling regimes are influenced by the mesoscale sea breeze circulation, which, in turn, will be perturbed by coastal upwelling. The results of a coupled sea breeze coastal upwelling model are discussed in the next section.

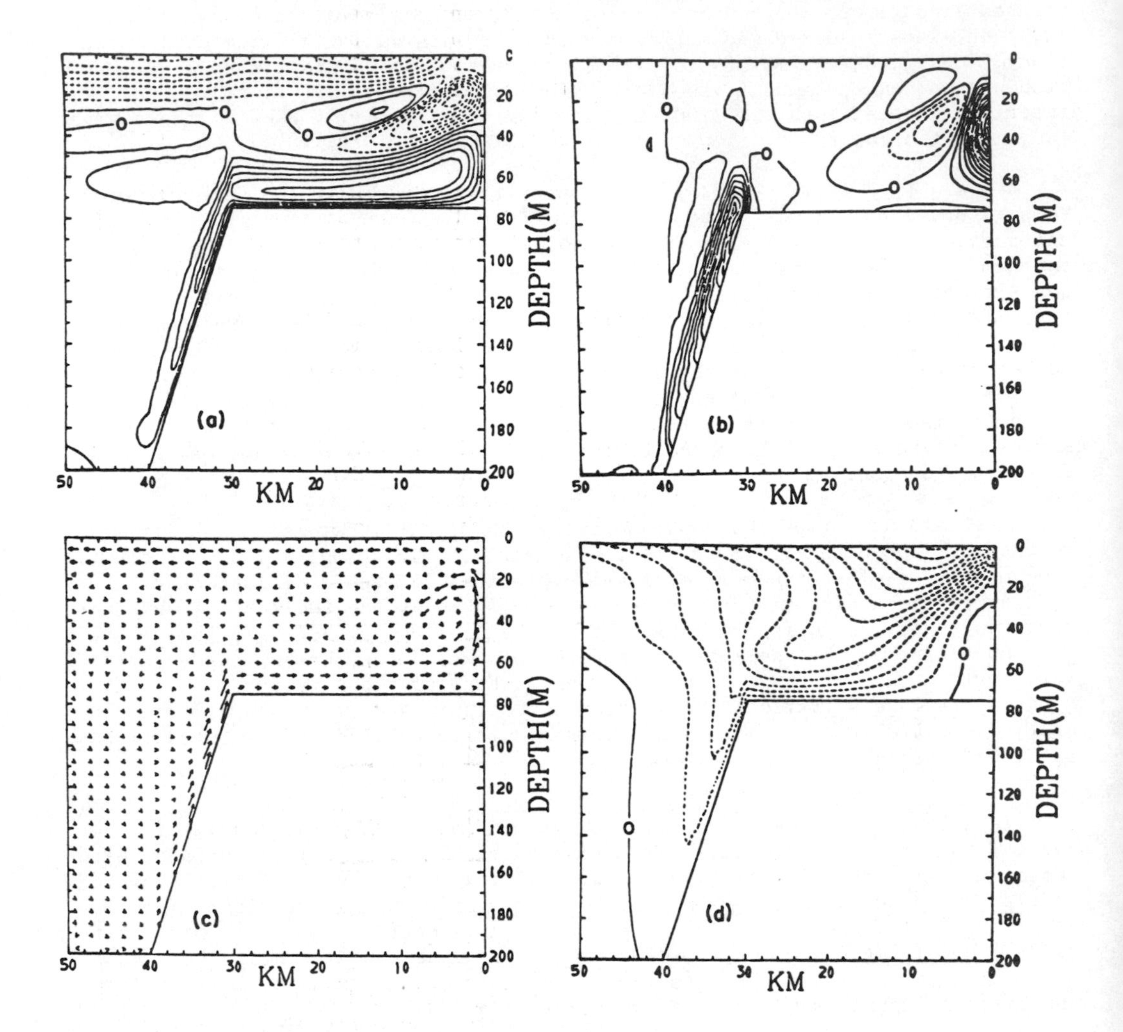

Fig. 11.3 - Velocity fields for the continental shelf event case at day 10; (a) contours of zonal current speed in 1 cm s^{-1} intervals, (b) contours of vertical motion in intervals of $.5x10^{-2}$ cm s^{-1}, (c) u-w vectors scaled to the maximum length in the field, (d) contours of longshore flow in 5 cm s^{-1} intervals. Dashed contours indicate negative values.

11.3 Air-sea interaction in an upwelling sea breeze regime

It is commonly assumed that ocean models and atmospheric models will eventually be coupled together to study the joint variability between the two geophysical fluids. For global models this is very difficult because the time scales for the two fluids are very different and there is an impedence mismatch. In upwelling regimes, strong sea breeze atmospheric motions occur. This affords an excellent opportunity to study the interaction of the two geophysical fluids on a mesoscale where the time scales are both order one to several days. A numerical model has been constructed which consists of coupling a two-dimensional, four-layer atmosphere to a two-dimensional, two-layer ocean through simple heat and momentum fluxes in order to study the mesoscale upwelling sea breeze regime.

The ocean model is described completely by Thompson (1974) and briefly in Section 11.2. It is capable of producing a realistic simulation of the time and space dependence of sea surface temperatures characteristic of coastal upwelling regimes.

The atmosphere model is described by Lee (1973). The model equations are those for the conservation of momentum, mass, and thermodynamic energy for a viscous, Boussinesq fluid in hydrostatic balance. The effects of moisture are not included in the model and the initial conditions consist of an Ekman balanced flow field.

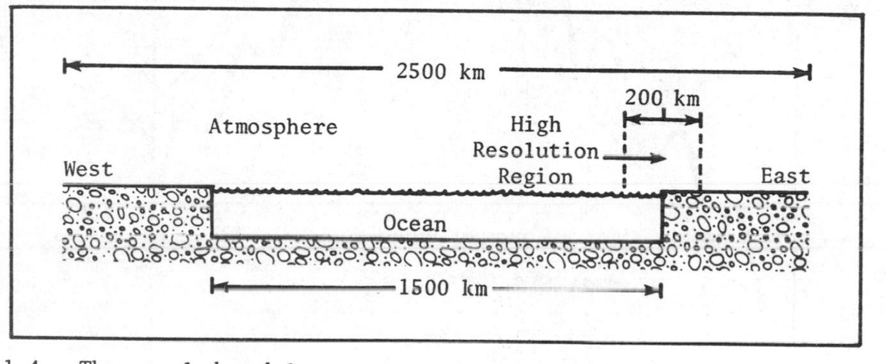

Fig. 11.4 - The coupled model geometry.

The coupled model geometry is shown in Fig. 11.4 and represents a 1500 km wide ocean bounded by two 550 km wide continents. Note the area labeled "high resolution region" at the eastern end of the ocean basin. Very high grid resolution is specified for both the ocean and the atmosphere (0(1.5 km)) in this region and the solutions there will be examined in detail.

Only the simplest possible coupling between the ocean and the atmosphere is considered. In particular, the atmosphere model predicts the wind stress which drives the ocean while the ocean model predicts sea surface temperatures. The atmosphere model is driven by a flux of sensible heat at the bottom boundary which, over the ocean, depends on the sea surface temperature and, over the land, depends on a prescribed time and space-varying land temperature. Since the predicted sea surface temperatures depend on the winds and vice versa, the coupled model is a true feedback system.

The following three questions can be examined with the model:

1. What is the oceanic response to sea breeze forcing and can this response play a role in coastal upwelling? This problem is of interest to oceanographers because it involves the economically important near shore waters and the biologically significant diurnal time scale.

2. What effect does coastal upwelling have on the sea breeze circulation? This problem is of interest to coastal meteorologists, especially those concerned with air pollution forecasts and forestry operations.

3. Can coastal upwelling be affected by a feedback loop with the atmosphere? This problem is of interest to both oceanographers and meteorologists.

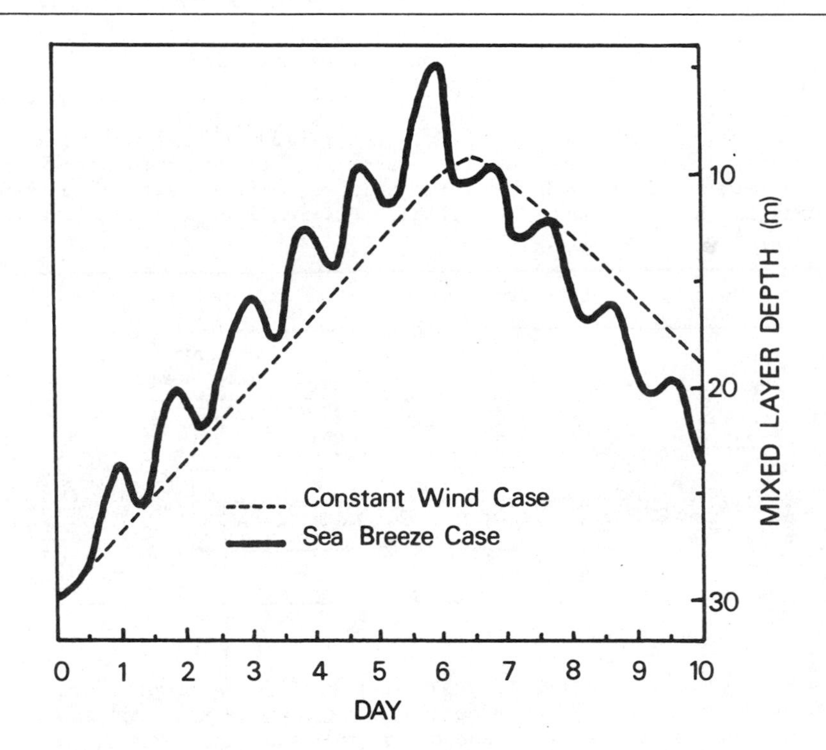

Fig. 11.5 - Time series of mixed layer depth in the ocean at the point 6 km offshore for a constant north wind case (dashed line) and for a case with a sea breeze superimposed on the same constant wind (solid line). Note the diurnal fluctuation for the sea breeze case. The interface rises more rapidly for the sea breeze case than for the constant wind case during the early stage of the integration when vertical advection is dominant. During the latter part of the integration when turbulent mixing dominates the effect of vertical advection, the mixed layer also deepens at a faster rate for the sea breeze case.

Only a brief summary of the model predicted answers to these questions will be given here. For a thorough discussion of the experiments and results see Clancy (1975). With regard to the first question:

1. The sea breeze forces a clockwise rotating, diurnal oscillation in the upper layer of the ocean near the coast which is superimposed on a slowly evolving motion field similar to that found by Hurlburt and Thompson (1973).

2. Since the angular frequency of the forcing is less than f, the oscillation does not take the form of a propagating wave.

3. Near the coast, the amplitude of the velocity perturbation is about 10-15 cm s^{-1}.

4. The periodic convergence and divergence associated with the oscillation and the kinematic boundary condition at the coast produces a diurnal fluctuation in the depth of the upper layer. The amplitude of this fluctuation is a maximum at the coast (being about 3 m there) but drops to zero beyond about 12 km from shore (see Fig. 11.5).

5. A diurnal sea surface temperature fluctuation with amplitude 0(1°C) is produced in the upwelling frontal region, primarily because of horizontal advection in the zone of large horizontal temperature gradients.

6. After the axis of the equatorward surface jet (Hurlburt and Thompson, 1973) moves seaward of about 10 km from shore, the sea breeze forcing causes a weak, poleward, surface counter-current to form diurnally within 10 km from shore. A poleward surface current also forms periodically seaward of the jet axis throughout the integration.

7. The sea breeze increases the upwelling rate by about 20 per cent over that of an experiment driven by a constant wind equal to the time average of the wind in the sea breeze forced experiment (see Fig. 11.5).

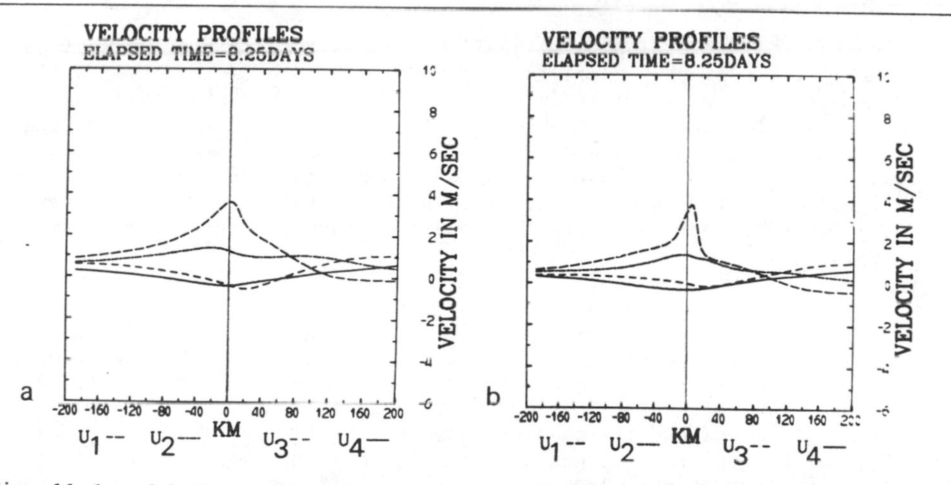

Fig. 11.6 - (a) The x dependence of the u components in the four layers of the atmosphere at 1500 LST for an experiment in which the atmosphere does not "see" the cold coastal waters. The coast is at the origin, the subscript 1 refers to the bottom layer and the thickness of each layer is about 500 m. (b) Same as (a) except that the cold water is allowed to affect the atmosphere.

The upwelling produces the following changes in the sea breeze circulation:

1. A narrowing of the width scale of the sea breeze resulting in a much sharper sea breeze front during the early stage of the cycle (see Fig. 11.6).

2. Faster inland propagation and longer persistence of the sea breeze front resulting in a threefold increase in the inland penetration of the sea breeze influence.

3. An overall strengthening of the sea breeze circulation with the highest speed occurring later in the day and further inland.

4. A weakening of the sea breeze seaward of about 6 km inland.

5. A significant increase in the magnitude of the vertical motions in the ascending branch of the circulation.

6. An increase in subsidence over the cold water near the coast.

7. A shift in the region of initial land breeze formation from about 10 km inland to a point near the upwelling front, about 20 km offshore.

8. A 75 per cent reduction in the maximum speed of the land breeze.

The cold sea surface temperatures near the coast do not produce any significant change in the mean longshore wind and hence there is no feedback to the upwelling. This result can be attributed to the fact that the width of the upwelling zone is small with respect to the radius of deformation in the atmosphere.

In conclusion, several points should be emphasized. First, it must be stressed that the present model is two-dimensional and, in the real world, three-dimensional effects are important. In addition, certain potentially important physical processes are absent from the atmosphere model. For example, cooling of the land due to inland advection of cold air from over the sea is not treated explicitly and the formation of clouds, fog, and precipitation is totally ignored. Yet, even the relatively simple physics retained in the model is sufficient to show a strong response of the sea breeze circulation to the upwelling.

Concerning the result of no feedback to the upwelling, it must be emphasized that only the simplest possible air-sea coupling is employed. Furthermore, a constant drag coefficient is used in all the experiments, while a stability dependent drag coefficient would be more appropriate. This is the case because a low level atmospheric stable layer is found over the upwelling zone in both the model and the observations and the effect of such a layer is to suppress atmospheric turbulence and, hence, the flux of momentum into the ocean (i.e., C_D decreases). As reported by Busch (Chapter 6) the drag coefficient for extremely stable conditions may be as small as one-half the value for neutral stability. A more complicated coupling taking this into account may yield a negative feedback to the upwelling rate.

The upwelling models discussed thus far have been two-dimensional and numerical. We now turn our attention to a three-dimensional analytical model of upwelling which illustrates a connection between coastally trapped long waves and classical Ekman theory.

11.4 Coastally trapped long waves and Ekman theory

Coastal upwelling is not two-dimensional as described in Section 11.2. The presence of the shore permits wave disturbances which may be barotrophic and/or baroclinic. These coastal waves have been reviewed recently by LeBlond and Mysak (1976). The irregularities of the coastline and bottom topography also force three-dimensional motions which are discussed in Section 11.5. Since the barotrophic motions do not alter the mass field in the ocean in a substantial manner, they are not discussed here (see LeBlond and Mysak, 1976). It is important, however, to review the theory of simple internal Kelvin waves and their potential role in coastal upwelling.

A simple explanation of coastal upwelling can be given by Ekman theory. Because of the earth's rotation, longshore winds can force an offshore Ekman transport. There can be no flow across the coastal boundary and so vertical motion is required near the coast. This theory is two-dimensional; upwelling at a given coastal location only occurs when longshore winds blow.

The two-dimensional theory described above does not always agree with observations. For example, in the summer of 1968, Walin (1972) obtained evidence of upwelling at a section of the Baltic coast which was at right angles to the wind (see Fig. 11.7). In this section an attempt is made to generalize Ekman theory so that longshore effects can be taken into account.

Consider the case of an ocean of uniform depth, D, with a straight coast line. Choose a right-hand system of axes x, y, z such that x is the distance seaward from the coast and z is the distance upward from the ocean surface. Consider the theory of upwelling for the case where only small perturbations from a state of rest and horizontally uniform stratification are produced. For this equilibrium state, the density is a function $\bar{\rho}(z)$ of depth only and so is the pressure $\bar{p}(z)$, the two being in hydrostatic balance,

$$\frac{d\bar{p}}{dz} = -g\bar{\rho}(z). \tag{11.16}$$

Let the perturbation pressure be p', the perturbation density, ρ', and the velocity components, u, v, and w. These satisfy the equations

$$\frac{\partial u}{\partial t} - fv = -\frac{1}{\rho_r}\frac{\partial p'}{\partial x} + \frac{1}{\rho_r}\frac{\partial X}{\partial z}; \tag{11.17}$$

$$\frac{\partial v}{\partial t} + fu = -\frac{1}{\rho_r}\frac{\partial p'}{\partial y} + \frac{1}{\rho_r}\frac{\partial Y}{\partial z}; \tag{11.18}$$

$$0 = -\frac{\partial p'}{\partial z} - g\rho'; \tag{11.19}$$

$$\frac{\partial u}{\partial x} + \frac{\partial v}{\partial y} + \frac{\partial w}{\partial z} = 0; \tag{11.20}$$

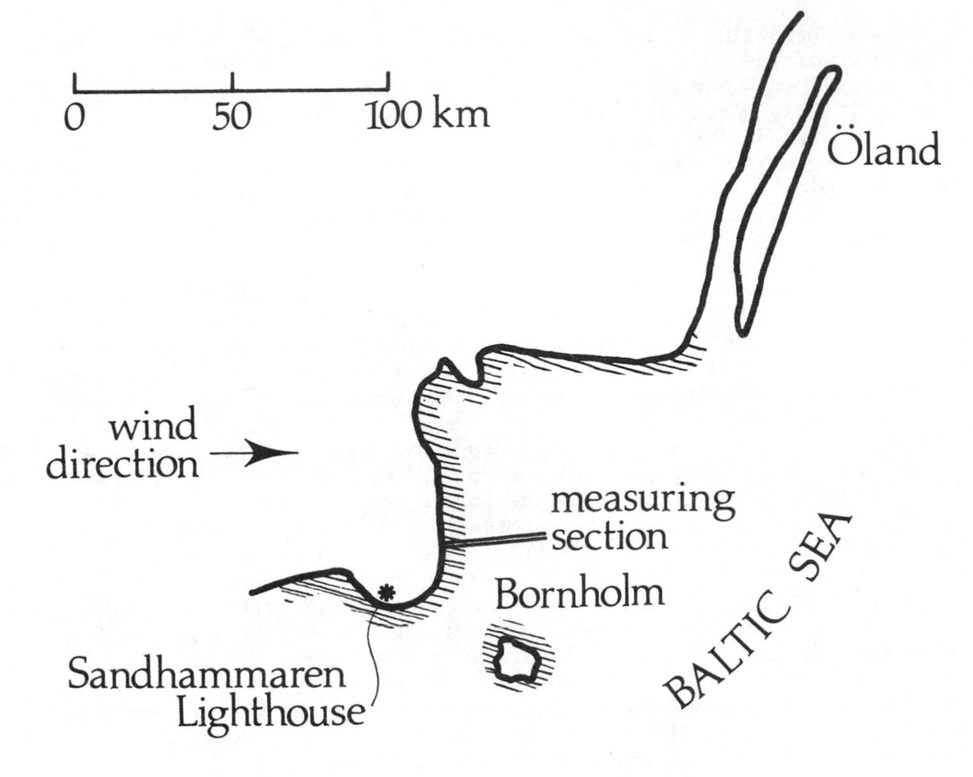

Fig. 11.7 - Map of the southern Baltic showing the position of the measuring section (from Walin, 1972). Wind measurements were made at Sandhammaren light house. Upwelling at the measuring section was often observed when the wind direction was at right angles to the shoreline as shown. The shaded region indicates the coastal water where strong internal movements are expected to occur - it is the region within the first baroclinic radius of deformation of the shoreline.

$$\frac{\partial p'}{\partial t} + w \frac{d\bar{\rho}}{dz} = 0. \qquad (11.21)$$

Density ρ_r is a constant which is a representative density, the Boussinesq approximation having been made. The vector (X, Y) is the tangential stress acting between horizontal planes, and takes a prescribed value (τ_0^x, τ_0^y), the wind stress, at the surface.

Solution of the above equations requires a knowledge of how (X, Y) varies with depth. There is some evidence that a reasonable approximation is obtained by assuming the currents are uniform over the upper mixed layer, of depth h_0. This is equivalent to assuming that (X, Y) are linear functions of z in the mixed layer, falling from their surface values to zero at the bottom of the mixed layer. For this model, therefore, the vertical derivative $\left(\frac{\partial X}{\partial z}, \frac{\partial Y}{\partial z}\right)$ of the stress is given by

$$\left(\frac{\partial X}{\partial z}, \frac{\partial Y}{\partial z}\right) = \begin{cases} \left(\tau_0^x, \tau_0^y\right) h_0^{-1} & -h_0 \leq z \leq 0 \\ 0 & z < -h_0 \end{cases} \qquad (11.22)$$

The problem can now be reduced to one involving horizontal variations only by the standard separation into normal modes. The modes are found by inquiring under what conditions, when X=Y=0, solutions can be separated into products of a function of z and a function of the other variables. The result is that there exist eigenfunctions $\hat{P}_n(z)$, $\hat{W}_n(z)$ and eigenvalues c_n such that u, v, p', w, ρ' can be expanded as follows

$$u = \sum u_n(x,y,t)\hat{P}_n(z); \qquad (11.23)$$

$$v = \sum v_n(x,y,t)\hat{P}_n(z); \qquad (11.24)$$

$$\frac{p'}{\rho_r} = \sum p_n(x,y,t)\hat{P}_n(z); \qquad (11.25)$$

$$w = \sum w_n(x,y,t)\hat{W}_n(z); \qquad (11.26)$$

$$g\rho' N^{-2} = \sum h_n(x,y,t)\hat{W}_n(z). \qquad (11.27)$$

The notation h_n is used because h_n is proportional to the displacement of the isopycnals. N is buoyancy frequency defined by

$$N^2 = -g\rho_r^{-1} \frac{d\bar{\rho}}{dz}. \qquad (11.28)$$

Because the density difference across the sea surface is so much greater than the density differences within the ocean, one mode has quite a different character from the others. This is the barotropic mode (n=0) for which, to a good approximation

$$\hat{P}_0 = 1 \quad ; \quad \hat{W}_o = z+D \quad ; \quad c_o = \sqrt{gD}. \tag{11.29}$$

This satisfies the conditions that $w=0$ at $z=-D$ and that $w=\partial\zeta/\partial t$ at the free surface $z=\zeta$. For the remaining baroclinic modes ($n=1, 2, 3, \ldots$) the separation of variables leads to the equations

$$-\hat{P}_n = \frac{d\hat{W}_n}{dz} ; \tag{11.30}$$

$$\frac{d\hat{P}_n}{dz} = \frac{N^2}{c_n^2} \hat{W}_n ; \tag{11.31}$$

which combine to give

$$\frac{d^2\hat{W}_n}{dz^2} + \frac{N^2}{c_n^2} \hat{W}_n = 0. \tag{11.32}$$

The eigenfunctions $\hat{W}_n$ and eigenvalues c_n are obtained by solving this subject to the conditions that $\hat{W}_n$ vanishes at $z=0$ and $z=-D$. The c_n have the dimensions of velocity and, for the non-rotating case, are the speeds at which long internal waves propagate. The modes are ordered so that c_n decreases as n increases.

To find the forced motions, it is necessary to expand the functions $\left(\frac{\partial X}{\partial z}, \frac{\partial Y}{\partial z}\right)$ in terms of the eigenfunctions $\hat{P}_n(z)$, e.g.,

$$\rho_r^{-1} \frac{\partial X}{\partial z} = \sum X_n(x,y,t)\hat{P}_n(z). \tag{11.33}$$

If the model of $\left(\frac{\partial X}{\partial z}, \frac{\partial Y}{\partial z}\right)$ described earlier is used, it is necessary to expand the step function

$$\left.\begin{array}{ll} 1 & -h_o<z<0 \\ 0 & z<-h_o \end{array}\right\} = \sum b_n \hat{P}_n(z), \tag{11.34}$$

in which case

$$(X_n, Y_n) = \frac{b_n}{\rho_r h_o} (\tau_o^x, \tau_o^y). \tag{11.35}$$

The equations for the modes now take the form

$$\frac{\partial u_n}{\partial t} - fv_n = - \frac{\partial p_n}{\partial x} + X_n ; \tag{11.36}$$

$$\frac{\partial v_n}{\partial t} + fu_n = -\frac{\partial p_n}{\partial y} + Y_n \, ; \qquad (11.37)$$

$$\frac{\partial p_n}{\partial t} + c_n^2\left(\frac{\partial u_n}{\partial x} + \frac{\partial v_n}{\partial y}\right) = 0. \qquad (11.38)$$

In addition, the functions h_n and w_n can be found from p_n using the following equations:

$$h_n = -\frac{p_n}{c_n^2}\, , \qquad (11.39)$$

$$w_n = \frac{\partial h_n}{\partial t}. \qquad (11.40)$$

The displacement of the isopycnals for the barotropic mode is of the same order as the displacement of the free surface, and so can be ignored. For the upwelling problem, therefore, it is necessary to solve only the equations for the baroclinic modes.

Form (11.36) and (11.37), u_n and v_n can be found in terms of p_n (the n subscript is dropped):

$$\left(\frac{\partial^2}{\partial t^2} + f^2\right) u = -f\,\frac{\partial p}{\partial y} - \frac{\partial^2 p}{\partial x \partial t} + \frac{\partial X}{\partial t} + fY, \qquad (11.41)$$

$$\left(\frac{\partial^2}{\partial t^2} + f^2\right) v = f\,\frac{\partial p}{\partial x} - \frac{\partial^2 p}{\partial y \partial t} + \frac{\partial Y}{\partial t} - fX. \qquad (11.42)$$

Substitution of (11.41) and (11.42) into (11.38) then gives

$$\frac{\partial}{\partial t}\left\{\frac{\partial^2 p}{\partial x^2} + \frac{\partial^2 p}{\partial y^2} - \left[\frac{\partial^2}{\partial t^2} + f^2\right]\frac{p}{c^2}\right\} = f\left(\frac{\partial Y}{\partial x} - \frac{\partial X}{\partial y}\right) + \frac{\partial}{\partial t}\left(\frac{\partial X}{\partial x} + \frac{\partial Y}{\partial y}\right). \qquad (11.43)$$

Upwelling is produced in a very narrow region adjacent to the coast, typically of order 20 km wide. Since wind patterns have scales much larger than this, the right-hand side of the above equation and $\frac{\partial^2 p}{\partial y^2}$ can be ignored (the forcing will re-enter the problem when the kinematic boundary condition is applied at the coast). Also, upwelling time scales are much greater than f^{-1} and so Eq. (11.43) can be simplified to

$$\frac{\partial}{\partial t}\left(\frac{\partial^2 p}{\partial x^2} - \frac{f^2}{c^2}\, p\right) = 0. \qquad (11.44)$$

Integration of this equation from an initial state of rest gives

$$\frac{\partial^2 p}{\partial x^2} - \frac{f^2}{c^2}\, p = 0. \qquad (11.45)$$

The solution which vanishes at large distances from the coast has the form

$$p = \phi \exp\left(-\frac{x}{a}\right) \tag{11.46}$$

where $a=\frac{c}{f}$ is the Rossby radius of deformation associated with the appropriate baroclinic mode. The c is taken to be positive in the northern hemisphere and negative in the southern hemisphere so that a is always positive. Equation (11.45) indicates that the density changes are confined to a coastal strip with a width on the order of the first baroclinic radius of deformation. The density changes are associated with a coastal jet, the velocity of which, by the approximations made earlier, is

$$v = f^{-1}\frac{\partial p}{\partial x} = -\frac{\phi}{c}\exp\left(-\frac{x}{a}\right) \tag{11.47}$$

(recall ϕ has units of cm^2s^{-2} from Eqs. 11.25 and 11.46).

It remains to apply the boundary condition that u vanishes at the coast. Using the approximations described immediately following Eq. 11.43 in Eq. 11.41, this condition is

$$u = f^{-1}\frac{\partial p}{\partial y} - f^{-2}\frac{\partial^2 p}{\partial x \partial t} + f^{-1}Y = 0 \quad \text{at } x = 0, \tag{11.48}$$

and so

$$-\frac{1}{c}\frac{\partial\phi}{\partial t} + \frac{\partial\phi}{\partial y} = Y. \tag{11.49}$$

This first order wave equation can easily be solved for given initial and forcing conditions by the method of characteristics. The complete solution then requires the solutions for all the baroclinic modes to be added. In practice, however, one could expect to get a reasonable approximation using only a few modes, perhaps only one or two.

The solution for the motion normal to the coast can be obtained by substituting Eq. (11.49) in the expression for u (see Eq. 11.48). This gives

$$u = \frac{Y}{f}\left[1-\exp\left(-\frac{x}{a}\right)\right]. \tag{11.50}$$

At large distances from the coast, this reduces to

$$u = f^{-1}Y; \tag{11.51}$$

i.e., the expression for the wind driven Ekman transport. It is this normal component of the Ekman transport which produces the upwelling as described earlier.

Equation (11.50) shows that the motion normal to the coast is locally determined, i.e., the motion for a particular coastal position, depends only on the wind stress at that coastal position. In contrast, the density changes and the variations in the strength of the coastal jet are not locally determined as (11.49) shows. To understand (11.49) physically, note that the solution in the absence of forcing is

$$\phi(y,t) = \phi(y+ct); \tag{11.52}$$

i.e., a freely propagating wave. In the absence of rotation, this wave is a long internal gravity wave. The effect of rotation is to trap the wave near the coast and to allow propagation only in one direction, namely with the coast on the right in the Northern Hemisphere and on the left in the Southern Hemisphere. This is a long internal Kelvin wave. Since Eq. (11.49) represents free wave propagation when there is no forcing, when there is forcing it describes a wind forced long internal Kelvin wave. To understand this wave, consider an observer moving along a path $y = y(t)$ such that he travels with the wave. Then

$$\frac{dy}{dt} = -c, \tag{11.53}$$

and so, using Eq. (11.49)

$$\frac{d\phi}{dt} = -cY. \tag{11.54}$$

Since the pycnoclinic amplitude is proportional to $-\phi$, Eq. (11.54) indicates that for an observer moving with the wave, the rate of change of pycnocline height has the same sign as cY. Since cY has the same sign as Y/fg from (11.50) it can be seen that the pycnoclinic height is increased when the Ekman transport is offshore and decreased when it is onshore. The upwelling at a given coastal position thus depends on how much the amplitude of the Kelvin wave has been increased or decreased by Ekman transport as it has propagated along the coast to the coastal position under consideration.

To illustrate this mechanism, consider again the observations of Walin (1972) in the Baltic Sea (see Fig. 11.7). The Kelvin waves travel with the coast on their right in the Northern Hemisphere so they propagate along the section of the coast where there is offshore Ekman transport and then turn southward and progagate along the section of the coastline perpendicular to the wind. The Kelvin wave amplitude is increased along the section parallel to the wind and is unchanged along the section perpendicular to the wind, so it is possible to have upwelling at the measuring section even though there is no offshore Ekman transport there.

The above theory has indicated how Ekman theory is modified when longshore and transient effects are taken into account. However, the theory did not include many effects which may be important. Curvature of the coastline, nonlinearity, longshore and shelf topography, and the β-effect are some of these. A discussion of how these factors modify the simple linear forced internal Kelvin wave theory described above will not be given here. Analytic models which consider the influence of coastline curvature, nonlinearity, and shelf and longshore topography are described by Clarke (1975), while results for the β-effect are given by Rowlands (1975). These complications have been studied in numerical models by Hurlburt (1974), Suginohara (1974), and Peffley and O'Brien (1976). This is the subject of Section 11.5.

11.5 Three-dimensional models of coastal upwelling

It has long been suspected that longshore variations in bottom topography and coastlines are an important factor in the development of upwelling flow patterns. However, there has been so little work in this area until recently that Mooers and Allen (1973) were led to comment that "There is, at the present, a lack of almost any guidance from theoretical studies on what the effects of longshore variations in topography may be". Among the few papers on this subject which were published before that time one might mention, in particular, Arthur (1965) and Yoshida (1967) who both discussed the effects of capes on steady state flows. They concluded that the strongest upwelling should occur on the south side of west coast capes.

Since 1973, the topic has attracted a great deal more attention. Suginohara (1974) studied the effects of longshore wind stress variability in a model with a straight coast and a bottom topography which did not vary in a coastwise direction. For the case of a flat bottom, he found that after the winds were shut off in the model, the upwelled portion of the pycnocline propagated poleward at the speed of an internal Kelvin wave. Along a north-south coast with a sloping shelf, Suginohara's model developed a poleward flow in the lower layer and northward propagating continental shelf waves. Kelvin waves and continental shelf wave dynamics were found to be important also by Gill and Clarke (1974) in a study which dealt with changes in the local upwelling intensity produced by longshore wind stress variations. The correlation between movements of the thermocline and changes in sea level at the coast, led them to suggest a possible working procedure for the prediction of upwelling (see Section 11.4). Pedlosky (1974a) also examined the relationship between local longshore currents, upwelling, and the stress distribution. In two other publications (Pedlosky 1974b, 1974c) he used a three dimensional, continuously stratified f-plane model which was integrated for cases of both steady and time dependent flows to investigate topographical effects. Shaffer (1974) found that longshore variations in the topography of the N. W. African shelf determine the distribution of upwelling for that area. To explain the observed "funnelling" of onshore transports, he proposed a homogeneous, f-plane model with a shelf which did not slope but which had an edge that varied in a longshore direction.

A numerical primitive equation model for upwelling situations was developed by Hurlburt (1974). Hurlburt's model is wind driven; it consists of two layers and involves variations along the coast and at right angles to it. The model studies show that the β-effect plays a fundamental role in the flow pattern associated with mesoscale longshore topographic variations. Though the model has been applied so far only to fairly simple or highly idealized topographies and coastlines, it has been found that longshore variations in the sea bottom configuration can produce barotropic perturbations which extend seaward far beyond the topographic features. Large amplitude internal Kelvin waves are excited in the model by coastline irregularities.

Hurlburt's model has been used by Peffley and O'Brien (1976) for several case studies, which deal with the onset and the relaxation of coastal upwelling under conditions which are meant to simulate the topography of the Oregon coast in a simplified way. They find that the sea bottom configuration tends to have a greater effect on the upwelling pattern in the model than do coastline irregularities. Two areas next to the coast develop a poleward undercurrent during spin-down. The topographic variations also favor local upwelling in specific regions along the coast. After the winds are shut off, a poleward flow is seen in some areas of the upper layer. Baroclinic continential shelf waves are observed at the same time.

Hurlburt's model does not allow for mixing across the layer boundaries. The layer densities ρ_1 and ρ_2 are therefore constant. Thermodynamics are not included in the model. These omissions imply the assumption that the time scale of vertical advection is short compared to that for vertical eddy diffusion. In actual fact, this assumption may not be very realistic, especially under conditions of strong surface turbulence, tidal mixing, or shear instability. There is observational evidence that, in the Oregon coastal upwelling region, periods of shear instability occur when the Richardson numbers drop below ten (Huyer, 1974). In a numerical model incorporating thermodynamics, Thompson (1974) has further shown that vertical mixing may be comparable to vertical advection during the week-to-ten-day upwelling cycle. However, for short period fluctuations and small interface displacements, Thompson (1974) found good agreement between a purely hydrodynamic model and one which includes thermodynamics. This was discussed in greater detail in Section 11.2. All of the cases to be discussed below have maximum interface displacements which are less than 30 m. Since a 50 m initial upper layer thickness is used in each case, this means the interface is always at least 20 m from the surface. It seems that thermodynamical considerations can be neglected in these circumstances. Hurlburt's model provides then a reasonably realistic simulation. Its basic features are contained in the purely hydrodynamic forms of Eqs. (11.1), (11.2), (11.5), and (11.7).

No slip conditions apply at the eastern and western boundaries. Northern and southern boundary conditions are the quasi-symmetric boundary conditions described by Hurlburt (1974). Peffley and O'Brien were interested in modelling the upwelling circulation of the Oregon shelf region, a region of approximately 500 km N-S extent. To model such a region adequately with a closed basin, it is necessary for the basin's horizontal dimensions to be 0(1000 km) for, as shown by Hurlburt and Thompson (1973), a Sverdrup interior will not develop in a closed basin of lesser dimensions. A Sverdrup interior is, however, one of the primary dynamical features of the upwelling circulation and must be preserved. Use of the quasi-symmetric boundary condition allows the problem to be solved with relative economy in an open basin of mesoscale N-S extent and still develop a Sverdrup interior in the model.

As an added complication to the necessity of 0(1000 km) width, the grid resolution in the eastern ocean must be fine enough to resolve boundary layer phenomena and the topographic forcing ($\Delta x < 5$ km). In an x-y model without a variable grid, this problem would be prohibitively expensive, requiring five times as many grid points. A coarse grid has been used in the western ocean since we are not interested in the solution there. Hurlburt and Thompson (1973) and Hurlburt (1974) have demonstrated that for a simple equatorward wind stress in the upwelling area, the solution for the eastern ocean is independent of that for the western boundary region provided the zonal* extent of the model is great enough to admit a Sverdrup interior. Discrete variation in x also allows the advantage of using the same value of Δx over all parts of a variable coastline.

Initial conditions for the model cases to be described below are either (*i*) rest or (*ii*) the quasi-balanced initial state prescribed by Hurlburt (1974).

---

*The terms "zonal" and "onshore-offshore" are used interchangeably herein to mean motion in the x or E-W direction. Flow perpendicular to isobaths is referred to as "cross-isobath" flow. The terms "meridional" and "longshore" are used interchangeably to mean motion in the y or N-S direction. Flow parallel to isobaths is referred to as "along-isobath" flow. Huyer, et al. (1975) found this convention for longshore flow to simplify the analysis of observational data without altering the qualitative results in any way.

The quasi-balanced initialization filters out the inertial oscillations, Rossby waves, and gravity waves caused by impulsive application of the wind stress at t=0, allowing an easier interpretation of the low frequency dynamics. In comparisons between cases started from rest and from a quasi-balanced state, Hurlburt (1974) found that the amplitude of external gravity waves excited by the impulsive application of the wind stress was reduced up to three orders of magnitude in the quasi-balanced initial state cases, without significant effect on low frequency aspects of the solutions.

The intent of Peffley and O'Brien (1976) was to put the Oregon bottom topography and coastline into the Hurlburt (1974) model and to study the resulting effects on the coastal upwelling circulation. The region of primary interest is the Oregon shelf, although, as noted by Huyer (1974), the northeastern Pacific coastal upwelling region stretches from Vancouver Island to Baja, California.

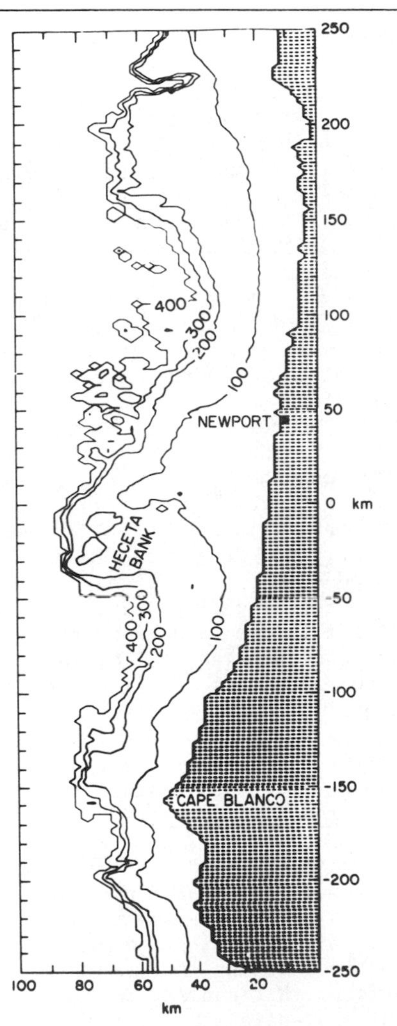

Fig. 11.8 - Bathymetric chart of the Oregon coastal region contoured by computer from the 2 km grid digitized bottom topography data. Depths are in meters. Isobaths below 400 m are not shown. Ordinate labels correspond directly to those of the x-y plots in subsequent sections. Although a right-handed coordinate system is used, positive abscissa values are used in this and other x-y plots for simplicity. References in the text to positions in x-y plots imply km.

A quick look at the almost meridional Oregon coastline gives the deceptively appealing impression of geometric simplicity. Bathymetric charts such as Fig. 11.8 however, reveal a complexity in the topography which will obviously make it more difficult to understand the observed ocean circulation. The most prominent coastline feature is Cape Blanco, north of which is an essentially meridional coast. Underwater topographic features include a small width scale O(10 km) Columbia River canyon (y=120 in Fig. 11.8). Another important part of the topography is a ridge system comprised of Stonewall and Heceta Banks (y=40 to y=-40 in Fig. 11.8).

Shelf widths in the region shown in Fig. 11.8 vary from 15 km off Cape Sebastian (y=-212) to 70 km for the area just south of the Columbia River canyon (y=225) and the area of Heceta Bank. Off Heceta Bank, the shelf edge may be in water as shallow as 100 m, while the shelf break off Cape Blanco is in much deeper water, O(300 m).

A smooth representation of the coastline, ignoring features with scales $\leq$ 20 km, was chosen for use in the model grid. The coastline used is depicted in Fig. 11.9. Notice that the 500 km region of Fig. 11.8 has been extended by adding two regions of y-independent coastline, 150 km in the north and 150 km in the south.

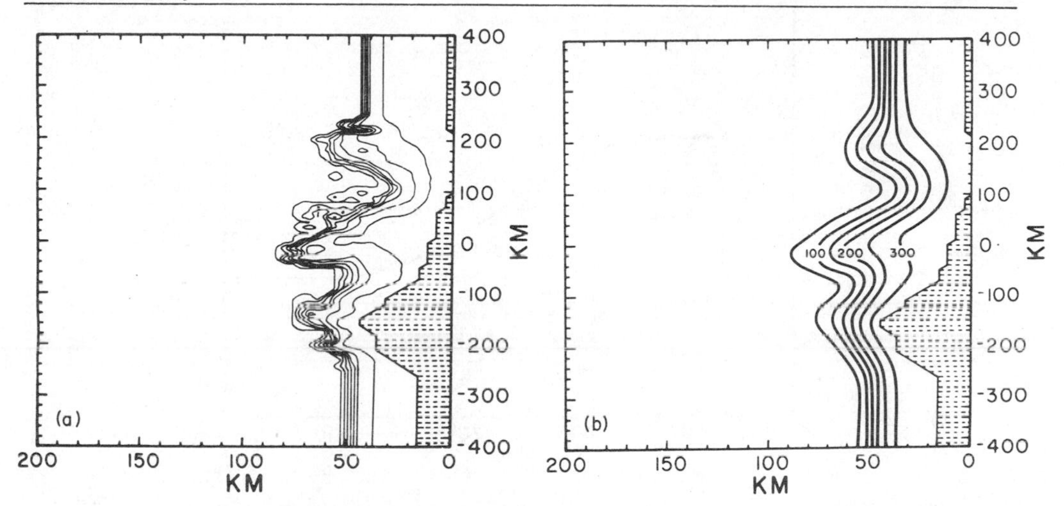

Fig. 11.9 - (a) Unsmoothed contours of bottom height shown on the model grid. Total depth of the basin is 400 m. The western-most height contour is at 50 m; the eastern-most is at 300 m. The maximum height of the topography is 345 m. Notice the lack of y dependence in the bottom topography and coastline in the northern-most 150 km and southern-most 150 km of the grid. The full 800 km N S extent of the model region is shown here. Although the total basin width is 5000 km, only the part within 200 km of the coast is shown in this and subsequent figures, this being the region of greatest interest. (b) Smoothed bottom topography heights in m obtained by smoothing the topography (a). This figure shows the coastline used in all model cases discussed herein, and the topography used in all cases discussed except the flat bottom case. The maximum height of the bottom is 345 m.

The topography shown in Fig. 11.9a produced severe instabilities in the model solutions. The steep slopes present in the actual topography demand extremely fine resolution, finer than was economically feasible in this study. Thus the topography was smoothed as shown in Fig. 11.9b. The actual near shore Oregon bathymetry was digitized and a Fourier analysis was made to determine the dominant longshore scales of variation; the smoothing scheme preserves both the qualitative aspects and the most important quantitative scales.

The permanent pycnocline is modelled as the layer interface, and upwelling is identified with displacements of that interface from its initial position. If the convention of using the 25.5-26.0 σ_t band to specify the permanent pycnocline is used, observations show that σ_t surface to be at a depth of 50-100 m off Oregon in the early summer. A density difference of 2 kg m^{-3} is used to approximate that between the upper wind mixed layer and lower layer. This value agrees with typical Oregon observations, e.g., Huyer (1974).

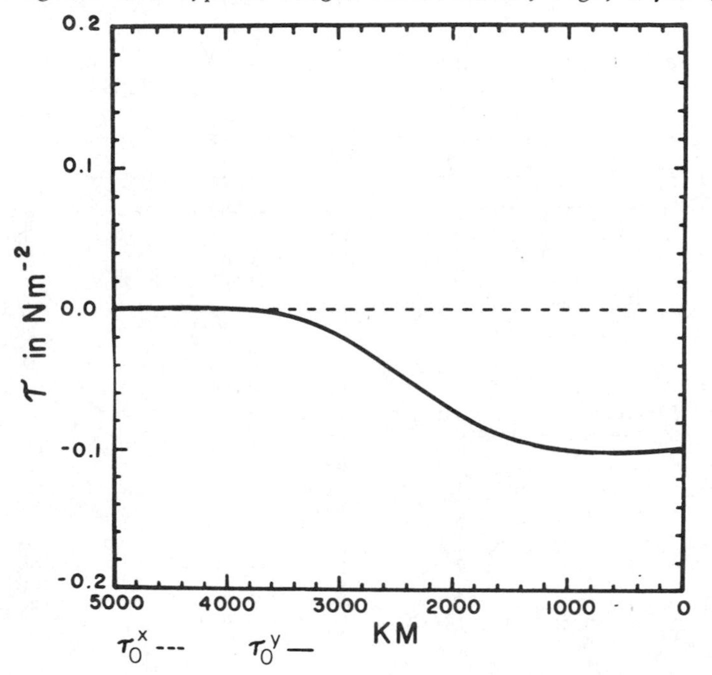

Fig. 11.10 - E-W profile of the y independent, time independent wind stress used in all spin-up cases described herein. The x component of the wind stress is zero everywhere.

The wind stress used is shown in Fig. 11.10. τ_0^x is zero everywhere; τ_0^y is y independent and time independent and constant over the 800 km nearest the coast. A simple wind stress with zero curl in the upwelling region was chosen to simplify interpretation of the Ekman dynamics. The maximum of $\vec{k}\cdot\text{curl}\ \vec{\tau}_0$ is 6×10^{-8} N m^{-3} at a distance of 2450 km offshore. The magnitude of τ_0^y in the upwelling area agrees with observations in the CUE-II experiment area, e.g., Halpern (1974).

Three model cases will be discussed. Case I used the coastline of Fig.

11.9b, but with a flat bottom, depth 400 m, and was started from rest. Case II used the topography and coastline of Fig. 11.9b and was also started from rest. Also, to study the relaxation of the upwelling, Cases I and II were repeated identically as described above except that the wind forcing was "shut off" at t=2.5 days and the integration continued for another 2.5 days with zero wind stress. In the subsequent discussion, we will refer to the initial period with wind stress forcing as the "spin-up", and the period with zero wind stress as the "spin-down". The wind forcing of the Oregon upwelling regime is highly variable and even changes direction during the upwelling season. Therefore, we feel that a study of the onset of upwelling and initial stages of decay is important to understanding the Oregon case.

We will examine several features of the model circulation which develop during spin-ups. Analysis of the model solutions will emphasize the role of longshore variability of the bottom topography and the coastline. Comparisons with Oregon observations and predictions of theory will be made when possible as an aid to understanding the dynamics.

As a review, we will first describe the essential dynamics of coastal upwelling for a β-plane case with equatorward wind stress, a meridional coastline and a flat bottom. The longshore flow is nearly geostrophic. A Sverdrup balance exists in the interior. Offshore flow in the upper layer is Ekman drift reduced by a longshore pressure gradient; u_2, the onshore flow in the lower layer, is geostrophic. Near the coast, the zonal flow must approach zero to satisfy the boundary condition, and geostrophy breaks down resulting in an equatorward jet in the upper layer and a poleward jet in the lower layer.

The longshore pressure gradient reduces the barotropic mode near the coast and permits the poleward undercurrent to develop. The cause of the longshore pressure gradient has been a subject of controversy. Hurlburt and Thompson (1973) considered an x-z model which neglected y derivatives except those of the pressure gradient terms and the Coriolis parameter. By comparing the upper layer y-momentum equation to the zonally-integrated upper layer vorticity equation, they derived the result

$$\int_{-L_x}^{x} \beta v_1 dx = g \frac{\partial}{\partial y}(h_1+h_2+h_b)\Big|_{x} - g \frac{\partial}{\partial y}(h_2+h_2+h_b)\Big|_{-L_x}, \qquad (11.55)$$

which shows how a N-S sea surface slope can be induced by the β effect.

When variable topograpy is considered, the situation becomes more complex, and as was noted by Sarkisyan and Ivanov (1971): ". . . the bottom relief competes successfully for significance with such an important factor as the β effect". Hurlburt (1974) listed three dynamical principles likely to aid explanation of dynamics associated with longshore variations in topography: (1) conservation of potential vorticity, (2) topographic β effect (β_T effect) and (3) joint effect of baroclinicity and bottom relief. He found the β_T effect to be important dynamically, even for mesoscale features, O(100 km).

How β_T becomes important can be illustrated by considering the barotropic vorticity equation (e.g., Hurlburt, 1974)

$$\frac{\partial}{\partial t}\left(\frac{\partial v}{\partial x} - \frac{\partial u}{\partial y}\right) = -f\left(\frac{\partial u}{\partial x} + \frac{\partial v}{\partial y}\right) - \beta v + \frac{1}{\rho_r}\, \underset{\rightarrow}{k}\cdot \text{curl}\left[\frac{\underset{\rightarrow}{\tau}_o - \underset{\rightarrow}{\tau}_b}{h}\right], \tag{11.56}$$

where the advective and diffusive terms have been neglected. Substituting from the continuity equation

$$\left(\frac{\partial u}{\partial x} + \frac{\partial v}{\partial y}\right) = -\frac{1}{h}\left(\frac{\partial h}{\partial t} + u\frac{\partial h}{\partial x} + v\frac{\partial h}{\partial y}\right), \tag{11.57}$$

and defining

$$\beta_T = -\frac{f}{h}\frac{\partial h}{\partial y} \tag{11.58}$$

yields

$$\frac{\partial}{\partial t}\left(\frac{\partial v}{\partial x} - \frac{\partial u}{\partial y}\right) - \frac{f}{h}\left(\frac{\partial h}{\partial t} + u\frac{\partial h}{\partial x}\right) + (\beta_T + \beta)v = \frac{1}{\rho_r}\, \underset{\rightarrow}{k}\cdot \text{curl}\left[\frac{\underset{\rightarrow}{\tau}_o - \tau_b}{h}\right]. \tag{11.59}$$

Thus, the N-S sloping topography yields an analogous term to the planetary vorticity advection in the integral of Eq. (11.56), and can therefore exert an effect on the longshore pressure gradient via Eq. (11.55).

Garvine (1974) has pointed out the importance of a realistic interior in coastal upwelling models. In Peffley and O'Brien (1976), the model interior approaches a Sverdrup balance

$$v_1 h_1 + v_2 h_2 = \frac{1}{\rho_r \beta}\, \underset{\rightarrow}{k}\cdot \text{curl}\ \underset{\rightarrow}{\tau}_o \tag{11.60}$$

but does not achieve that balance by day 5. All cases exhibit barotropic Rossby waves. This is expected for the cases started from rest due to impulsive application of the wind stress curl.

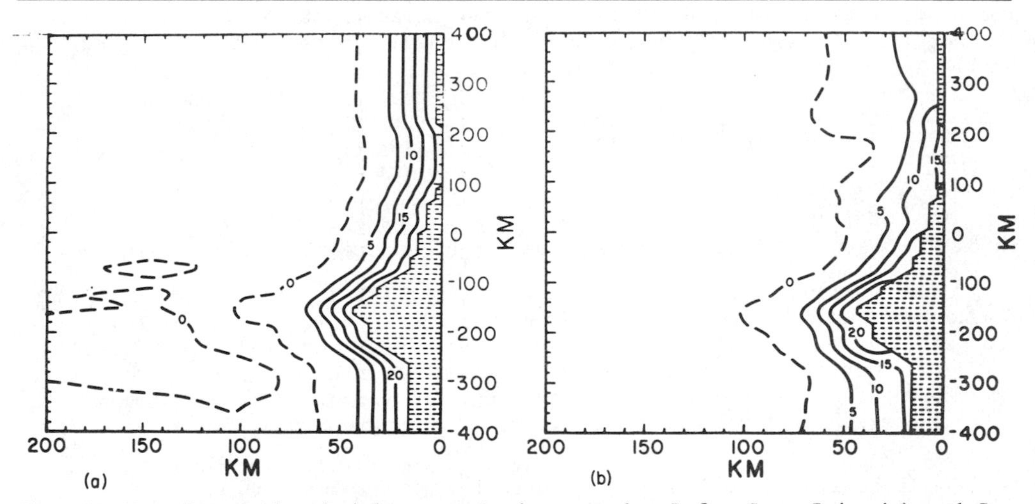

Fig. 11.11 - Pycnocline height anomaly in m at day 5 for Case I in (a) and Case II in (b). The pycnocline height anomaly is the displacement of the layer interface from its initial position of rest. In this and subsequent figures, positive contours are indicated by solid lines, the zero and negative contours by dashed lines.

Upwelling at day 5 for the flat bottom (I) and topography (II) cases is compared in Fig. 11.11. Quite striking is the longshore variability in the upwelling pattern for the case with topography compared to the flat bottom case. Generally weaker upwelling near the coast in the topography case is due to upwelling being induced further offshore by E-W bottom slope. The layer interface of Case II exhibits vertical velocities at the coast of approximately 6×10^{-5} m s^{-1} for the cape region and 4×10^{-5} m s^{-1} for the CUE study area ($y \simeq 100$). Halpern (1974a), estimating vertical velocities in the CUE area during two periods of southward winds, found values of 6.6×10^{-5} m s^{-1} and 1.25×10^{-4} m s^{-1}. Downwelling is seen in areas west of the zero contour.

The favored areas of upwelling are the cape and the head of the mesoscale canyon (axis at y=120). An interesting comparison can be made with Fig. 1 of Smith et al. (1971), a synoptic mapping of sea surface temperatures for a large part of the Oregon coast for a typical day of the coastal upwelling season. It shows pronounced upwelling south of Cape Blanco and another area of relatively strong upwelling between 44°40' and 45°20' (y=50 to 120).

The intensity of upwelling shows an exponential decrease with distance offshore with an e-folding width roughly 15 km. Studies of sea surface temperatures in the CUE-II area (y=50 to 100) led Holladay and O'Brien (1975) to conclude that the mean isotherms tended to parallel isobaths. Due to the difference in scales, no very meaningful comparison can be made with Fig. 11.11b, except that for the mesoscale, the pattern of upwelling exhibits significant departures from the simple picture of upwelling paralleling the isobaths.

The role of coastline irregularities in influencing upwelling has been studied by Hurlburt (1974). For a flat bottom case with a cape roughly resembling that in the present study, he found: ". . . the upwelling follows the coastline, but otherwise this cape exhibits no dramatic effect on the pattern of vertical motion. Upwelling is greatest near the point of the cape and slightly greater on the southern side than on the northern side". This conclusion is borne out in Fig. 11.11a.

Several concepts shed light on explaining the pattern of longshore variation of upwelling in Fig. 11.11b. In addition to Shaffer's (1974) study mentioned above, Hurlburt (1974) also studied effects of a symmetric mesoscale (half width ~100 km) canyon on the upwelling pattern. He found: (1) decreased upwelling north of the canyon axis, (2) increased upwelling south of the axis and (3) enhanced upwelling on the axis near the shore. For a symmetric ridge, he found effects opposite to those of the canyon. Hurlbut found these effects to be consistent with the β_T effect. In both cases, the dominant longshore scale appeared to be that of the topographic disturbance. These basic results are seen to explain the observed pattern of upwelling. In particular, note enhanced upwelling south of the canyon (axis at y=120) and at the coast near the canyon axis. Deflection of the zero contour at y=180 (cf. flat case) is a persistent feature explained by the local "ridge canyon" system. The main ridge (axis at y=0) and the canyon to the south act in the same manner to divert the height anomaly contours. The conclusion is that the influence of topography is dominant over the coastline variations in determining the longshore pattern of upwelling for the Oregon coast.

Figure 11.12 shows contours of the zonal component of upper layer velocity for the flat bottom and topography cases. Although the flat case is

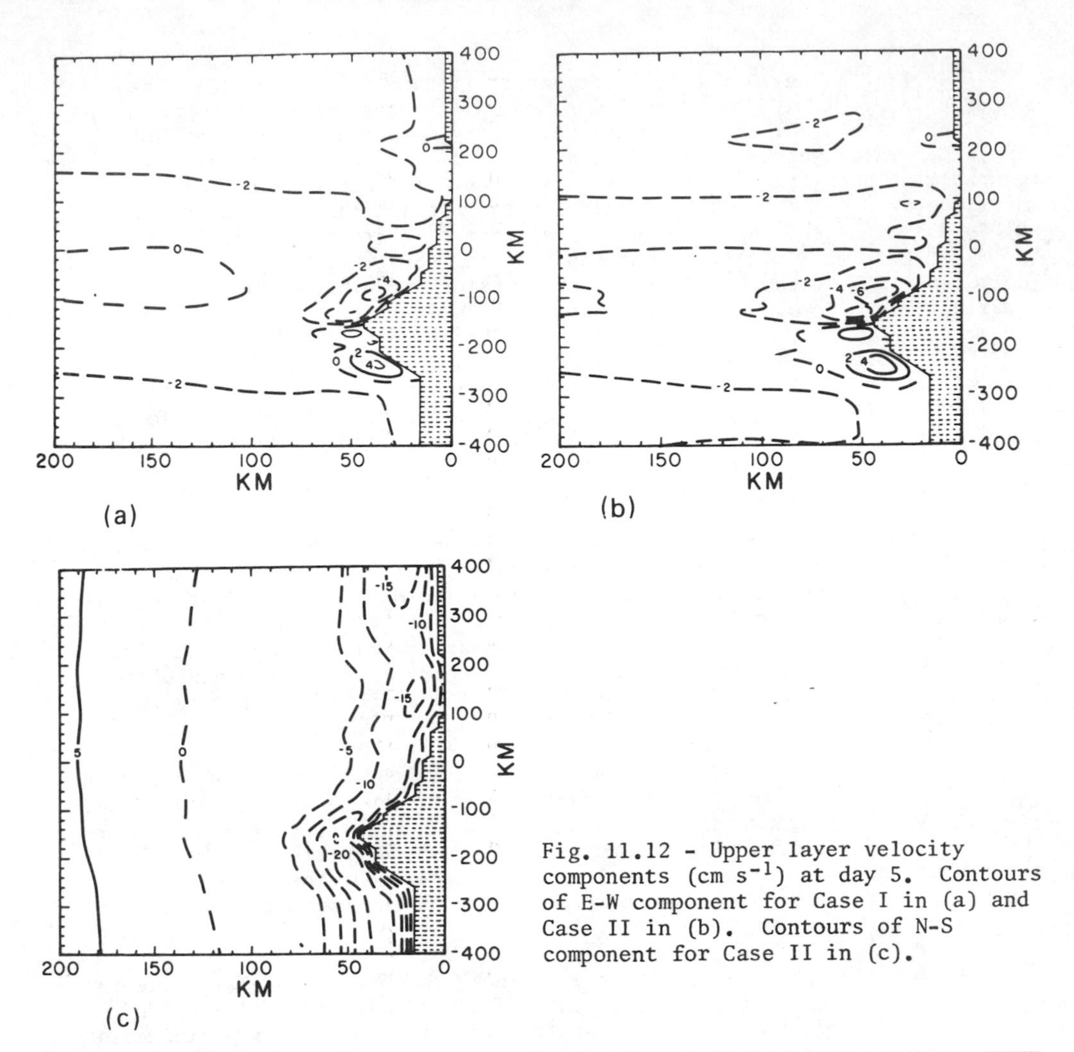

Fig. 11.12 - Upper layer velocity components (cm s^{-1}) at day 5. Contours of E-W component for Case I in (a) and Case II in (b). Contours of N-S component for Case II in (c).

---

relatively contaminated by inertial oscillations, the close qualitative similarity between Figs. 11.12a and 11.12b suggests the preliminary conclusion that topography has little effect on the upper layer zonal flow.

For an upper layer thickness of 50 m, Ekman dynamics predicts $u_1 = \tau_0^y (\rho_r h_1 f)^{-1}$ or $u_1 \simeq -0.02$ m s^{-1}. Figure 11.12 shows offshore speeds in good agreement with this. Mooers and Allen (1973) reported offshore velocities for the Oregon region to be 0.10 to 0.30 m s^{-1} in the upper Ekman layer. According to R. L. Smith (1974), the surface Ekman layer is apparently restricted to the upper 20 m. Thus, the model predicts transports in good agreement with the observations, but the artificially thick layer of the model causes a decrease in zonal velocity.

The longshore dependence of u_1 appears to be dominated by coastline variations, and their associated length scales. Hurlburt (1974) has explained the existence of strong zonal flow along the relatively E-W portions of a coastline for the eastern ocean circulation.

Figure 11.12c shows contours of the upper layer N-S component of velocity for the topography case. It is essentially similar to the flat bottom case (not shown) but exhibits a stronger poleward flow west of x=120. The equatorward surface jet in the topography case has somewhat greater longshore variability, and the jet maximum is slightly further offshore and stronger than in the flat bottom case. O'Brien and Hurlburt (1972) first modelled the equatorward jet in a x-z, f-plane model explaining its existence in terms of conservation of potential vorticity.

The existence of the jet can also be understood in terms of upper layer momentum balance (Hurlburt and Thompson, 1973). Consider the linearized upper layer momentum equation, neglecting viscous terms

$$\frac{\partial v_1}{\partial t} + fu_1 = -g \frac{\partial}{\partial y} (h_1+h_2+h_b) + \frac{\tau_0^y}{\rho_r h_1} . \tag{11.61}$$

In the interior, a Sverdrup balance exists and $\partial v_1/\partial t$ is negligible. The right side of Eq. (11.61) has negative sign since the geopotential gradient, although positive, is smaller in magnitude than the Ekman drift term. In the upwelling boundary layer $u_1 \to 0$ over the scale of the baroclinic radius to satisfy the kinematic boundary condition, but the right side of Eq. (11.61) changes relatively little, forcing $\partial v_1/\partial t<0$. No slip at the coast and the viscous boundary layer (width 5 to 7 km for the present study) result in the development of a jet structure in the longshore flow. The existence of the jet is well documented (e.g., Mooers, et al., 1976; Stevenson, et al., 1974; Huyer, et al., 1975) with equatorward speeds 0.20 to 0.30 m s^{-1} in the jet maximum located 15 to 20 km offshore.

Theory predicts an e-folding width of the baroclinic radius of deformation for the equatorward jet. The offshore side of the jet in Fig. 11.12c has an e-folding width of 12 to 14 km, in good agreement with the observations of Huyer (1974). The e-folding width of the jet displays relatively little longshore variability.

Huyer (1974) compared geostrophic calculations from hydrographic observations to v-component time series from current meters off Newport. She concluded that ". . . the observed currents were mainly geostrophic during the period 5 to 20 July (1972) even though the winds were variable."

It is in the lower layer that bottom topography should exert its greatest influence. Figure 11.13 compares zonal flow in the lower layer of Case I to that for Case II. Obviously, the topography has a strong effect on the offshore and longshore structure of u_2. There are several main features of the onshore-offshore flow pattern. Each develops before day 1 and persists qualitatively, although affected by oscillations of an inertial time scale. Notice the region of strong onshore flow of $\sim$0.03 m s^{-1} just south of the cape. Although this feature of the flow appears to be associated with the coastline variations, the flat bottom case exhibits no such feature. In fact, flow south of the cape in Case I is directed offshore! In the case with bottom topography, onshore flow is relatively strong, $\sim$0.01 m s^{-1}, near shore for most of the N-S extent of the basin. There is a region of strong offshore

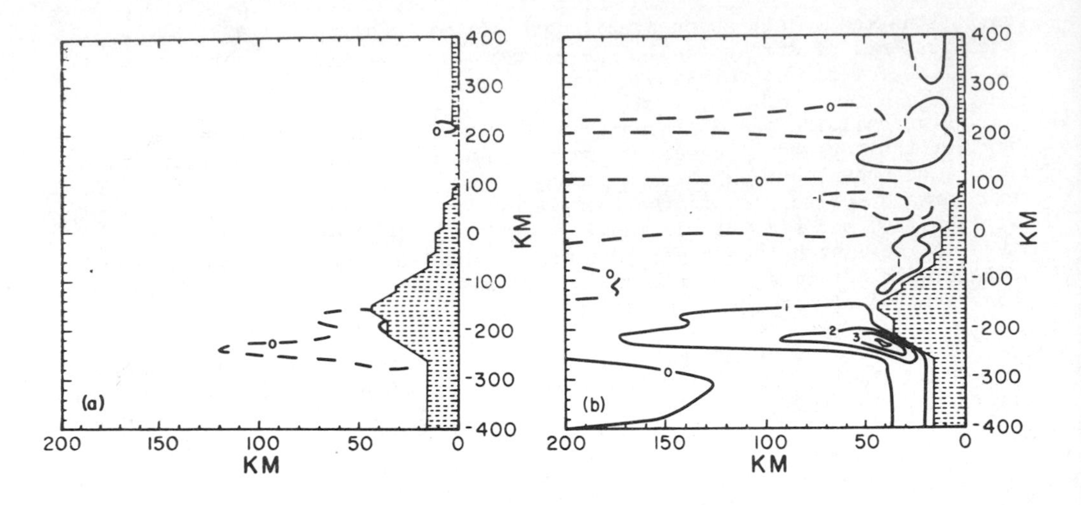

Fig. 11.13 - Lower layer E-W velocity component (cm s^{-1}) at day 5 for Case I in (a) and Case II in (b). The same contour interval is used in both figures.

flow with axis at $y \simeq 50$.

Mean velocities from the CUE-I moored current meter array, as tabulated by Huyer (1974) for a period of generally southward winds, indicate mean zonal velocities for deep flows to be generally ≤ 0.01 m s^{-1}. The deep current meters at NH-10 and NH-15 ($x \simeq 25$ to 30, $y \simeq 50$) show qualitative agreement with Fig. 11.13b, but the NH-20 deep flow is directed onshore, contradicting the model results.

Zonal flow in the lower layer can be understood in terms of the linear y-momentum equation (Hurlburt and Thompson, 1973)

$$\frac{\partial v_2}{\partial t} + fu_2 = -g \frac{\partial}{\partial y} (h_1+h_2+h_b) + g' \frac{\partial h_1}{\partial y} - \frac{\tau_b^y}{\rho_r h_2}. \qquad (11.62)$$

In the interior a geostrophic balance holds. The kinematic boundary condition forces u_2 to zero in the upwelling region and, for the flat bottom case, $\partial v_2/\partial t>0$. For the case with E-W sloping topography, offshore transport over the shelf is approximately the same as in the interior. Thus, mass continuity requires the onshore flow over the shelf to become supergeostrophic, forcing $\partial v_2/\partial t<0$ until a frictional balance is achieved in Eq. (11.62). The generally stronger onshore flow nearshore noted above can thus be understood as a consequence of mass continuity and the rising topography.

Figure 11.13b demonstrates that topographic variations affect the pattern of onshore flow well beyond the region of sloping topography, producing barotropic flows which seem to be governed by the barotropic radius of deformation. In particular, the strong onshore flow just south of the cape appears to be due to a divergence in v_2.

Longshore variations in deep onshore flow have been observed (Shaffer, 1974) in the NW Africa upwelling region. For canyons of approximately half the width of those in the Oregon area, Shaffer observed neutrally buoyant floats to move shoreward along the canyon axis. He concluded: "The main "onshore" compensation flow takes place along the axis of the canyon". This is not observed in Fig. 11.13b. Instead, a pattern of offshore transport north of a ridge, and onshore transport south of a ridge is seen, with the exception of the ridge at $y \simeq -150$. This pattern agrees with the results of Hurlburt (1974) for a simple symmetric ridge. The NW Africa observations and the area of onshore flow north of the ridge at $y \simeq -150$ in Fig. 11.13b lead us to speculate that the longshore distribution of deep zonal flow is quite sensitive to the longshore scale of the topographic variation.

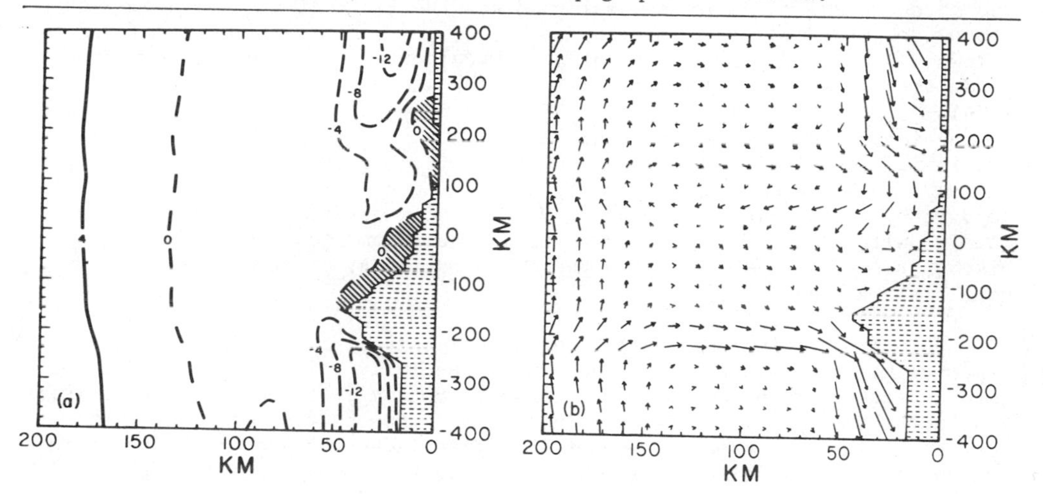

Fig. 11.14 - Lower layer N-S component of velocity (cm s^{-1}) at day 5 for Case II in (a). Lower layer velocity vectors (b) for Case II at day 5. The arrows define streamlines, not geographical direction, i.e., the x component is multiplied by 800/200=4. The length of the arrows is scaled by the longest arrow in the frame.

Figure 11.14a presents contours of the lower layer N-S component of velocity for the topography case at day 5, and Fig. 11.14b the lower layer velocity streamline vectors for the same case. For the flat bottom case (not shown) meridional flow at day 5 is everywhere weakly poleward in the lower layer, with a jet structure developing in the upwelling zone similar to that seen by Hurlburt (1974) for a β-plane flat bottom, meridional coast case. But the picture in Fig. 11.14a is quite different. Topography induces distinct variations in both the offshore and longshore structure of the flow. In particular, the equatorward jet structure of the y-independent north and south boundary regions implies the dominance of a strong barotropic mode there. In the region of longshore varying topography, the barotropic mode is much reduced; in fact, two regions of poleward flow develop (shaded areas of Fig. 11.14a). The more southern region appears during the first day of integration developing first just north of the main cape. The other area of poleward flow develops first at $y \simeq 210$ at about day 1.25. Topography clearly has a significant effect on the longshore flow.

A poleward undercurrent, coincident with the upwelling season has been observed in Oregon (Mooers, et al., 1976; Huyer, 1974). Huyer has observed statistically significant deep poleward flow at several CUE-I moored current meter stations, no poleward flow at others. The CUE-I data do not provide a detailed description of the offshore dependence of the deep flow; however, its strength is not observed to decrease with distance offshore. In the upwelling boundary layer the E-W barotropic and baroclinic pressure gradients compete in driving the geostrophic longshore flow. Huyer concluded that there is a persistent vertical shear in the longshore flow during the upwelling season which varies with location and time. The shear was observed to obey the thermal wind (geostrophic) relation.

Hurlburt and Thompson (1973) have reviewed theoretical explanations of the poleward undercurrent and explained how the β-effect can produce a longshore pressure gradient sufficient to reduce the barotropic mode, allowing poleward flow to develop. The poleward undercurrent observed in the present study is similar to that which developed for sharp shelf cases in the x-z models of Hurlburt and Thompson (1973) and Thompson (1974), being restricted to the region immediately adjacent to the coast. As explained in the discussion of Eq. (11.62), u_2 becomes supergeostrophic over an E-W sloping bottom in response to requirements of continuity, forcing $\partial v_2/\partial t<0$. Over the flat topography, a moderately increasing N-S pressure gradient in the lower layer coupled with a decrease in other terms of the y-momentum equation in the viscous boundary layer can allow the N-S pressure gradient to dominate very near the coast and $v_2>0$.

In a linear time-dependent, continuously stratified f-plane model with E-W sloping, but y-independent topography, Pedlosky (1974c) has found a poleward undercurrent to develop in response to bottom friction and the vorticity constraint imposed by the E-W sloping bottom. He observed that the undercurrent begins as a broad flow, but for times large compared to the barotropic spin-up time and small compared to the diffusive time scale, shrinks to a flow resembling that found by Hurlburt and Thompson (1973).
Pedlosky concluded that this inshore counter-current gives way over diffusive time scales to a <u>steady</u> counter-current discussed in Pedlosky (1974b).

The feature of the undercurrent of most interest in the present study is its longshore variability. Comparison of Fig. 11.14a with the upwelling contours of Fig. 11.11b shows a tendency for areas of poleward (or weaker equatorward) flow to coincide with areas of stronger upwelling, i.e., an increased baroclinic mode.

Hurlburt (1974) has investigated the effects of N-S sloping topography on the development of the poleward undercurrent. Case two, using wedge shaped topography with N-S slopes of opposite sign, showed the following. Topography sloping upward towards the north ($\beta_T>0$) resulted in an enhanced poleward flow near the shore. The opposite was true for $\beta_T<0$. As previously discussed, the topographic β-effect can act in Eq. (11.56) to affect the strength of the poleward undercurrent. To summarize, regions of $\beta_T>0$ would be expected to exhibit stronger poleward flow than regions for which $\beta_T<0$. Hurlburt states that the fundamental scale of influence seems to be the barotropic radius of deformation.

There is general agreement with the predictions of the β_T effect. It should also be noted that the undercurrent is more likely to develop in areas where the shelf is wider. The equatorward barotropic mode is more effectively reduced by $\int \beta v_1 dx$ in these areas with wider shelves.

Several interesting features of the flow are apparent in Fig. 11.14b. A strong divergence occurs in the longshore flow near the cape with an associated response in the onshore flow dictated by continuity. There is a general tendency to follow the isobaths, but regions of cross-isobath flow do occur. Oregon observations (Huyer, 1974) indicate that the orientation of the major axis of flow is determined by the bathymetry, tending to be oriented along the local bottom contours. However, mean cross-isobath flow has been observed at all levels at the DB-7 current meter string ($x=18$, $y=70$), and the model results indicate cross-isobath flow there. An anticyclonic gyre (axis at $y \simeq 100$) with scale governed by the barotropic radius seems to be associated with the main canyon. A weaker gyre exists at $x \simeq 130$, $y \simeq 240$. Their positions remain quite stationary during spin-up.

The efforts in modelling the onset of coastal upwelling discussed above suggest several ways in which significant improvements can be made. Inclusion of a more realistic wind stress, with time dependency in a three-dimensional model has been left for further work. A careful treatment of the problem of initialization is needed. The model of Oregon upwelling is an idealization of the actual underwater topography. Although the principal scales of variation have been preserved, bottom slopes are too gradual in comparison to actual Oregon topography, and no realistic shelf break is included. Finer resolution must be used in x and y to include the shelf break and steep bottom slopes, requiring a larger usage of computer core storage and a smaller time step.

The non-linear numerical model developed by Hurlburt (1974) has been used to investigate the effects of Oregon-like bottom topography on the onset and decay of the coastal upwelling circulation. Variations in bottom topography are found to play a very important role in explaining observed mesoscale features of the Oregon upwelling circulation. The region near Cape Kiwanda is shown to be an area of favored upwelling due to presence of an underwater mesoscale canyon. In general, for an equatorward wind stress, the pattern of longshore variation in upwelling intensity is explained by the rule: greater upwelling equatorward of (and lesser upwelling poleward of) a canyon axis, consistent with the topographic β-effect. Topographic variations are found to dominate over coastline irregularities in determining the longshore structure of upwelling. Specifically, the above studies indicate that the stronger upwelling observed near Cape Blanco is mainly due to local bottom topography rather than the cape itself. Zonal mass balance is not seen for the Oregon upwelling regime, with net zonal transports always following the pattern: offshore transport north of a ridge and onshore transport south of a ridge under forcing by southward blowing wind. We conclude that the inclusion of realistic bottom topography is essential to an understanding of local upwelling and the longshore current structure of the Oregon region.

Our review up to this point has been concerned only with coastal upwelling. In the next sections we will discuss three types of upwelling which occur in the open ocean.

11.6 Ice edge upwelling

Open ocean upwelling is believed to play an important role in bottom water formation. If dense water is raised to the surface in high latitudes it will be exposed to cold air and may become even denser and eventually sink to the bottom. In this section we present an interesting case where a long ice edge acts as a boundary.

Open ocean upwelling is usually due to Ekman pumping induced by an atmospheric cyclone (Section 11.7). There is the possibility of another type of open ocean upwelling, namely, the wind driven upwelling near an ice edge. It is interesting to note that large populations of euphausiids (krill) are often found along the edge of the ice in the Gulf of St. Lawrence (Platt, personal communication). Development of these herbivorous populations would be consistent with the occurence of upwelling at the ice front pumping nutrient rich water to the illuminated surface layer of the ocean and permitting growth of phytoplankton.

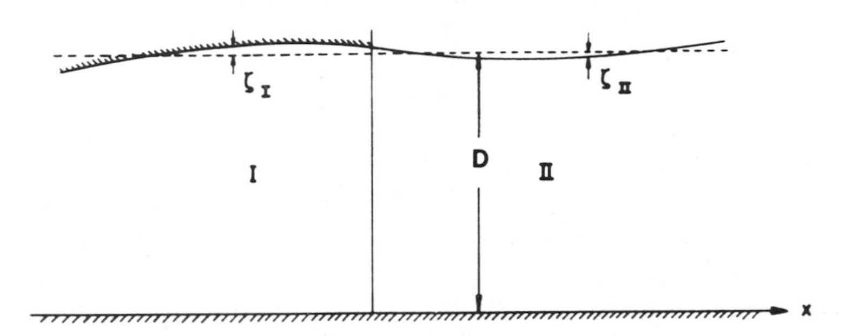

Fig. 11.15 - Model geometry and variables.

The ice is modelled as a mathematically thin surface, free to move in the vertical direction, but not in the horizontal. As will be seen, the upwelling is induced because the ice shields the water in region I from the wind (Fig. 11.15).

As a first step towards an understanding of the dynamics of the upwelling at an ice edge, consider a linear, constant density model. This does not mean that stratification can be neglected in the real oceanic case, although this model may give a good description of the actual conditions when the water is nearly homogeneous (as is the case when the bottom water formation takes place).

We have also neglected the horizontal diffusion terms. Therefore, the model will give a poor description of the details near the ice edge. For instance, we expect to find discontinuities in the surface slope there, but this will not affect the intensity of the upwelling.

From the equations of motion it is possible to derive an expression for the surface displacement ζ. Application of the kinematic boundary conditions at the surface and at the bottom yields the following governing equation:

$$\left(\frac{\partial}{\partial t} + d\right)\nabla_H^2\zeta - \frac{1}{gD}\left(\frac{\partial}{\partial t} + f\right)\frac{\partial\zeta}{\partial t} = \frac{f}{\rho_r gD}\underset{\to}{k}\cdot\text{curl}\,\underset{\to}{\tau}_o$$

$$+ \frac{1}{\rho gD}\frac{\partial}{\partial t}\left(\nabla_H\cdot\underset{\to}{\tau}_o\right), \qquad (11.63)$$

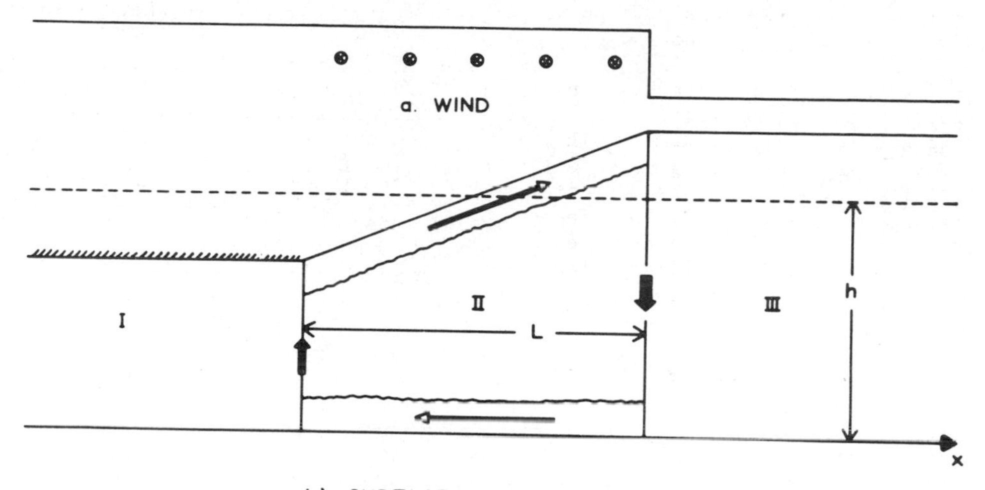

Fig. 11.16 - The steady state solution for a two-dimensional model with a northward wind stress confined to a belt of width L. Both surface elevation and circulation scheme are shown.

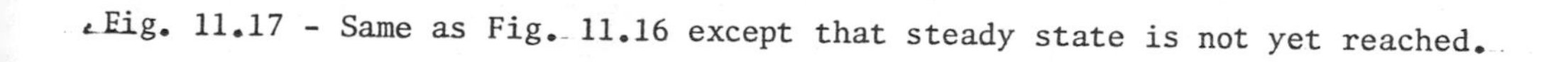

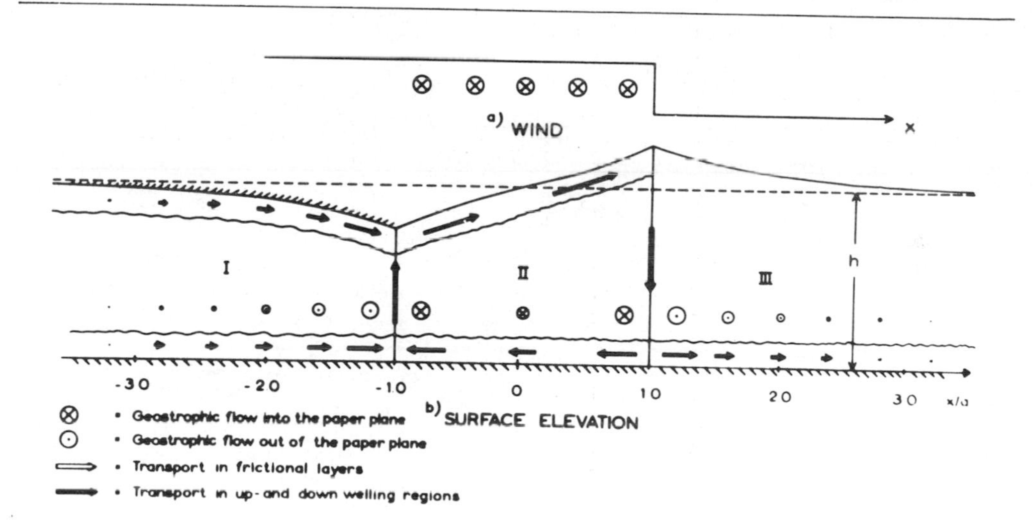

Fig. 11.17 - Same as Fig. 11.16 except that steady state is not yet reached.

where d is a friction coefficient. For further details see Gammelsrød, Mork, Røed (1975), (henceforth called GMR).

An instructive example is the steady state case with no variation of wind stress along the ice edge. It turns out that the Ekman transports in the boundary layers in the ice covered region have the same direction. Therefore to have finite displacement when $x \rightarrow \infty$, the only possible solution is a horizontal ice surface. This means that the water under the ice is quiescent and all possible circulation is confined to the open ocean region. With a constant wind stress applied in a limited belt region, we obtain the circulation picture as shown in Fig. 11.16. This steady state may be reached from an initial undisturbed surface in the manner as sketched in Fig. 11.17.

The transport of water from the ice covered region is determined by the difference between the wind driven surface current and the boundary current near the bottom in region II (see Fig. 11.17). When the surface slope in region II becomes steeper, the geostrophic induced boundary current at the bottom becomes greater and the transport of water from the ice covered region decreases. Therefore, the ice surface will flatten out. The "wave" shape of the surface in region II will also disappear due to the divergent transport in the bottom layer. Thus the picture we have in Fig. 11.17 will, in the limit of large time, be the steady state shown in Fig. 11.16.

The present brief discussion clearly demonstrates the possibility of upwelling at an ice edge. In the steady state model the ice edge acts more or less as a coast. It is further shown in GMR that if the wind stress is a harmonic function along the ice edge, the upwelling may be stronger in confined regions that the corresponding simple coastal upwelling.

A major shortcoming of the model is the constant density assumption. The case with a continually stratified ocean has been solved by Allan Clarke (private communication) during the NATO conference using a different approach. His results indicate that the magnitude of the vertical motion in the ice edge upwelling is half the corresponding coastal one. The area in which the upwelling takes place, however, is twice as large so that the total amount of upwelled water is the same in both cases.

11.7 Storm-induced upwelling in the open ocean

In the previous sections, the presence of a coast or long ice edge provided a mechanism for a one-sided divergence and subsequent vertical motion. It is also possible that the shape of atmospheric storms can provide differential motion which will lead to vertical motion in the ocean. The influence of hurricanes (typhoons) on the ocean is discussed in this section.

There appear to be two characteristic domains of storm induced effects in the ocean. Geisler (1970) used an idealized stress distribution and a linear, two layer ocean model to demonstrate that the primary response will be an upwelling when the storm translation speed is less than the baroclinic long wave speed. However, for rapidly moving storms the predominant response will be an oscillatory vertical displacement of the thermocline, referred to as a wake. Pollard (1970a) has shown that the energy input to the ocean will be absorbed primarily in inertio-gravity waves for storms moving faster than the baroclinic wave speed. For the same stress input, the rate of working or the amount of energy absorbed depends crucially on the time during which the wind acts (Veronis, 1965). Either rapid increases in wind speed, or changes in wind direction, having time scales less than one inertial period will induce inertio-gravity modes. The horizontal convergence and divergence associated with the inertial rotation of the currents can lead to significant vertical displacements of the thermocline along the path of storms. A greater amplitude may be expected for shallow mixed layers and for intense storms.

One implication of these theories is that significant upwelling due to extra-tropical cyclones is unlikely, because the stress input is on a large space scale and on a longer time scale than the inertial period. Possible exceptions may be found in the regions of the surface low center or near particularly intense fronts where the wind direction changes rapidly. A second implication is that upwelling may be expected near the center of slowly moving tropical cyclones. In these cyclones the wind speed increases to a maximum perhaps 20-60 km from the center. The subsequent decrease in wind speed within the "eye" of the storm, and the reversal of wind direction on the opposite side, produces a tremendous imbalance in the upper layers of the ocean. On the other hand, for slowly moving storms in which simple Ekman reasoning can be applied, the outward deflection of the surface layers must be accompanied by an upwelling of the subsurface waters on the scale of the baroclinic radius of deformation as shown by O'Brien and Reid (1967).

Because tropical cyclones tend to occur in late summer and early autumn when the mixed layer depth is a minimum, one may expect significant changes in the upper layers of the tropical oceans. This will be particularly important for acoustical propagation if either the upwelling penetrates to the surface, or if the amplitude of the inertio-gravity oscillations on the thermocline are a significant fraction of the mixed layer depth. The wake left in the path of tropical cyclones represents a local perturbation within the relatively homogeneous thermal field away from current boundaries. Examples of the thermocline being brought to the surface, with sea surface temperature decreases of 5-6°C as reported by Leipper (1967), are relatively rare. Typical decreases are about 2-3°C, because most tropical cyclones move faster than the baroclinic wave speed. Although the effects of the tropical cyclone are restricted to a relatively narrow region (perhaps within five times the radius of maximum wind speed), the baroclinic perturbations may persist for time scales of weeks. This may have biological implications in situations when the upwelled water is brought to the surface (Iverson, 1976).

The extent of which the movement or intensity of a tropical cyclone is affected by crossing the wake of a previous storm is uncertain. Brand (1971) deduced indirectly a bias in the operational forecasts made by the Joint Typhoon Warning Center (Guam), which he attributed to wake-crossing effects. However, Ramage (1974) has found cases in the South China Sea in which three successive tropical cyclones passed over the same region. Ramage demonstrated that the intensity changes of these cyclones were more affected by the passage of upper level troughs than by the sea surface temperature. This is not surprising considering that the energy supply from the ocean to the tropical cyclone occurs over a much larger region than the scale of the wake left by previous storms.

Because of the importance of open ocean upwelling in the tropical oceans, we will focus on model simulations and the limited data available for validation of tropical cyclone-induced effects. Tropical cyclones produce one of the best examples of the importance of vertical transports in the upper layers of the ocean. The combination of strong upward heat flux and large downward momentum fluxes may be expected to result in significant mixed layer depth and temperature changes. Parameterization of the mixing processes, which forms the basis of the one-dimensional models discussed earlier in this volume, will play a crucial role in the modelling of the oceanic response to the tropical cyclone. However, the large horizontal gradients in the thermal response, and the importance of upwelling previously discussed, prohibit the application of a one-dimensional model.

Previous models of O'Brien and Reid (1967), O'Brien (1967), (1968), Geisler (1970) and Gilbert (1974) have focused on the dynamical response of the upper ocean to stationary or slowly moving stress patterns. These layer-type models have illustrated the role of the upwelling region near the center, but did not treat the exchange of mass between layers due to the convectively and mechanically generated turbulent mixing. By contrast, a recent model of Elsberry, Fraim, and Trapnell (1976) emphasized the role of the vertical fluxes by incorporating a mixing layer model similar to Kraus and Turner (1967) and Denman (1973). For the stationary, symmetrical storm case, the horizontal currents were derived from the Ekman transport. However, results similar to those shown here were also obtained with dynamical equations of the type used for the two-dimensional coastal upwelling models (Section 11.2). The predicted thermal structure as a function of radius is shown in Fig. 11.18. Each profile represents a simulated temperature-depth sounding at increasing radii from the center of the storm (labelled as zero). In this case the radius of maximum wind speed (RMW) was at 10 per cent (45 km) of the entire domain of the model storm. The initial thermal structure was represented as a mixed layer temperature (MLT), depth (MLD), and thermocline gradient equal to 30°C, 30 m, and 0.1°C cm^{-1}, respectively. The deviations of the predicted temperatures from these initial conditions are shown in Fig. 11.19. Three significant domains of thermal structure variations can be identified in Figs. 11.18 and 11.19. Near the center (r=0) the isotherms within the thermocline are deflected upward in response to the upwelling. The upwelling does not penetrate to the surface because of the opposing effect of mixing processes. Temperature profiles within the RMW tend to have a smooth transition between the surface and the thermocline layers, as was found by Leipper (1967) in the observations after the passage of Hurricane Hilda. Subsurface temperature decreases of several degrees are predicted at the original MLD in this region (see Fig. 11.19). In the second domain, extending outward from the radius of maximum wind speed, the change in thermal structure is dominated by mixing processes. The temperature "ledge" separating the surface waters from the thermocline in the model appears to correspond to the region of "thermoclines having unusually steep gradients" noted by Leipper (1967) in bathythermographs taken nearly a week after storm passage. This mixing dominated region, which was not included in the previous dynamical models, has important consequences for thermal structure changes. First, there is a decrease in sea surface temperature of more than 1.5°C, which is due to heat loss to the storm and to the downward heat flux as the mixed layer deepens. As a result, a second layer is produced that has temperatures warmer than the initial values, as shown by the shading in Fig. 11.18, and the positive deviations in Fig. 11.19. Because the mixed layer penetrated into the region of general upwelling within the thermocline, one can identify this as a third layer, with negative temperature deviations. Support for this three layer vertical structure, and the magnitude of the temperature deviations for the mixing dominated regime, may also be found in the time series of thermal changes observed at the ocean weather ship "Tango" (Fedorov, 1973). It is difficult to compare the model predicted changes with other existing measurements because the pre-existing conditions were not observed. One exception is some measurements by Leipper before and after Hurricane Betsy, but the data was affected greatly by the presence of

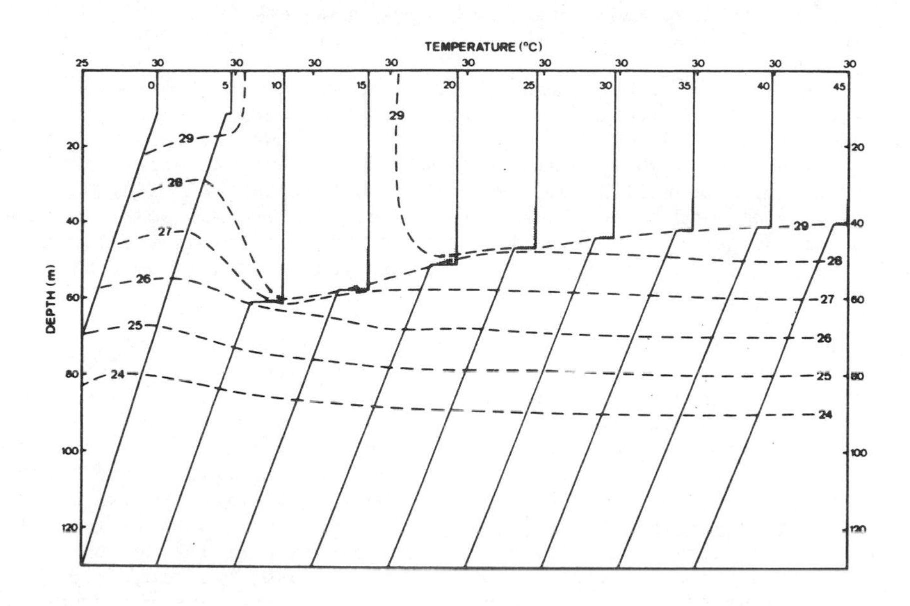

Fig. 11.18 - Simulated temperature-depth soundings after 24 h for the standard experiments are shown at intervals of 5 grid points (22.5 km) within the inner half of region. Shading indicates the layer of warmer temperatures. The depth of several isotherms is shown by dashed lines.

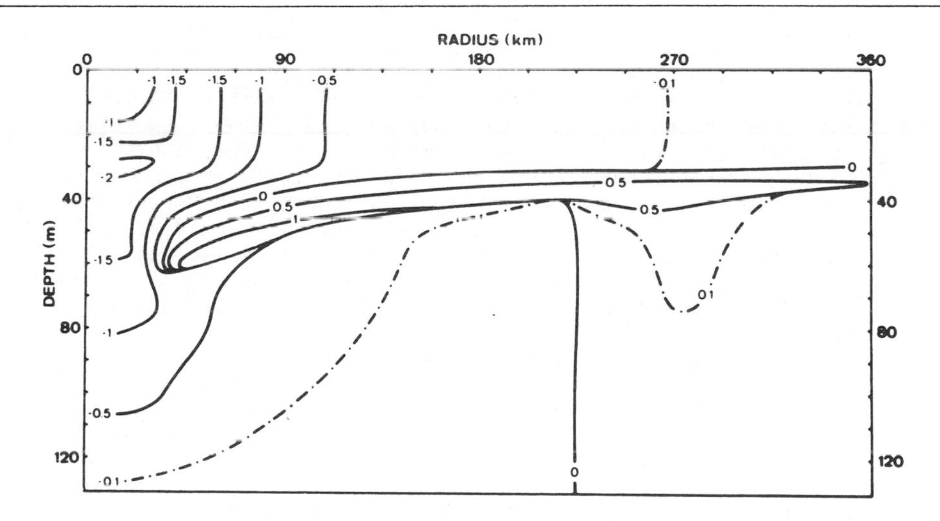

Fig. 11.19 - Distribution of temperature deviations from initial conditions for the standard experiment.

the Gulf of Mexico loop eddy during the nearly two week interval between the ship cruises.

Fedorov (1973) attributed the intermediate warmer layer to downwelling, which is inconsistent with the negative temperature anomalies in the lower layer. The region of downwelling in the model is shown in Fig. 11.19 to occur at larger radii, where positive temperature anomalies exist throughout the thermocline. Downwelling occurs in the model at the radius at which the radial profile of wind speed changes curvature. There is some evidence that similar changes of wind profiles in hurricanes determine the location of downwelling zones. This downwelling domain is generally well separated from the mixing dominated regime.

The forcing caused by the atmospheric pressure minimum excites a barotropic mode in the ocean. Kajiura (1956) showed that the associated deformation in the free surface height is translated with the storm and leaves no ridge behind. As the deformation is distributed over an area having a radius of the order of the barotropic radius of deformation, this effect is not included in the numerical models. The more important response is the baroclinic effect due to the mixture of upwelling and the inertio-gravity wake extending to the rear of the storm (Geisler, 1970). An appreciable wake is generated only if the region of strong wind stress curl passes over the region in less than half a pendulum day (Veronis, 1965; Geisler, 1970; Pollard, 1970a). Geisler notes that the steady state amplitude of the baroclinic response was so large for all storm speeds that a non-linear theory was required. One example of a numerical experiment to simulate the response in the open ocean will be discussed below. We will not review the effects of tropical cyclone passage on the coastal shelf flow (Forristall, 1974) or on the western boundary currents (Suginohara, 1973).

Elsberry and Grigsby (1976) have examined the oceanic response in a two-dimensional (x-z) cross section oriented along the storm motion vector. The dynamical equations and the semi-implicit solution technique for the two layer model follow those used by O'Brien and Hurlburt (1972) for a two-dimensional coastal upwelling simulation. However, the continuity equation was modified to include the mixing between layers due to convectively and mechanically generated turbulence. The model was started with constant MLD, MLT and thermocline gradient fields, so that the fluid was initially at rest within the 1500 km domain of integration. A model hurricane (Elsberry, Pearson and Corgnati,1974) of constant intensity was translated from the right hand side of the basin to the position shown in Fig. 11.20. The translation speed of 3.25 m s^{-1} exceeded the baroclinic long wave speed so that an oscillatory motion of the thermocline was expected. Radial (i.e., along the translation vector) components averaged over the upper layer were alternately directed toward and away from the storm with a wavelength of slightly more than 300 km. The corresponding vertical velocity at the interface height represents the type of wake predicted by the linear theory of Geisler (1970), but with realistic amplitudes. One of the effects of including turbulent mixing in the dynamical model was to increase substantially the depth over which the stress was distributed. Consequently the response amplitude should be less than in the linear theory. The first upwelling peak trails the storm center by about 100 km. A similar result (unpublished) has been obtained by Rangarao Madala of the U. S. Naval Research Laboratory using a three-dimensional ocean model. During the period of integration the inertio-gravity mode shown in Fig. 11.20 did not decay significantly as the model contained no dispersive mechanism for propagating the energy to deeper layers.

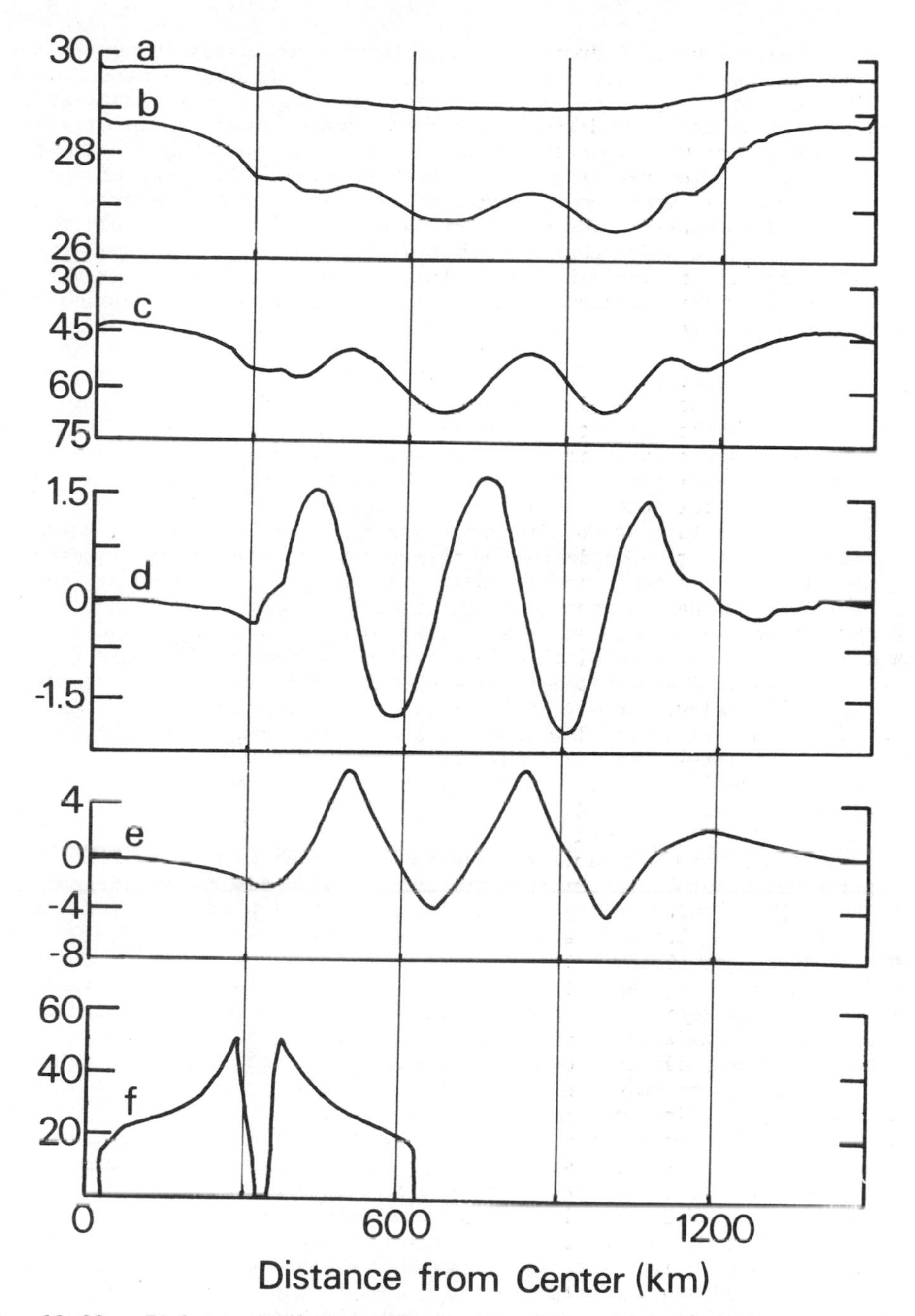

Fig. 11.20 - 72 hour predicted values with the storm moving at 6.5 kt. Surface stress is portioned 70% to current production and 30% to mixing. Shown are: (a) mixed layer temperature (°C); (b) thermocline temperature (°C); (c) mixed layer depth (m); (d) ocean vertical velocity ($m\ m^{-1}$); ocean radial velocity ($m\ s^{-1}$); and (f) tangential wind velocity ($m\ s^{-1}$).

The corresponding mixed layer depth and temperature predicted by Elsberry and Grisgby (1976) are also shown in Fig. 11.20. Peak-to-peak amplitude of the MLD oscillations in the wake are about 20 m and propagate at the same speed as the storm. The mixed layer temperature decreases steadily as the storm approaches, levels off and then continues to decrease. Behind the storm the rate of upward heat flux is greatly diminished, and the turbulent generation is insufficient to cause further mixing between the layers. As the vertical fluxes tend to zero, the MLT remains constant, even though the MLD is oscillating in response to the inertio-gravity wave. Both Pollard (Chapter 8) and Woods (Chapter 9) have suggested that such inertio-gravity modes will produce mixing, but these effects have not been parameterized in the present model.

These model experiments illustrate the importance of vertical fluxes in altering the thermal structure of the upper oceans during tropical cyclone passage. The predicted thermal changes are generally in agreement with the data of Leipper (1967), which is still the most extensive set available. Until recently it has not been possible to obtain thermal structure data on time scales shorter than the week or so necessary to traverse the region using research ships. P. Black of the National Hurricane and Experimental Meteorology Laboratory has surveyed regions within two hurricanes using airborne expendable bathythermographs (AXBT). Black was the first to document (see Sheets, 1974) the existence of thermocline oscillations similar to those predicted by Geisler (1970) and simulated in the model results discussed above. Although one must be rather cautious in interpreting such a limited set of data, it appears that the wavelength is much shorter than either the analytical or the numerical results would predict for that storm movement. An oscillation in MLT of about 0.25C with the same wavelength as that of the thermocline oscillations was also indicated. Black attributed this MLT variation to mixing caused by shearing instability near the wave crests on the thermocline.

One of the shortcomings of Black's data set, which is also true for nearly all other case studies of tropical cyclone induced effects, is the lack of an adequate specification of the pre-storm conditions. Consequently, it has not been possible to do heat budget calculations without making rather arbitrary assumptions regarding the initial conditions. Furthermore, the thermal structure has not been sampled successively to estimate the transient response rate of the ocean surface layers. In an attempt to overcome these data shortcomings, W. Schramm of the U. S. Naval Postgraduate School collected ocean thermal data immediately before and after the passage of Typhoon Phyllis in August, 1975. A series of pre-calibrated AXBT's were dropped in advance of the storm along a section normal to the forecast track. Phyllis, moving at 9 m s^{-1}, attained maximum intensity of 55 m s^{-1} in the region of the AXBT section. Those sounding locations that were close to the actual storm track were resampled about 10 h after the storm passage, and again after 48 h. It is hoped that the differences between these soundings will provide an adequate specification of the oceanic thermal structure change for calculating the heat budget. This should indicate the applicability of the numerical models discussed above. In addition to the section normal to the path, a series of AXBT drops were made approaching the storm from the rear, and along a parallel track. These data should permit additional estimates of the wavelength and amplitude of inertio-gravity modes excited by an intense, fast moving tropical cyclone. Collection of data sets such as this are essential to improve our ability to model storm-induced effects in the open ocean.

11.8 Modelling of fronts in the upper ocean

Fronts in the upper layers of the open ocean and the upwelling associated with them are of considerable interest to oceanographers, meteorologists and biologists. The large temperature gradients across them may exert some influence on mesoscale atmospheric motions and the observed vertical oceanic circulations associated with these fronts are important to biological productivity. Woods (1974b) has also suggested that these motions might be responsible for a significant fraction of the vertical heat transport through the seasonal thermocline.

The amount of data available on fronts in the open ocean is small and in particular there are very few observations of the formation of such fronts. We know, however, that there exist large-scale eddying motions in the ocean. In sub-regions of such an eddy field, we expect the large-scale flow to approximate a horizontal deformation field. In the atmosphere such deformation fields are thought to be an important frontogenetic mechanism and Hoskins and Bretherton (1972) have simulated atmospheric frontogenesis with a model forced by horizontal deformation. In this section we describe an attempt to see whether oceanic fronts can be simulated by such a model, to make an estimate of the time scale of the frontogenesis and thus to suggest whether horizontal deformation might be an important frontogenetic mechanism. The model is highly idealized and produces fronts which are independent of the y coordinate from an initial state which is a baroclinic density field not varying in the y direction and with a very large horizontal scale in the x direction. The action of the deformation field ($u=-ax$, $v=ay$) on such a distribution of a passive scalar would be to concentrate the x gradients around $x=0$, where they are asymptotically infinite. Density is not a passive scalar but while gradients are small, the density field is approximately passively compressed. As the gradients increase, however, non-linear effects, which create a vertical circulation, become important and rapidly sharpen the front near the ocean surface. The model does not include any mechanism to oppose increasing gradients at the surface and they become infinite there after a finite time. This final state is unrealistic but it can be shown that the assumptions of the model are violated locally before this stage is reached. The important point is that the model is capable of predicting in a physically consistent manner the sharp horizontal gradients which are often observed.

We will present a brief outline of the mathematical model. The starting point is the set of primitive equations for an incompressible, inviscid fluid in hydrostatic balance. The Coriolis parameter is assumed constant. Radiation effects and other heat sources and sinks are neglected. The vertical coordinate used is $z=(1-p/p_r)p_r/\rho_r g$, where p_r and ρ_r are constant reference values of the pressure and density. We denote physical height by h_1, the geopotential by $\phi=gh_1$, and the specific volume by $\alpha=\rho^{-1}$. Then the equations governing the system are:

$$\frac{\partial\phi}{\partial z} - \frac{g}{\alpha_r}\alpha = 0; \qquad (11.64)$$

$$\frac{\partial \underset{\rightarrow}{v}}{\partial t} + (\underset{\rightarrow}{v}\cdot\nabla_H)\underset{\rightarrow}{v} + w\frac{\partial \underset{\rightarrow}{v}}{\partial z} + f\underset{\rightarrow}{k}\times\underset{\rightarrow}{v} + \nabla_H\phi = 0; \qquad (11.65)$$

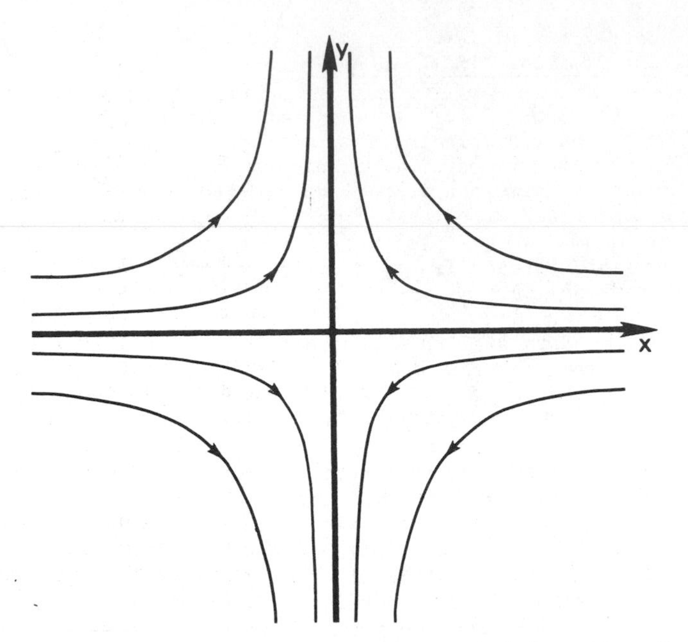

Fig. 11.21 - Typical streamlines of the horizontal deformation field.

$$\frac{\partial \alpha}{\partial t} + \underset{\rightarrow}{v} \cdot \nabla_H \alpha + w \frac{\partial \alpha}{\partial z} = 0; \qquad (11.66)$$

$$\nabla_H \cdot \underset{\rightarrow}{v} + \frac{\partial w}{\partial z} = 0. \qquad (11.67)$$

We let $u=-ax+u'$, $v=ay+v'$, $w=w'$, $\alpha=\alpha'$, $\phi=faxy-(a^2+\frac{da}{dt})\frac{y^2}{2} - (a^2-\frac{da}{dt})\frac{x^2}{2} + \phi'$. Typical streamlines of the horizontal deformation field are shown in Fig. 11.21. If we then assume that the primed quantities are independent of y and that a is independent of z, we may obtain a consistent set of equations in only two space dimensions. On the basis of a scale analysis we make the major simplifying approximation of neglecting the acceleration terms in the x momentum equation. Thus we assume geostrophic balance between the pressure gradient across the front and the velocity along it. We also make a transformation of coordinates to (χ,z,t) where $\chi=x+f^{-1}v'$. After some analysis it may be shown that the solution at any desired time can be obtained without integrating in time by solving the elliptic problem:

$$f^{-2} \frac{\partial^2 \phi}{\partial \chi^2} + \left[N_I^2\left(\frac{\chi}{L}, \frac{\partial \phi}{\partial \chi}\right)\right]^{-1} \frac{\partial^2 \phi}{\partial z^2} = 1,$$

$$\text{where } fv' = \frac{\partial \phi}{\partial \chi}; \; \frac{g}{\alpha_r}\alpha' = \frac{\partial \phi}{\partial z},$$

$$\text{and } N_I^2 = \frac{g}{\rho_r}\frac{\partial \alpha'_I}{\partial z}. \qquad (11.68)$$

The subscript I denotes values in the initial state and $\alpha_r=\rho_r^{-1}$. N_I^2 is a known function analogous to the Brunt-Väisälä frequency in the initial state.

Time enters the problem only through $L(t)=L_I\exp(-\int_{t_I}^{t} a(t)dt)$ and the boundary conditions. L is the length scale of the solution in the χ direction; $L=L_I$ in the initial state, where L_I is so large that gradients of v' are negligible. Thus the time origin $(t=t_I)$ is rather arbitrary and we choose a new origin $(t=t_1)$ such that the solution at this time is not significantly different from a passive compression of the true initial state by the deformation field. The times given below are all relative to this new origin, but it should be remembered that they depend on the value of the deformation rate. The value used here is 10^{-5} s^{-1}, based on some observations of Saunders (1973b). Very few observations suitable for calculating a in the open ocean are available and thus the value used may not be typical.

The governing equations can be solved analytically on the boundaries to provide the boundary values required in the solution of Eq. (11.68) for which a finite difference scheme is used. The solution thus obtained provides the v' and α' fields at any time; the u' and w' fields may be obtained indirectly from such solutions at two times since the same fluid particles may be identified at each time and thus their trajectories calculated.

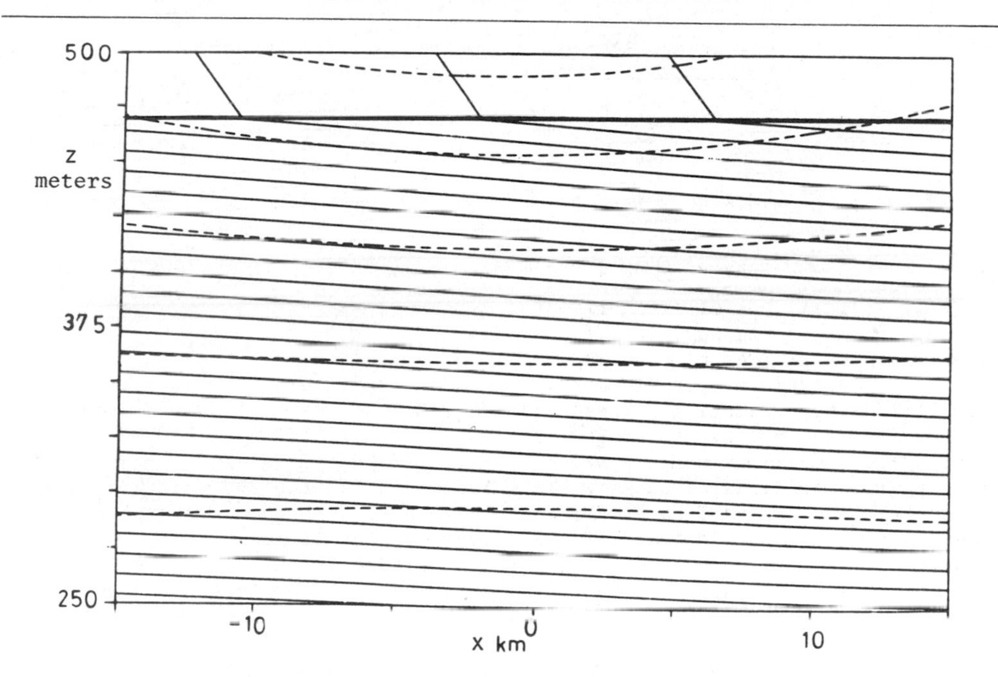

Fig. 11.22 - Solution for σ_T (solid lines) and v' (broken lines) at 0 h. Velocity contour interval is 0.04 m s^{-1}; bottom-most contour is 0.0 m s^{-1} and the velocity increases upwards. σ_T contour interval is 0.1; bottom-most contour is 28.05. The boundary between the mixed layer and the thermocline is shown as a heavy line.

The initial conditions for the solutions described here are intended to model a 500 m deep ocean with a shallow mixed layer overlying a thermocline. The boundary between the two regions is a discontinuity in vertical density gradient, the mixed layer being characterized by a small constant value of N_I^2, while the thermocline is represented by a relatively large value of N_I^2 which varies only with x/L. The initial surface σ_T field has an inverse tangent profile with a value of 25.96 at $-\infty$ and 25.09 at $+\infty$. The lower boundary, $z=0$, is taken as a surface of constant $\sigma_T(=30.86)$ in order to avoid frontogenesis there.

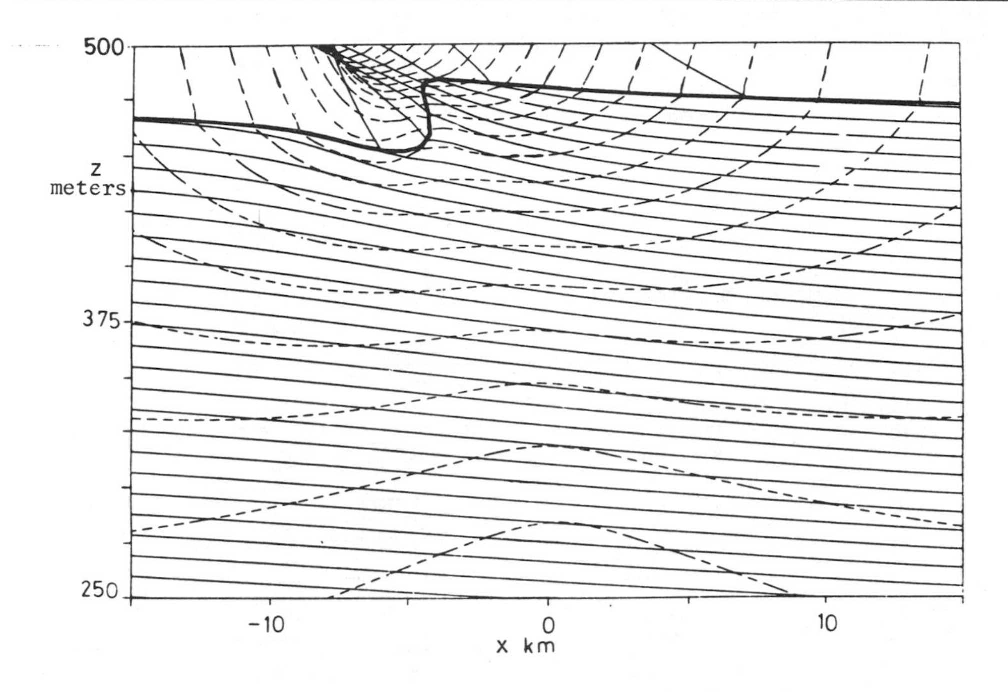

Fig. 11.23 - As for Fig. 11.22 but at 77 h. Bottom-most velocity contour is -0.04 m s^{-1}.

Figures 11.22 and 11.23 show the solution in the top 250 m at 0 h and 77 h. By 77 h the gradients at the surface have already become infinite while those on the mixed layer thermocline boundary are still finite. The model is theoretically capable of predicting infinite gradients on both these surfaces but the internal boundary is free to move in order to counteract the increasing gradients along it. Thus instead of infinite gradients we see in Fig. 11.23 a remarkable distortion of the internal boundary. The mixed layer, originally 30 m deep everywhere has shallowed to a minimum depth of 17 m in the less dense water and deepened to a maximum depth of 49 m in the denser water. There is a well defined frontal region running through the mixed layer and this can also be traced into the thermocline but there it loses its identity with increasing depth. The main features of the V' field are an intense jetstream with a maximum velocity of 0.8 m s^{-1} centered at the surface front but extending over the upwelling and downwelling tongues of water and a broad slow return flow in the lower half of the region. A typical value of the mean surface slope required to drive such a flow pattern is 0.5 cm km^{-1}. Examination of the

Richardson number in these solutions suggests that in reality mixing should have become important in a very localized region around the front before infinite gradients are predicted and we speculate that such local mixing might lead to a front with no singularity at the surface but with strong gradients extending to greater depth. Unfortunately, it is not possible to verify this with our present model. The neglect of mixing seems to be the first limitation on the model; the neglected acceleration terms do not become significant until later in the front's development.

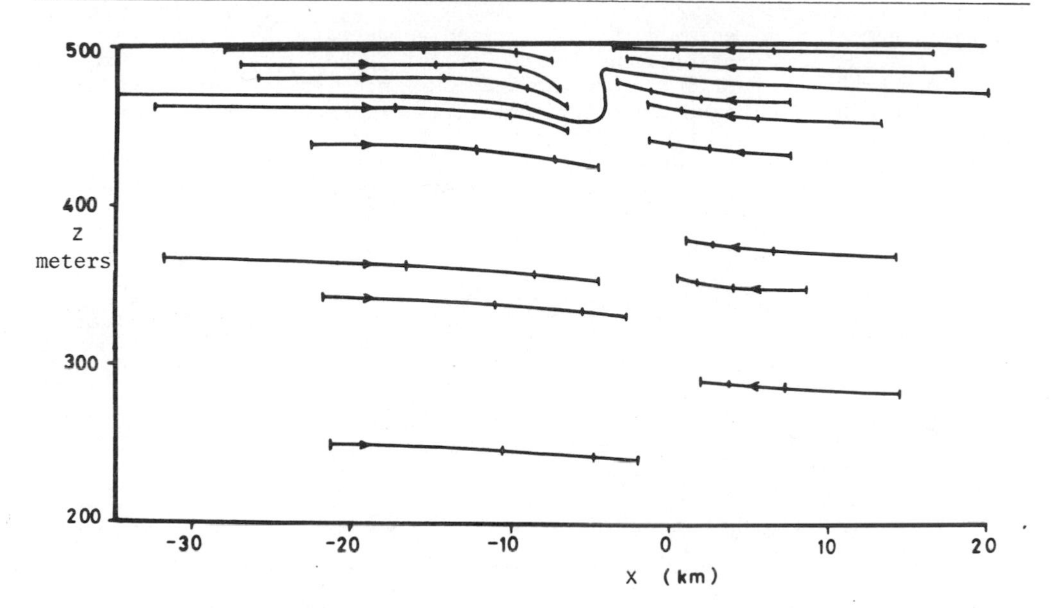

Fig. 11.24 - Trajectories of selected fluid particles. The positions of the particles at 19.25 h, 38.5 h, 57.75 h and 77.0 h are marked by ticks. The position of the mixed layer thermocline boundary at 77 h is shown.

Figure 11.24 shows the u' and w' velocities in the form of the trajectories of selected fluid particles, the position of each particle at four equally spaced times being shown. In the less dense water we have upward motion and in the denser water downward motion. The magnitude of the vertical velocities increases over the period shown and is a maximum in the tongues, the upwelling velocity reaching 29 cm h^{-1} while the downwelling velocity reaches 66 cm h^{-1}.

It has been shown that a horizontal deformation field can produce very sharp fronts in the ocean in much the same way as in the atmosphere. The vertical motions are crucial in the sharpening of the fronts and also have a significant effect on the mixed layer depth. In the example shown the mixed layer depth is altered by 10 m or more over a region about 10 km wide. This feature is one we might expect observational studies to show and which has indeed been observed by McLeish (1968), Voorhis and Hersey (1974), and Levine and White (1972). McLeish notes that the contrast in mixed layer depth across fronts is often considerable and that on occasions the mixed layer may not exist on the warm side.

It is clear that to make further, more detailed comparisons of models of ocean frontogenesis with relatively high resolution, simultaneous observations of temperature salinity and velocity are required. It would also be highly desirable to have more observations from which the deformation rate could be calculated. If the value used here is of the correct magnitude then the time scale of frontogenesis by horizontal deformation is about 3-4 days. This would suggest that horizontal deformation could be an important frontogenetic process in the upper ocean since this time scale is much smaller than the lifetime of the eddies which produce the deformation field.

Acknowledgements: We enjoyed the meeting and appreciate the kindness of Eric Kraus, Roberto Frassetto and Enrico Cattani on our behalf. We thank the other members of our working group who made our intellectual pursuit so enjoyable. These are Dominique Begis, France; Chris Brockmann, Germany; Kaye Burnett, U.S.A.; Armando Fiuza, Portugal; and Edward Horne, Canada. It is unfortunate that we were not able to drink 20 liters of that excellent local wine. Space does not permit us to list the many organizations which support our research in upper ocean circulation.

Chapter 12
THE SEA SURFACE

O. M. Phillips

12.1 Introduction

From the point of view of oceanic modelling, the sea surface is simply the upper boundary across which the atmospheric forcing is communicated, the variations in atmospheric pressure, the tangential stress resulting from the wind, and the buoyancy flux resulting from the exchange of sensible heat and surface evaporation. But as a result of the wind, the surface is disturbed by waves that range in scale from capillary components with wavelengths of a few millimeters to swell with a wavelength of perhaps several hundred meters. What fraction of the wind stress, the momentum flux from the atmosphere to the underlying water, is available to drive the oceanic circulation and what fraction is radiated away as waves? How are the heat transfer characteristics of the ocean surface influenced by the wave field, with its sporadic breaking and locally intense mixing? To what extent do the high frequency wave motions, through non-linear interactions, influence the mean height of the sea surface? Can non-linear processes, whether associated with surface waves or tidal components, generate steady circulations driven only indirectly by the wind field? In storm surges, how much of the change in sea level can be attributed to the varying tangential wind stress and how much to non-linear wave effects?

The purpose of this paper is to provide a framework in which questions like these can be addressed. Underlying them all are both the general characteristics of wave motions and the particular attributes, such as wave breaking and the nature of energy input from wind to waves, that are peculiar to surface waves. In the next section, therefore, a brief account is given of some of the general characteristics of waves, followed by a consideration of the particular processes - input from the wind, wave-wave and wave-current interactions, and losses through wave breaking. Finally, the implications of these processes for modelling will be described to the extent that they are understood at the present time.

12.2 Wave conservation equations

When a wave train of any kind propagates through a medium that is otherwise at rest, the energy density of the wave field E is specified by

$$\frac{\partial E}{\partial t} + \nabla \cdot (\underset{\to}{c}_g E) = 0, \qquad (12.1)$$

where $\underset{\to}{c}_g = \nabla_{\underset{\to}{k}} \omega$ is the group velocity, the velocity of energy propagation, and ω the wave frequency. If the medium is uniform and constant in time, both the wave frequency and the wave number are constant, the relation $\omega = \omega(\underset{\to}{k})$, the dispersion relation, being characteristic of the type of wave under consideration.

If the medium is in motion or if its properties (in the present context, generally the water depth) vary, the scales of variation being large compared with the wavelength, some new effects arise. In the first place, the waves are convected by the current field $\underset{\to}{u}$ so that the apparent frequency, the wave frequency observed at a fixed point, is modified by the advection of waves by the current:

$$\upsilon = \omega + \underset{\to}{\kappa} \cdot \underset{\to}{u}. \tag{12.2}$$

The distribution of the local wave number vector in the wave train is specified by the 'conservation of waves',

$$\frac{\partial \underset{\to}{\kappa}}{\partial t} + \nabla \upsilon = 0, \tag{12.3}$$

as given, say, in Phillips (1969). This simple kinematical modification is accompanied by more profound dynamical effects. The velocity of energy propagation is $(\underset{\to}{c}_g + \underset{\to}{u})$ and in a varying current, energy is exchanged between the the wave train and the current by the working of the radiation stresses $T_{\alpha\beta}$. The wave energy balance becomes

$$\frac{\partial E}{\partial t} + \frac{\partial}{\partial x_\alpha}[(c_{g_\alpha} + u_\alpha)E] + T_{\alpha\beta}\frac{\partial u_\alpha}{\partial x_\beta} = 0. \tag{12.4}$$

The radiation stress is given by Bretherton (1971) as

$$T_{\alpha\beta} = \frac{E}{\omega}\kappa_\alpha c_{g_\beta}. \tag{12.5}$$

For ocean waves in deep water,

$$T_{\alpha\beta} = \begin{bmatrix} \frac{1}{2}E & 0 \\ 0 & 0 \end{bmatrix} \tag{12.6}$$

where the one direction is taken parallel to $\underset{\to}{\kappa}$. The interaction between a mean and a fluctuating motion is familiar in other contexts: in turbulent motion, the Reynolds stress is precisely of this kind and the working of the Reynolds stresses against the mean velocity gradient provides the source for turbulent energy in shearing flows. The Reynolds stress also appears in the mean momentum balance for turbulent flow as an additional stress, generated by the fluctuating motion, acting on the mean flow. In the same way, the gradient of radiation stress provides a body force per unit volume acting on the medium which can, in turn, lead to steady motions. The circulations described by Ronday following this paper provide a very interesting example of this in a field situation.

The dynamics of wave trains in slowly varying media is simplified considerably by considering the wave action density N, the energy density divided by the intrinsic frequency. Bretherton and Garrett (1968) have shown than in a moving medium, the local rate of change of wave action is balanced by the convergence of the flux of action:

$$\frac{\partial N}{\partial t} + \nabla \cdot \{(\underset{\to}{c}_g + \underset{\to}{u})N\} = 0 \quad ; \quad N \equiv \frac{E}{\omega}. \tag{12.7}$$

This action conservation principle is considerably simpler than the energy balance (12.4) to which it is completely equivalent, because of the disappearance of the radiation stress interaction term. With the kinematical conservation equation (12.3) and the dispersion relation $\omega=\omega(\underset{\to}{\kappa})$, it provides a complete set of equations from which the action density of the wave field can be found. If one is concerned with the influence of the wave train upon the ambient medium (such effects as wave set-up or set-down, or the generation of wave-induced currents), the radiation stress, given by Eq. (12.5), is included in the mean momentum balance.

Oceanic surface waves do not, however, occur as series of wave trains but as a whole spectrum of components with a wide range of scales and directions of propagation. For surface waves, the wave number spectrum is defined by

$$\Psi(\underset{\to}{\kappa}) = (2\pi)^{-2}\iint \overline{\zeta(\underset{\to}{x},t)\,\zeta(\underset{\to}{x}+\underset{\to}{r},t)}\,\exp[-i\underset{\to}{\kappa}\cdot\underset{\to}{r}]d\underset{\to}{r}, \tag{12.8}$$

the normalization factor being chosen such that the mean-square surface displacement is simply the integral over the spectrum:

$$\overline{\zeta^2} = \iint \Psi(\underset{\to}{\kappa})\,d\underset{\to}{\kappa}. \tag{12.9}$$

Since the mean potential energy of the wave field is $\frac{1}{2}\rho g\overline{\zeta^2}$ and since there is equipartition between mean potential and mean kinetic energy to sufficient accuracy, the spectrum of wave energy is simply $\rho g\Psi(\underset{\to}{\kappa})$.

More limited information is contained in the frequency spectrum of the surface displacements at a fixed point. This is defined by

$$\Phi(\upsilon) = \frac{2}{\pi}\int_0^\infty \overline{\zeta(\underset{\to}{x},t)\,\zeta(\underset{\to}{x},t+t')}\cos(\upsilon t')dt', \tag{12.10}$$

the cosine factor being used since the covariance is an even function of t'. The frequency υ is ordinarily taken as positive and the normalization is such that

$$\overline{\zeta^2} = \int_0^\infty \Phi(\upsilon)d\upsilon. \tag{12.11}$$

The frequency spectrum contains no directional information and, with the use of the dispersion relation, can be expressed as an integral of $\Psi(\underset{\to}{\kappa})$ over all directions of a fixed wave number magnitude $|\underset{\to}{\kappa}|$. The details are given, for example, in Phillips (1969).

The spectral density of wave action is now defined as

$$N(\underset{\to}{\kappa}) = \frac{E(\underset{\to}{\kappa})}{\omega} = \frac{\rho g\Psi(\underset{\to}{\kappa})}{\omega}, \tag{12.12}$$

and, corresponding to Eq. (12.7), the conservation of spectral density of wave action is expressed by

$$\{\frac{\partial}{\partial t} + (\underset{\to}{c}_g + \underset{\to}{u})\cdot\nabla\}N(\underset{\to}{\kappa}) = 0. \tag{12.13}$$

The spectral action density is thus conserved following a wave group. Note the slight differences in form between Eqs. (12.7) and (12.13); these arise since, in a moving medium, an element of area in wave number space is not constant but satisfies a conservation equation that can be derived from Eq.

provide a balancing energy supply, its spectral density would be expected to decrease subsequently before attaining an equilibrium level. This overshoot phenomenon was in fact originally observed in the field by Barnett and Wilkerson (1967). At larger wave numbers, significantly greater than those associated with the spectral peak, the net transfer of wave action is very small; if the spectral density is in a state of saturation, then Eq. (12.14) reduces simply to an approximate balance between input from the wind at these wave numbers and loss by wave breaking. The input from the wind is determined by the detailed dynamics of air flow over the waves. Under saturated conditions, the frequency and intensity of wave breaking adjusts itself so that the rate of energy dissipation is equal to the rate at which energy is supplied from the wind.

The problem of calculating the rate of energy transfer from wind to waves is an old and difficult one. The air flow over waves produces variations in surface stress, both normal stress (pressure) and tangential stress. The normal energy flux from air to water is the mean value of the scalar product of the applied stress and the velocity of the water surface, and this can be expressed as the product of the separate means of the two quantities, together with the average of the product of the fluctuations:

$$\overline{\vec{\tau}\cdot\vec{u}} = \overline{\vec{\tau}}\cdot\overline{\vec{u}} + \overline{\vec{\tau}'\cdot\vec{u}'} . \qquad (12.15)$$

This separation corresponds to the partition of the total energy flux into the part which produces mean currents and the part that supports the wave motion, a fact that is not obvious but was established by Davis (1972). The energy flux to the wave motion then involves the fluctuations in surface stress associated with air flow over waves which depends on the response of the turbulence near the water surface to the undulatory mean flow. The primary source of the energy flux appears to involve the component of the pressure fluctuations in phase with the wave slope and the first attempt to calculate this was by Miles (1957) who supposed that the influence of the turbulence was negligible except insofar as it is involved in the maintenance of the logarithmic profile for the mean flow. However, the neglect of the atmospheric turbulence in this context is hardly realistic. Bryant (1966) was the first to consider the question at all seriously and a rather unsuccessful attempt to account for the influence of the turbulence was presented by Phillips (1969). The problem was re-formulated by Davis (1972) who used several different closure approximations to calculate the energy flux from wind to waves. Closure schemes in turbulent shear flow are still rather ad hoc (Bradshaw, Ferris, and Atwell, 1967; Hussain and Reynolds, 1970; Lumley, 1970) and different methods, which may be reasonably satisfactory in other flows, were found by Davis to give very different results when applied to this problem. The situation is not one in which firmly established methods lead to results that one might seek, with some confidence, to verify experimentally. On the contrary, because of the sensitivity of the results to the assumptions made, the air flow over waves appears to provide an ideal context to test the theories of turbulent stress generation themselves.

Davis found that the rate of energy input to the waves is also very sensitive to the assumed profile of the mean flow very close to the water surface, a disconcerting finding since, in nature, this depends crucially on the distribution of small-scale roughness. The problem is avoided in the recent calculations by Gent and Taylor (1976) who suppose that the surface has an assigned aerodynamic roughness which may either be uniform over the surface

(12.13) (see Phillips, 1969, p. 147).

12.3 The balance of action spectral density in a surface wave field

Equation (12.13) is linear in the action spectral density $N(\underset{\rightarrow}{\kappa})$ and ignores both input to, and losses from, the wave components. The non-linear wave interactions do provide a slow interchange among the wave components; the wind, blowing over the water surface, provides an input of wave energy and wave action; while dissipative processes, principally wave breaking, extract energy and action from the wave field. The resonant interactions among surface waves occur among groups of four components and in these interactions, both wave energy and wave action are conserved. When integrated over all wavenumbers, the net rate of change of action density resulting from wave interactions therefore vanishes so that, for a particular component, the rate of change of wave action can be represented as the divergence (in wave number space) of the spectral flux of wave action density $F(\underset{\rightarrow}{\kappa})$. If $W(\underset{\rightarrow}{\kappa})$ represents the rate of increase of action spectral density associated with energy flux from the wind, and $L(\underset{\rightarrow}{\kappa})$ the rate of loss, primarily from wave breaking, then the balance of action spectral density can be represented as

$$\{\frac{\partial}{\partial t} + (\underset{\rightarrow}{c}_g + \underset{\rightarrow}{u})\cdot\nabla\}N(\underset{\rightarrow}{\kappa}) = -\nabla_{\underset{\rightarrow}{\kappa}} F(\underset{\rightarrow}{\kappa}) + W(\underset{\rightarrow}{\kappa}) - L(\underset{\rightarrow}{\kappa}). \qquad (12.14)$$

If the nature of the terms on the right-hand side is known, this equation, together with the kinematical conservation equation and the dispersion relation, provides a complete specification of the wave field.

The statistical mechanics of non-linear interactions in gravity waves have been studied extensively by Hasselmann (1962, 1963), and the net transfer of action spectral density to a particular component can be represented as a Boltzmann-type integral involving multiple integrations over all sets of wave numbers that contribute to the wave number $\underset{\rightarrow}{\kappa}$. The evaluation of the coupling coefficient as given by Hasselmann is no easy task, involving long and heavy algebraic manipulations followed by numerical computations and the computational error in the end result appears to be substantial. One qualitative result is, however, of great significance. It is that most of the energy transfer occurs among groups of almost identical wave numbers: the spectral fluxes of action and energy density are local in wave number space so that these quantities at any given wave number depend on the spectral densities near this wave number, rather than at wave numbers significantly larger or smaller. Very recently, a greatly simplified method of calculation has been given by Longuet-Higgins (1976) and developed by Fox (1976). Fox calculates both the rate of redistribution of action in an initially concentrated spectrum and in a distributed spectrum, the results being probably much more reliable than the earlier calculations based upon the Boltzmann equation. The original calculations suggested that the sharpness of the spectral peak might be enhanced by wave-wave interactions but Fox's work indicates that this is not so. In a distributed spectrum, energy and action flow away from the spectral peak, primarily towards slightly lower wave numbers, i.e., slightly longer waves. The growth of an individual long wave component is therefore influenced by the flux of energy and action from higher wave numbers and by direct energy input from the wind; by the time this particular component has grown to the extent that it occupies the spectral peak, the net action flux from it is negative so that, unless the wind is sufficiently vigorous to or which may vary with respect to phase of a long wave. The results of these calculations seem much more encouraging, the order of magnitude variations found by Davis for different closure models being reduced to variations by only a factor of two. Clearly, our ability to calculate the rate of energy transfer to the waves leaves a great deal to be desired.

Direct experimental measurement of the energy flux is also very difficult. Measurement of the surface stress involves positioning of an instrument very close to the water surface as it heaves up and down. The instrument must be as nearly flush with the water surface as possible, which renders it liable to be swamped. It must record reliably when subjected to random vertical accelerations and, if used at sea, must be capable of withstanding exposure. Finally, the aim of the measurement is to determine a rather small phase difference from 180° in a pressure signal whose magnitude may have only the order of the hydrostatic head produced by a column of <u>air</u> of the height of the waves. It is little wonder, then, that the experimental results available show considerable scatter - the fact that there is any consistency at all is a tribute to the experimental skill of those involved.

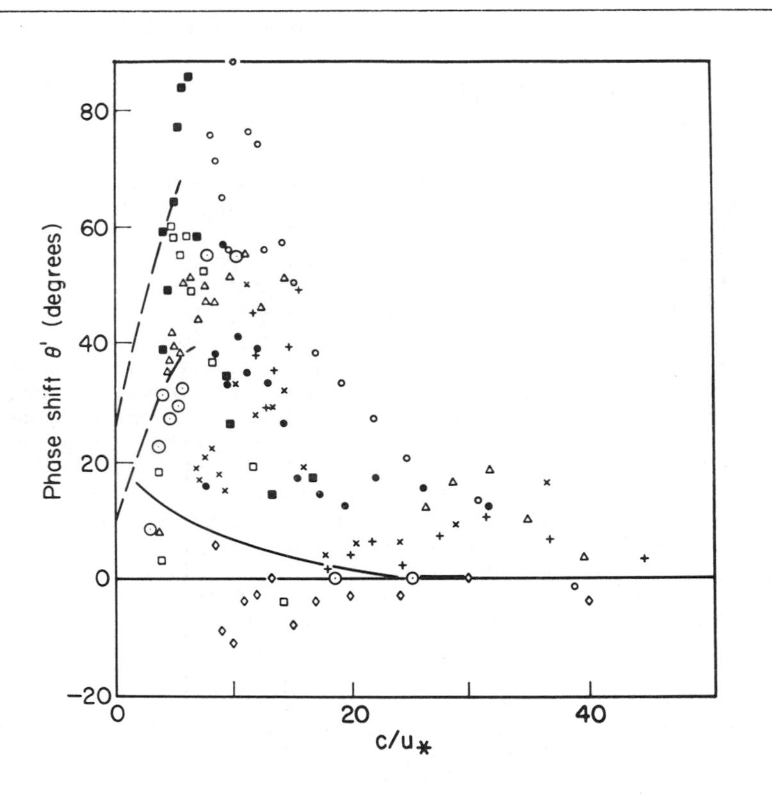

Fig. 12.1 - Phase shift of pressure signal from 180 degrees plotted against c/u_*: +, run 1; ○, run 2a; ●, run 2b; ×, run 3; □, run 4a; ■, run 4b; ▲, run 5; △, run 6; ◊, Longuet-Higgins et al. (1963); ⊙, Shemdin and Hsu (1967); ----, envelope of Kendall's (1970) wind tunnel results; —, Miles's (1959) theoretical prediction.

The first measurements were made by Longuet-Higgins, et al. (1963) using a rather large, flat buoy in the North Atlantic. Though other observations taken at the same time were outstandingly successful, the pressure measurements were inconclusive in this regard - no significant phase differences from 180° were found. Shemdin and Hsu (1967) studied the air flow over mechanically generated water waves in the laboratory and Dobson (1971) using a very small buoy in the shallow water of Burrard Inlet, Vancouver, made an extensive series of observations and found consistently high coherences between the pressure signal and the surface displacement over a considerable range of frequencies. The results of these measurements are summarized in Fig. 12.1. In spite of the large scatter, it is evident that the largest phase differences from 180° occur when c/u_* is of the order 10. The solid curve represents the contribution calculated by Miles on the assumption that the profile is logarithmic right down to the linear region. Some further progress in the development of a properly argued theory of the response of turbulent flow to wave deformations has been made by Davis (1974) and by Gent and Taylor (1976) but until these theories are firmly based, the measurements, although they are scattered, will likely provide a better test for the theories than the theories provide an explanation for the measurements.

Energy is lost from the wave field predominantly as the result of wave breaking. In many circumstances, wave breaking is far more widespread than the occurrence of highly visible white caps. Individual wavelets, with wavelengths of the order 10 centimeters, usually break without air entrainment. At these scales in the presence of an active wind, the water surface is frequently characterized by a series of small, irregular steps where the fluid at the surface is being turned under by micro-scale breaking. The conditions of incipient breaking have been discussed by Banner and Phillips (1974) who show that the maximum amplitude that a wavelet can attain without breaking is dependent sensitively on the surface drift induced in the top few millimeters by the wind drag. This surface drift is augmented near the crest of a longer wave so that the maximum amplitude of wavelets, riding on top of the longer wave, is reduced further. In fact, Phillips and Banner (1975) show that freely travelling short wave components can be erased altogether by a longer wave in the presence of wind.

The turbulence generated in micro-scale breaking is itself of very small scale. Flow visualization methods, described by Banner and Phillips, show that a turbulent wake trails behind a small-scale breaking region, the depth of which is comparable to the wave amplitude. This turbulent motion can be expected to dissipate rather rapidly, so that much of the energy input into small-scale components from the wind is lost from the waves by breaking and then dissipated into heat close to the water surface. However, as the waves lose energy by breaking, they also lose momentum and this is communicated by breaking to the water close to the surface. As far as the generation of ocean currents is concerned, then, it matters little whether the momentum is first transferred to short wave components or whether it is communicated to the water by tangential shear. Only in the case of momentum flux to longer waves which are unsaturated and which radiate away, is the momentum from the wind not available to support the mean current.

12.4 The implications for modelling

This interplay of rather complex physical processes occurring at the water surface has some important consequences for oceanic modelling. The most direct application is to the technique of wave forecasting. Wave forecasting methods still rely heavily on empirical classification of observed data but more recently, serious attempts have been made to incorporate some of the physical effects summarized by Eq. (12.14). The most recent of these is by Hasselmann, Ross, Muller, and Sell (1976). There are, however, a number of problems which remain. The processes of energy input from the wind and dissipation from the waves are certainly very non-linear in the ratio u_*/c so that the mean energy input into the wave field is not a function simply of the mean wind speed but also of the statistics of fluctuations about this mean. The gustiness of the wind does not seem to have been included in any of the models developed so far; it is likely to be important but we have as yet no assessment of it.

Wind-generated waves, encountering shallow water or currents, can generate a set-up or set-down of the mean water level. If the current velocity is uniform with depth, the mean momentum balance can be expressed as

$$\frac{\partial \widetilde{M}_\alpha}{\partial t} + \frac{\partial}{\partial x_\beta}\{\widetilde{U}_\alpha \widetilde{M}_\beta + T_{\alpha\beta}\} = Z_\alpha + \tau_\alpha \; ; \quad \alpha, \beta = 1,2, \tag{12.16}$$

where $\widetilde{M}_\alpha$ is the total momentum of waves and current, $\widetilde{U}_\alpha$ the total volume transport of the two, $T_{\alpha\beta}$ is the radiation stress, and τ_α is the surface wind stress. The quantity

$$Z_\alpha = \rho g(h+\bar{\zeta}) \frac{\partial \bar{\zeta}}{\partial x_\alpha} \tag{12.17}$$

is the force per unit area horizontal surface area associated with the maintenance of a slope in mean water level. The wave field can be calculated from the action conservation principle (12.7) and the radiation stress is then given by Eq. (12.5). Gradients in radiation stress are therefore associated with changes in mean water level as pointed out originally by Longuet-Higgins and Stewart (1962). In considerations of storm surges, this effect is not usually considered; the magnitude of the effect is generally of the order of the root-mean-square wave slope.

In oceanic modelling, one is concerned with the fraction of the wind stress that is available to support the sub-surface motions. Fortunately, the considerations of energy and momentum flux described earlier indicate that the bulk of the surface stress is supported by the short wave components which, in a spectrum that is saturated or nearly saturated, lose momentum at a rate that is, on average, the same. The effective stress, to be taken as the upper boundary condition in a numerical model, is equal to the wind stress minus the rate at which energy is radiated away by longer waves, a quantity proportional to the mean square slope of these components. In all except the most extreme circumstances, this reduction is likely to be a small fraction of the wind stress.

A more important consequence is likely to be found in the parameterization of the heat flux. The heat transfer coefficient commonly used is the one developed for turbulent flow over a flat plate. On the ocean surface, there are many additional processes, notably small scale wave breaking, that can influence the heat transfer and suggest a parameterization that should also involve the characteristics of the wave field. Kitaigorodskii, in his monograph (1970) gives an excellent critical review of observational data and present methods of parameterization; though none can be said to be satisfactory, his book certainly gives as good an account as is possible at this stage.

Acknowledgement: I wish to thank the U. S. Office of Naval Research for its support under contract No. N00014 67A 0163 0009.

Chapter 12a

NOTE ON MATHEMATICAL MODELS OF RESIDUAL CIRCULATION IN THE NORTH SEA

François C. Ronday

The North Sea is characterized by strong tidal and storm currents superimposed on a slowly varying residual flow created by the in and out-flows of two branches of the North Atlantic current and by non-linear interactions of tides and surges.

As the residual circulation pattern evolves very slowly with time it can be regarded as a steady flow or, at most, as a succession of - say, seasonal - steady states. This can be expressed in mathematical form by considering a time T sufficiently large to cover at least one or two tidal periods - to a large extent tidal oscillations and transitory wind currents thus cancel over T - but smaller than the characteristic time of residual changes (T can go from a few weeks to a few months). The residual circulation is then defined as the mean motion over T.

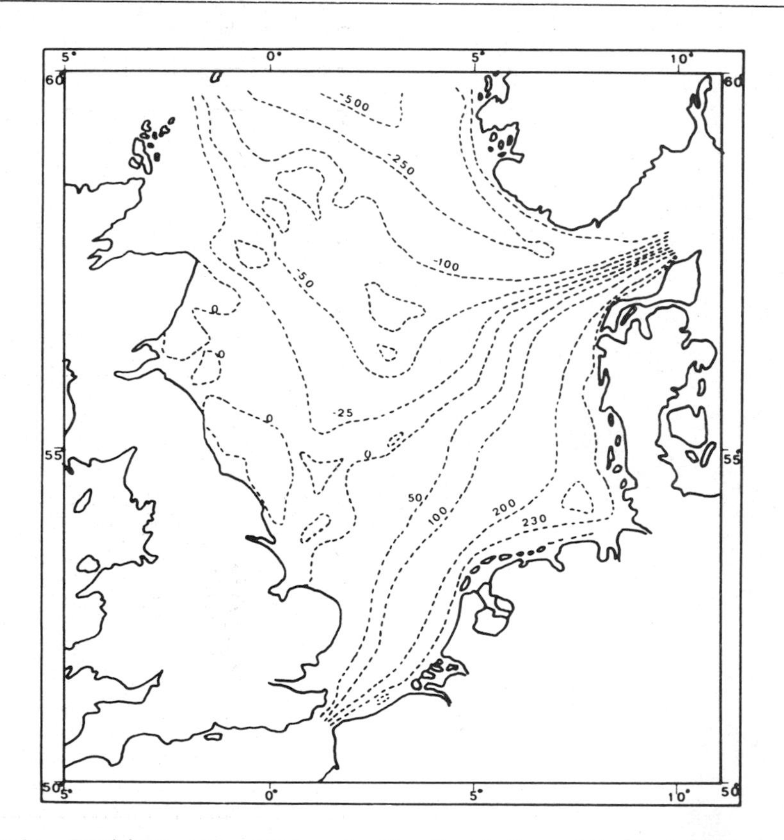

Fig. 12a.1 - Residual circulation in the North Sea with the tidal stress. Streamlines ψ=const (in $10^3 m^3 s^{-1}$); bottom friction coefficient $K_b = 3.\ 10^{-3} ms^{-1}$

In principle, one could, with appropriate boundary conditions, solve the long wave equations for the actual flow and separate the residual currents from tides and surges by averaging the solution over T.

In practice, this method must be rejected. Indeed, currents due to tides and storm surges are one or two orders of magnitude larger than the residuals. Thus the solution of the long wave equations is essentially the storm and tidal contribution; the residual part being negligible, of the same order as the error on the transient motion. By averaging the solution one would get the residual flow with a 100% error (Nihoul and Ronday, 1975).

To overcome this difficulty and reduce the error to 10%, one must (Ronday, 1975):

1) solve the long wave equations for the transient motions,

2) average these equations over T and solve the steady state resulting equations for the residual flow.

In the averaged equations, the transient motions still appear in the non-linear terms producing the equivalent of an additional stress on the mean motion. But this stress can be calculated explicitly using the results of the preliminary long wave model. (This stress has been called, in brief, the

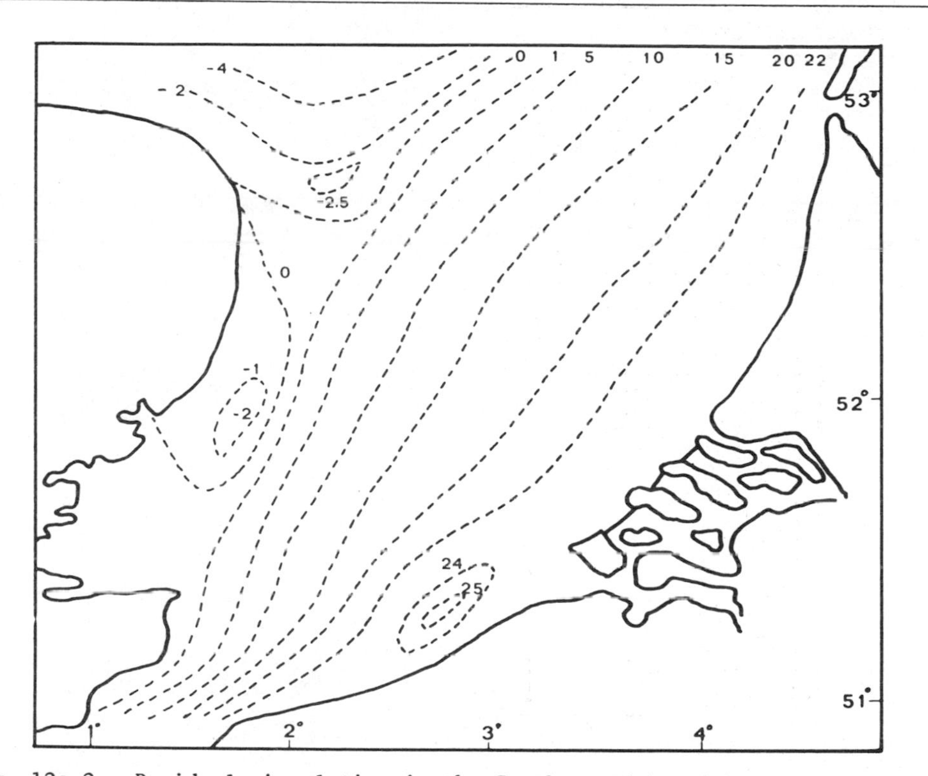

Fig. 12a.2 - Residual circulation in the Southern Bight with the tidal stress. Streamlines ψ=const (in $10^4 m^3 s^{-1}$); bottom friction coefficient K_b=3 10^{-3}m s^{-1}.

"tidal stress" to emphasize that, although wind conditions may vary from one season to the next, the tidal contribution is always present in the additional stress).

The results of these models have been described in previous publications (e.g., Math. Modelsea, 1975; Nihoul and Ronday, 1975; Ronday, 1975) and it has been shown that they are the only ones to reproduce in detail the observations.

Figure 12a.1 shows the residual circulation in the North Sea computed by the "tidal stress" model. Residual gyres, unrevealed with classical models, can be seen in various places.

Figure 12a.2 shows the circulation in the Southern Bight in more detail. The presence of a residual gyre off the Northern Belgian coast is confirmed by measurements of turbidity, sedimentation and pollutant concentrations (Math. Modelsea, 1975). It predicts a current going south along the Northern Belgian coast. This result is supported by long series of observations.

Part IV

BIOLOGICAL MODELS OF THE UPPER OCEAN

Chapter 13

ECOLOGICAL MODELLING OF THE UPPER LAYERS

John Steele

13.1 Introduction

In physical oceanographic models there are certain well-defined starting points such as the Navier-Stokes equations. In ecological modelling the major problem is that there are too many possible formulations of the basic equations for the interactions of organisms. There tend to be two methods of escape from this diversity. The first is towards very general mathematical models where only the simplest characteristics of broad classes of animals are considered (May, 1973; Maynard Smith, 1974). The second approach is to set up computer simulation models of a few particular species or trophic groups in one specific environment*. The disadvantages of both methods are obvious; the testability of the former is extremely limited; generalizations from the latter are not intended.

The major difficulty in testing theories is that even more than in physical oceanography the variance associated with any set of data is usually very much greater than the mean. By averaging of large sets of data it is possible to present a picture of mean conditions at several trophic levels; for example, the changes in nutrients, phytoplankton and herbivores during the spring outburst in the North Sea (Steele and Henderson, 1976). This is comparable to considering only the average temperature gradients from north to south in the ocean or atmosphere. It ignores the variability which, in the ecological as in the physical aspects, is of considerable practical as well as theoretical interest. It appears more and more likely that organisms, particularly fish, depend on the spatial variability of their food supplies as much as on the mean levels (Steele, 1976). Thus, sampling programs and theoretical models are becoming more concerned with the analysis and simulation of variability, than with prediction of average conditions. (See Denman and Platt, Chapter 14, for a detailed consideration of this problem in relation to phytoplankton).

13.2 Components of a model

Since there are so many possible ways of constructing ecological models, this paper will deal mainly with the building blocks and discuss the design problems, rather than give examples of the completed fabrication.

The classical problem in ecology concerns the factors determining the long term stability of populations. There have been two main components to this discussion which, although never treated in isolation, receive different degrees of emphasis. Firstly, this long term stability may depend on the cleverness of animals whose feeding or reproductive behavior produces density dependent responses which, at the population level, induce stability (e.g., Holling, 1965). Secondly, the complexity of the physical environment may play a critical role in preventing instabilities getting out of hand (Huffaker, 1958) either through mixing or dispersion of populations over a range of

---

*The Sea Vol. VI, (ed. Goldberg, McCave, O'Brien, and Steele, 1976) contains examples of the uses of simulation models in biological oceanography.

habitats, or (the other side of the coin) through relative isolation of portions of each component population.

In considering the theoretical simulation of these problems, there are, in fact, a pair of equations which can form a starting point but their main relevance is that they are unsatisfactory and the questions concern how they may be reformulated. These equations are the well-known Lotka-Volterra, prey-predator, relations (Pielou, 1969):

$$\dot{P} = aP - bPH; \tag{13.1}$$

$$\dot{H} = cPH - dH; \tag{13.2}$$

where, for this discussion, P represents phytoplankton and H the herbivorous zooplankton which grazes on the phytoplankton; a is a growth rate, d a mortality rate, and ($\dot{\ }$) indicates a time rate of change. The grazing on P by H has the simplest possible functional form, bPH, and the growth of H is some fraction of the food intake ($c<b$).

Equations (13.1) and (13.2) are unsatisfactory because they generate nuetrally stable cycles in P and H. Thus, they give the illusion of cycling in the populations but random variations introduced stochastically in this model (corresponding to environmental or population variability) will cause a random walk to extinction of either P or H (Bartlett, 1957). For this reason modifications and expansions are concerned with inducing either stable steady state solutions or limit cycles.

There are four possible types of complications which can be introduced for this purpose:

1. each of the functional relations (the four terms on the right hand sides of Eqs. (13.1) and (13.2)) can be made more realistic;

2. other equations can be added, particularly the nutrients on which the phytoplankton depend;

3. the phytoplankton or the herbivores can be considered as made up of numbers of individuals with varying responses.

Finally, and perhaps most relevant here,

4. the spatial characteristics of the physical environment can be introduced.

Each of these will be considered separately to indicate the nature of the problems arising in the construction of a simulation model. Of course any model is likely to involve all of these with varying degrees of complexity.

13.3 Functional relations

The relation of grazing by H on P given in Eq. (13.1) as $H \cdot f(P) = H \cdot bP$ is certainly too simple. From experimental work it is known that, as P increases, grazing rate by a unit of H (e.g., a copepod) tends to some upper limit. Thus the term bP could be replaced by $f(P)=bP/(\delta+P)$ as a representation of this effect.

The new pair of equations

$$\dot{P} = aP - bPH/(\delta+P) \tag{13.3}$$

$$\dot{H} = cPH/(\delta+P) - dH \tag{13.4}$$

will be unstable to small perturbations near the steady state values of P and H and the same random walk to extinction would be expected. This, in turn, can be overcome if a threshold grazing level is introduced, giving the functional response for $P>P_0$ as

$$f(P) = b(P-P_0)/(\delta+P-P_0),$$

when for P below a certain concentration the pair of equations becomes positively stable damping out small perturbations. There is, however, considerable argument about the magnitude of P_0 and thus the possible significance of this effect. Here, as for the other functional responses, the local stability or instability of the steady state $P=P^*$ does not depend on the exact nature of the functional response f(P) but only on the slope at $P=P^*$ through the relation (Steele, 1974)

$$f'(P^*) > f(P^*)/P^* \text{ for stability.} \tag{13.5}$$

For the grazing relation there is, at least, a considerable body of experimental work on which this argument can be based. The validity of the predation term dH is much more available to guesswork. We are concerned here with closing what is essentially an open system. In reality the term should be dFH where F represents the carnivores eating H. There would then need to be an equation for F and so on. Thus the problem is to parameterize F in terms of H. In the equations used so far this is done by taking F as constant and assuming it can prey proportionately on all concentrations of H. Taking F as constant could be justified for certain time scales on the assumption that, in fact, F grows very much more slowly than H. However, as with H, it could be reasonable to assume that there is some upper limit to its predation rate and so, again, this could be represented as

$$dH/(\delta'+H). \tag{13.6}$$

But again this would lead to an unstable system. Further, there are problems about the relation of P to H at steady state (P^*, H^*). The simplest herbivore Eq. (13.2) has the implication that $P^*=d/c$ and so is independent of, say, variations in its own growth rate coefficient, a. Also using Eq. (13.6) would imply that as P^* increases, H^* decreases. This might be acceptable in considering possible relations within one small environment. It goes counter to the evidence from a comparison of a range of quasi-steady environments in subtropical areas (Blackburn, 1973) where P^*, H^* and F^* increase in phase. What may appear as a minor problem of closure of the upper end of the model has a critical effect on its structure in relation to the parameterization of predation and thus to the time and space scales which the model is intended to represent.

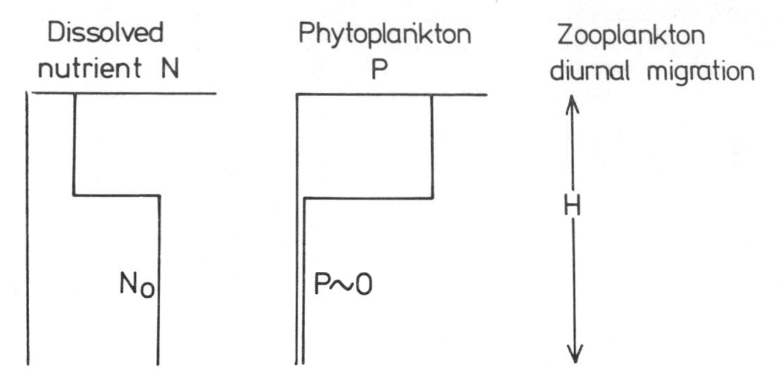

Fig. 13.1 - A simple representation of a two-layered ecosystem.

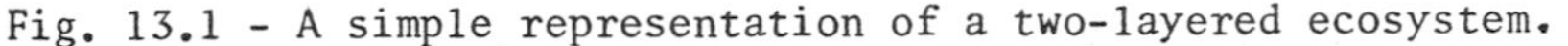

13.4 Extensions of models

The main extension required when dealing with the upper layers of the ocean is the introduction of a limiting nutrient. For simplicity, one nutrient, N, only will be considered here. This requires also the introduction of a very much simplified structure, a two-layered ocean, Fig. 13.1. Assume that primary production occurs only in the upper layer, that mixing between the layers is described by an exchange rate k for the upper layer, and that in the lower layer N is constant at N_0 and P=0. This is equivalent to assuming that the upper mixed layer is of constant depth and that at the bottom of this layer the light intensity is so low that growth of phytoplankton by photosynthesis cannot occur in the lower layer.

We know that the herbivorous zooplankton are capable of moving actively from the upper to the lower layer. They tend to do this in a diurnal cycle being mainly in the upper layer at night and the lower layer in daylight. It is assumed that they do this in a regular manner so that their grazing of the phytoplankton in the upper layer is proportional to their population density, H. Then the simplest representation following from Eqs. (13.1) and (13.2) is

$$\dot{N} = -aNP + (b-c)PH + k(N_0-N); \qquad (13.7)$$

$$\dot{P} = aNP - bPH - kP; \qquad (13.8)$$

$$\dot{H} = cPH - dH; \qquad (13.9)$$

It is assumed here that the common unit in which N, P, H are measured is nutrient content (e.g., mg nitrogen/m^3). Also the nutrients grazed by H but not used for growth are returned to the water in the upper layer. The main feature of these equations is that the steady state is now stable to small perturbations. It is of interest that the existence of positive steady state values for N, P and H depends on the condition $N_0>k/a+d/c$ involving the ratio of mixing to phytoplankton growth rate.

It is possible to extend this approach to multi-layer models (e.g., Radach and Maier-Reimer, 1975; Steele and Mullin, in press) which can be used to investigate the conditions for development of such observed features as the mid-water maximum of chlorophyll (the pigment in phytoplankton which can be used as an index of their concentration). For comparison with observation such models would require estimates of mixing coefficients (or vertical eddy diffusivity) for each layer. The difficulty (or impossibility) of obtaining such estimates is a major constraint on the usefulness of such models.

13.5 More realistic organisms

The use of simple state variables for the phytoplankton and zooplankton are obvious over-simplifications. As an example, the zooplankton should be considered as being composed of cohorts of copepods with numbers Z_i and individual weights W_i, Fig. 13.2. Thus

$$H = \sum_i W_i Z_i;$$

$$\dot{H} = \sum_i Z_i \dot{W}_i + \sum_i W_i \dot{Z}_i;$$

and this separates the herbivore equation more realistically into its growth and mortality components. In turn this allows more of the experimental information on growth to be used, but it also requires reproduction to be specified explicitly. The consequences of this complication, in a simulation model, is that it can introduce limit cycles into the response of P and H (Steele, 1974). Essentially, this is similar to the effects of the use of time delays in other ecological models.

In this example, variability in time is introduced by the basic biology of the organisms. For a single cohort there is an obvious dominant frequency. For groups of cohorts under certain simulation conditions lower frequencies can arise at approximately twice the period of the copepod's life cycle (Steele and Mullin, in press). In a model with both space and time one might expect that such temporal scales would induce dominant spatial scales. The present problem is whether such complicated space-time models are conceptually manageable.

These illustrations of the details of model building are intended to focus attention on the effects, intentional or inadvertent, which each particular functional form used for a biological component can have on the general nature of the response of the whole. These must be considered in relation to the physical factors which are also to be introduced.

13.6 Spatial effects

Thus, for the moment, I shall return to the simplest representation as a guide to the character of the responses.

$$\frac{\partial P}{\partial t} = aP - bPH + \frac{\partial}{\partial x}\left(K_x \frac{\partial P}{\partial x}\right); \tag{13.10}$$

$$\frac{\partial H}{\partial t} = cPH - dH + \frac{\partial}{\partial x}\left(K_x \frac{\partial H}{\partial x}\right); \tag{13.11}$$

where a very simple one-dimensional form for horizontal diffusion with coefficient K_x is used. The effect of diffusion is to make these equations

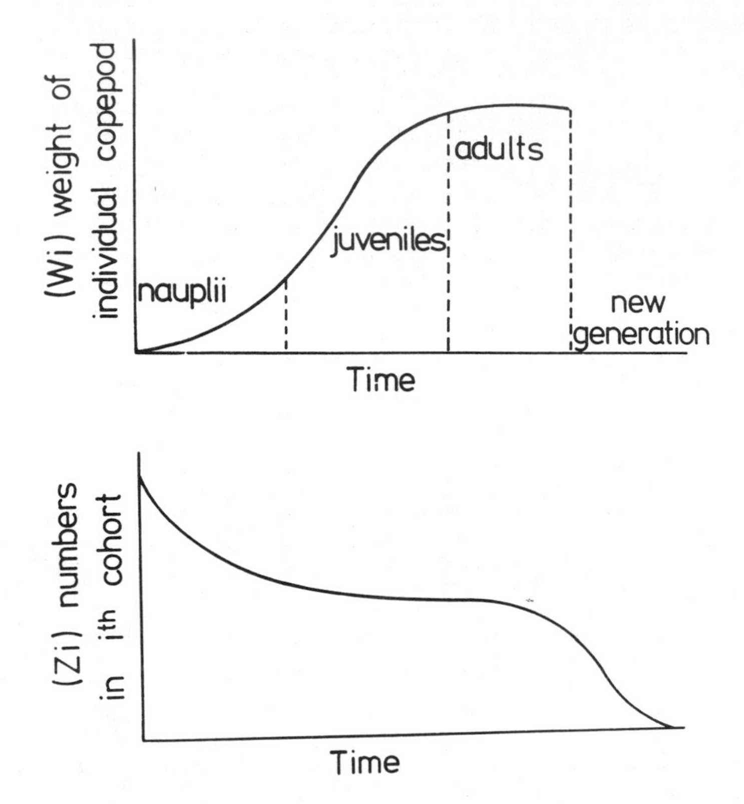

Fig. 13.2 - Schematic representation of changes with time in weight of individuals, and numbers in a cohort.

stable to any single oscillatory perturbation (Murray, 1975). In other words, the mean values of P and H over an interval (0,L) in x will still display neutrally stable cycles but, unlike the case for P on its own, when there is a critical wavelength (Denman and Platt, Chapter 14), a perturbation at any non-zero wavelength is damped out.

If we introduce a destabilizing term from the simple non-spatial form, Eqs. (13.3) and (13.4), then the two variable system has the same general character as that for P alone with a critical wavelength dividing a stable (low wavelength) region from a region of longer wavelengths where the system is unstable to small perturbations. Okubo (1974) has shown, using a mathematical approach derived from chemical kinetics (Turing, 1952) that if (a) one of the equations (say for P) has a stabilizing functional form and the other (for H) has a destabilizing form but the equations without diffusion are stable; and also, (b) there are different K's for P and H, say K_P, K_H, then, for K_P sufficiently greater than K_H, the resulting equations with diffusion are unstable for a small perturbation and, as Okubo points out, could induce patchiness. The question is whether this effect could operate on relevant spatial scales. The ecological basis for a possible difference in apparent

K's is that, whereas the plants are confined to the upper layers, the copepods migrate diurnally between the upper and lower layers.

At a very simple level, if horizontal turbulence is thought of as "real" and mainly wind induced, then turbulence could be greater in the upper layer and effectively $K_P > K_H$. If, however, one thinks of the shear and vertical mixing as dominant in the description of horizontal dispersion (Kullenberg, 1972) then it is likely that $K_H > K_P$. In this way the alternative concepts of turbulence interact with the earlier question of whether any "stabilizing" behavior of animals is at the hervibore or carnivore level.

This discussion of Eqs. (13.10) and (13.11) has been concerned solely with the linear, small perturbation aspects, as a development of the earlier concern with stability. However, the other main, and probably more important, factor is that in these equations, as in all fully developed ecosystem models, the nonlinear terms representing interactions between trophic levels are a crucial component. As a result of this, one can expect that perturbations in any small range of wavelengths will produce responses at all other wavelengths. Also, nonlinear effects will occur if one of the coefficients is perturbed at a range of wavelengths. The questions arising concern the quantitative nature of such effects. By developing the solution of Eqs. (13.10) and (13.11) in Fourier series,

$$P = \sum_{n=0}^{N} A_n \cos n x \quad ; \quad H = \sum_{n=0}^{N} B_n \cos n x ;$$

and expressing $K = K^* n^{-1}$ to give the scale effect on the diffusion coefficient, it is possible to explore the spectral character of solutions to Eqs. (13.10) and (13.11) when stochastic perturbations are included (Steele and Henderson, in press). These spectral distributions can then be compared with observations such as those of Denman and Platt (Chapter 14).

The ecological problem concerns the variety of ways in which perturbations can be introduced into the model. As mentioned earlier, the combination of vertical migration of the herbivores and horizontal velocity shear will redistribute the animals relative to the phytoplankton. Also the coefficient, a, for phytoplankton growth depends on the depth of the mixed layer and so can vary with any oscillatory motion or eddy structure imposed on the thermocline. These factors can be parameterized as inputs over a range of wave numbers with the amplitudes varying randomly with certain limits. The artificiality of this procedure is apparent and emphasizes the two inadequacies of Eqs. (13.10) and (13.11). Firstly, the representation of water motions in the upper layer as simple horizontal diffusion is not appropriate for the biological processes. Secondly, the implicit restriction to a single horizontal layer is also inadequate for the representation of the interactions between water and organisms.

The same kind of problems exist for purely vertical models with the interactions of biological and physical parameters confined to this one dimension (Steele and Mullin, in press). The particular problem which is relevant here concerns the representation of the vertical structure of the water column. Any multi-layer numerical model taking average conditions in each layer is an approximation to an environment supposed to have smooth vertical gradients in physical factors such as diffusion coefficient or phytoplankton concentration. We know that vertical structure is not smooth and three rather different questions arise: (1) could one simulate vertical structure more realistically; (2) how is this structure related to horizontal motions; and (3) how important are such structures to the plant and animal populations?

A quite different approach to these problems can be developed through large-scale experimental modelling. In Scotland and Vancouver Island, ecosystem experiments have been carried out using plastic enclosures 3 m diameter by 17 m deep to study the response of planktonic food webs in such restricted conditions.* Implicit in these experiments is the aim of finding the response when horizontal motions on scales greater than 3 m are removed; in other words, to investigate whether large-scale horizontal diffusion is an essential process in population stability. However, the restrictions imposed by the walls of the bag introduce other simplifications. From thermistor chains hung inside and outside the bags it is observed that the walls of the bags act as a low pass filter to heat exchange, damping out effectively all temperature fluctuations with frequencies less than a few hours but equilibrating inside and outside temperatures with a lag of 4-12 hours (Farmer and Henderson, personal communication).

It seemed likely that similar damping effects would occur with vertical temperature distributions. Observations using a thermistor probe have shown that, whereas a layered structure is observed outside the bags, the inside has a smooth temperature gradient, reminiscent of the curves drawn from a few discrete observations. Thus the environment inside is simplified not only horizontally but also vertically and so many approximate more closely than the outside world to the simple theoretical models. The ecological results obtained so far (e.g., Davies, Gamble, and Steele, 1975) suggest that plants and animals can survive in this simple environment with population structure not too different from the outside.

---

*The results of work in 1974 are described in the Bulletin of Marine Science (in press).

Chapter 14

BIOLOGICAL PREDICTION IN THE SEA

Kenneth L. Denman and Trevor Platt

14.1 The sea as an ecological environment

There is not the slightest doubt that physical modelling of the mixed layer is crucial to biological prediction in the sea: over most of the ocean the upper mixed layer overlaps almost exactly the layer of primary biological activity. Modelling of the mixed layer is therefore synonymous with modelling the layer of fundamental biological interest.

For the purposes of this discussion, the most significant characteristic of the upper layer of the ocean as a biological environment is its variability in space and time. We could represent this environmental variability in wave-number or frequency space extending over a very broad spectrum. Of course, the biological properties in the sea could also be represented in wave number or frequency space extending over an equally broad spectrum; we are interested then in the coupling mechanisms between the processes represented by these two spectra. But if we chose to analyze the marine ecosystem in terms of individual organisms we would find that the size spectrum of the organisms themselves extends over many orders of magnitude: the corresponding spectrum of characteristic times (generation times) for the organisms is rather less broad than the size spectrum, but intimately related to it. The inverse linear relationship between size and growth rate (adjusted for temperature) is so fundamental (Sheldon, et al., 1973) that it might be used as the basis for a predictive theory of particle growth in the sea, and provides the rationale for the hypothesis, provoked by the suggestive data of Sheldon, et al., (1972), that the concentration of biological material in the open ocean is approximately constant irrespective of size. In other words for quasi steady-state conditions in the upper layer $N_i W_i = \mu$ where N_i is the number density in the ith size category, W_i is the mean weight of a particle in the ith size category, μ is a constant to first order and the increase in size between the categories i=1, 2, ---, is geometric. The hypothesis treats particles with a range of characteristic sizes extending over eight decades. This working hypothesis would probably not be applicable (nor has it been tested) for the upper layer of the coastal zone.

In this paper, we propose to focus most of the attention on that portion of the size spectrum represented by the primary producers or phytoplankton (∿ 5 -100 μm). By virtue of its generation time (∿ 1 day) this is the group that is most closely coupled to the typical response scales of the upper layer of the ocean. And there are other reasons which justify the emphasis on this band of the size spectrum: it is the group for which we have the most information on distribution and growth rate; the short generation time is readily measurable; the growth is of the simplest possible kind (conversion of solar energy); the concentration field can be investigated by automatic measuring techniques; and finally, theoretical descriptions of the relationship between physical and biological systems in the ocean, although still rudimentary, are more developed for the phytoplankton than for any other group.

While we intend to concentrate on the size group that we believe is the most fruitful to study, we anticipate that results found for the phytoplankton will be generalizable in some way to other parts of the size spectrum, so long as we recognize that organisms of different sizes will interact with the physical environment on different time scales.

The central theme of this paper, then, will be the importance of the space-time variability of the upper layer for biological prediction, in particular for prediction about the phytoplankton community. We shall, however, make some reference to the problem of prediction for other size groups.

14.2 The general prediction equation

Biological prediction may refer to the estimation of either the biomass (abundance) of organisms in a given place at some future time or their growth rate at a given time and place. Prediction of the biomass requires a predictive equation for the growth rate:

$$B(t) = B(o) + \int_0^t P(t')dt', \tag{14.1}$$

where $B(t)$ is the estimated biomass at the future time t, $B(o)$ is the initial (measured or given) biomass and $P(t')$ is the estimated growth rate at time t'.

In practice, we estimate growth rate from the product of biomass and specific production rate:

$$P = B \cdot P^B. \tag{14.2}$$

Here P^B is the calculated growth rate normalized to the phytoplankton chlorophyll biomass B. In what follows, we consider the effect of the space-time variability of the upper layer on these two quantities B and P^B.

14.3 The growth term

All models of specific photosynthesis are empirical. The most important independent variable is the flux of available light. Since irradiance decreases exponentially with depth, the growth of phytoplankton is strongly depth-dependent. A recent study (Jassby and Platt, 1976) has shown that the most consistently useful representation of the relationship between specific growth rate and light (at light intensities below the threshold of photoinhibition) is the following:

$$P^B = P_m^B \tanh\left(\frac{\alpha I}{P_m^B}\right). \tag{14.3}$$

In this equation, α is the slope of the curve for values of irradiance near zero, and P_m^B is the specific production rate at optimal light intensity. It can be shown (Platt, et al., 1975) that all existing empirical models of phytoplankton growth may be recast in terms of these same two common parameters α and P_m^B.

The effect of other environmental variables on growth may be expressed through their influence on α and P_m^B.

It is therefore relevant to discuss the constancy of α and P_m^B. We have to remember that since they represent physiological parameters of organisms, they are capable of adaptation to changing conditions. In practice we find that, in addition to a dependence on depth, α and P_m^B change seasonally (with the succession of species), and that the seasonal change is modulated by systematic changes on a 24 hour time scale and by response to environmental changes such as those associated with passing weather systems. We do not know exactly how fast the physiological parameters can change in response to meteorological transients (e.g., the sudden and substantial reduction in irradiance due to passage of a fog bank), but experimental work now in progress promises to resolve the question.

It is worth mentioning here a mechanism of feedback between the biological and physical systems. We can write the production rate at any depth z (positive upwards) as

$$P_B(z) = P_m^B \tanh\left[\frac{\alpha I_o}{P_m^B} \exp(\gamma z)\right], \qquad (14.4)$$

where I_o is the photosynthetically active light intensity penetrating the surface and γ is an average attenuation coefficient for visible light which is depth dependent. Since the phytoplankton themselves make a substantial contribution to the attenuation coefficient, instantaneous growth is modified strongly by the products of previous growth:

$$P_B(z) = P_m^B \tanh\left\{\frac{\alpha I_o}{P_m^B} \exp\int_z^o [\gamma_W(z') + \gamma_S B(z')]dz'\right\}, \qquad (14.5)$$

where γ_W is the extinction coefficient for sea water and γ_S is the extinction coefficient per unit of chlorophyll biomass. Finally, physical models of the upper layer (e.g., Denman and Miyake, 1973b) may be quite sensitive to the choice of $\gamma=\gamma_W+B\gamma_S$, higher values of γ increasing the capacity to absorb solar radiation and thereby enhancing the stability of the layer.

In addition to temporal changes in α and P_m^B, spatial variations are also important. Both α and P_m^B decrease with depth (Platt and Jassby, 1976). Horizontal variations in P_m^B have been observed on a relatively small scale (hundreds of meters) in in-shore water (Platt and Filion, 1973). The occurrence of such variations is thought to depend on the extent to which the upper layer has been mixed by turbulent motions over the previous few generations. Fluctuations on this scale have not been observed in the open ocean, but both α and P_m^B should show variations on a scale $\sim$ 10 km in areas of intermittent upwelling, for example.

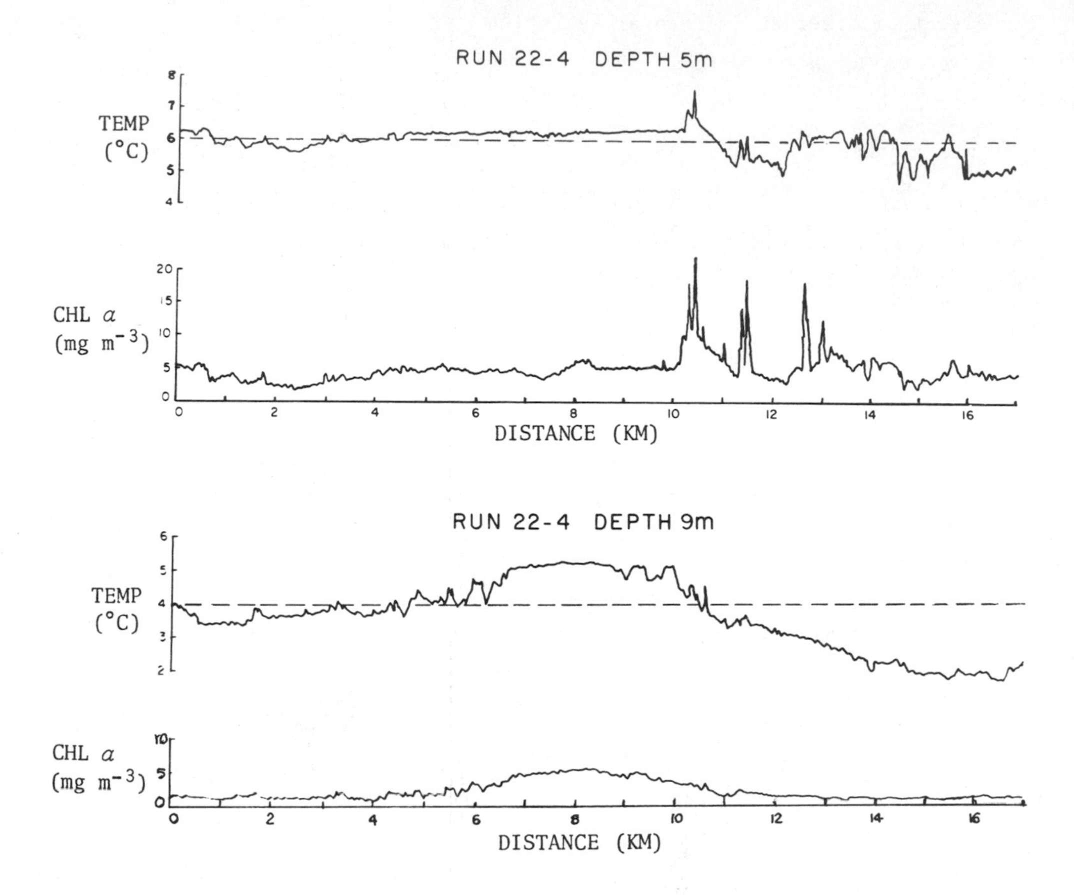

Fig. 14.1 - Series of chlorophyll *a* and temperature obtained simultaneously at depths of 5 m and 9 m along a 17 km horizontal transect (Denman, 1976)

14.4 The biomass term

There is a considerable body of evidence which indicates that the spatial distribution of phytoplankton biomass is highly non-uniform (Platt, et al., 1970; Platt, 1972; Denman and Platt, 1975; Powell, et al., 1976; Fasham and Pugh, 1976; Denman, 1976) (Fig. 14.1), over a range of length scales from a few meters up to tens of kilometers. To describe this variability mathematically requires that we formulate a differential equation for the biomass B containing, as a minimum, the effects of physical transport and phytoplankton growth. That is, we may regard the phytoplankton as a non-conservative contaminant of the diffusive-advective physical field. The main problem to resolve is how to parameterize the physical transport. All previous attempts have used the concept of a turbulent diffusion coefficient. Thus, in one dimension:

$$\frac{\partial B}{\partial t} = K_H \frac{\partial^2 B}{\partial x^2} + \kappa B, \qquad (14.6)$$

where K_H is the horizontal coefficient of turbulent diffusion and κ^{-1} is the generation time of the organisms. Such an equation leads to the formulation of a critical length scale L_c for which the opposing effects of diffusion and growth are just balanced (Kierstead and Slobodkin, 1953). Fluctuations in B on scales smaller than L_c would be damped out by diffusion while fluctuations on scales larger than L_c would persist. Using literature values for the quantities involved gives estimates for L_c from 10^2 to 10^3 m (Okubo, 1974; Platt and Denman, 1975).

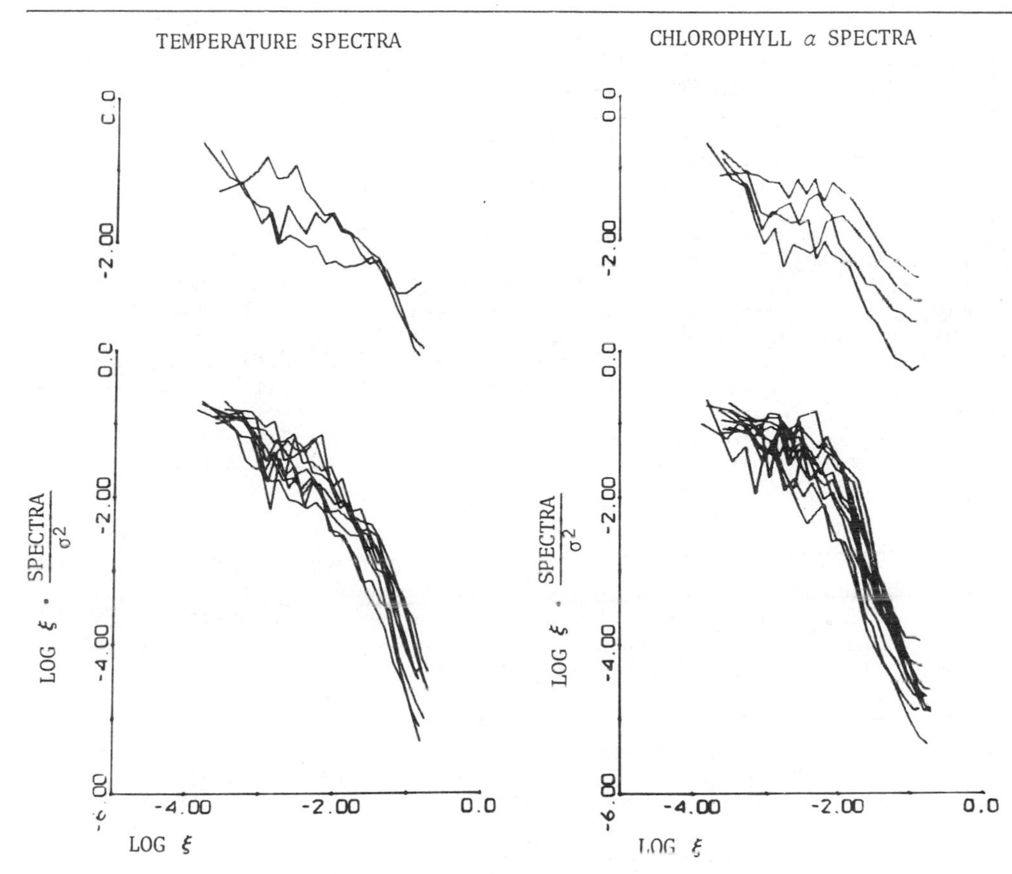

Fig. 14.2 - Normalized power spectra for temperature and chlorophyll *a* plotted against the inverse wavelength ξ, both on log scales. Each spectral estimate has been multiplied by ξ and divided by the total variance under the spectrum. Chlorophyll spectra for four runs, those with the lowest signal variances, did not exhibit the characteristic shape and were plotted separately above the rest (as were the corresponding temperature spectra) (Denman, 1976).

This approach leaves much to be desired. It precludes the possibility of growth outside a domain of size L_c, thus underestimating L_c. It would be more realistic to allow growth outside the fluctuation, but at a different rate than that inside. Again, the effect of all mortality due to grazing by herbivorous zooplankton has been neglected, leading also to an underestimate of L_c. But a more fundamental difficulty with the Kierstead and Slobodkin method is that the turbulent diffusion coefficient K_H parameterizes only those effects at length scales smaller than the one of interest. In the real world there is a continuous distribution of variance among scales (Fig. 14.2). If we express the diffusion in terms of ε, the rate of turbulent dissipation of kinetic energy (and the analogous expressions for contaminants in the flow), then we do include the cascade of kinetic energy from larger to smaller scales, and hence the deformation of fluctuations of scale L_c and larger by the large-scale features of the flow field.

However, reliable experimental data on the energy dissipation ε in the upper mixed layer are sparse (Stewart and Grant, 1962; Belyaev, et al., 1975). For prediction purposes, one would want to know how ε depends on wind speed, sea state and surface currents. Even then, it would be fortuitous if a single parameter would be sufficient to describe the dispersive characteristics of the upper layer, except in a gross, average sense.

Whatever way we may finally select to parameterize the spatial inhomogeneity of phytoplankton biomass, its study is of considerable significance to the problem of predictability. It can help to estimate the reliability of our sampling in the real world and thus to specify the probable magnitude on the error of the initial biomass condition; it can help in the formulation of the nonlinear interactions between the organisms and the medium, and between different size categories or trophic levels of organisms; it can help in the selection of the fundamental space scales and in the parameterization of spatial fluctuations for the construction of grid models for biomass; and, also in grid models, it indicates the effects on prediction at the largest scales through aliasing of fluctuations at scales smaller than the mesh size.

14.5 Predictability times

Biological prediction has many features in common with meteorological prediction. Both involve noisy non-conservative, nonlinear systems with very many degrees of freedom. In such cases, our ability to predict variance is, for practical purposes, as important as our ability to predict means.

Following the work of Lorenz (1969) and Robinson (1971) concerning fundamental limits on weather prediction, we can begin to discuss the limitations on biological prediction in the ocean using forward integration of a differential equation in the biomass. Briefly, the Lorenz and Robinson method consists in calculating the length of time taken for two states of the system, which differ initially by an amount equal to the typical observational error, to evolve into states which are as dissimilar as would be two states chosen randomly from among all possible states of the system. This is a measure of predictability time, and it is limited by the nonlinear transfer of error variance from the smallest scales via the intermediate scales to the largest scales of motion.

The predictability estimates of Lorenz and Robinson for the atmosphere cannot be carried over directly to oceanography because of differences in the shape of the kinetic energy spectrum and in the rate of dissipation of kinetic energy. Also, we would require an estimate of the spectrum of kinetic energy per unit wavenumber, but energy spectra in the ocean are usually measured in frequency space and it is not often justified to derive a wavenumber spectrum from them.

But an alternative method of Robinson (1971) is applicable to our problem. It consists in finding the time Δt taken for 50% dilution of matter contained in a circle of radius 1/2 l by diffusion through its circumference:

$$\Delta t = l^2 K_H^{-1} (\sqrt{2} - 1)^2/8, \qquad (14.7)$$

where K_H is a horizontal eddy coefficient of turbulent diffusion. These predictability times are about 50% larger than those calculated by the first method since they consider the effect of the diffusion only for scales smaller than the scale of interest (l), ignoring the transfer of kinetic energy from larger scales through the scale of interest to smaller scales.

We can make a similar calculation for the upper layer of the ocean using the empirical equation of Okubo (1971) which relates K_H to the length scale l:

$$K_H \cong 0.01\ l^{1.15}. \qquad (14.8)$$

This gives us the following estimates of Δt as a function of l:

SCALE l	PREDICTABILITY TIME Δt
10 m	10 min
100 m	1.5 hr
1 km	10 hr
10 km	3 days
100 km	3 weeks

Such an extremely simple-minded approach cannot be expected to tell the whole story, even for the physical system. In addition, for a biological population we should take into account that the reproduction of the organisms extends the lifetime of the fluctuations. Clearly, new ways should be sought to parameterize this problem.

14.6 More general equations for the biomass

The foregoing discussion has emphasized that nonlinear effects set finite limits on predictability, even if the equations of "motion" are exact. In practice, the pelagic ecosystem is so complex, even in the open ocean, that we could not pretend to be able to make an exact mathematical description. Even quite general biomass equations, then, are of limited use for prediction beyond a few days.

For theoretical and pedagogical purposes, however, generalized equations for phytoplankton biomass do have some utility. Among the most recent and best of this genre are the models of Iverson, et al. (1974), Radach and Maier-Reimer (1975) and Winter, et al. (1975). The following equation, adapted from Platt and Denman (1975), illustrates the problem:

$$\frac{\partial B}{\partial T} + \underset{\to}{V}\cdot\nabla_H B + W\frac{\partial B}{\partial Z} = K_H\nabla_H^2 B + K_V\frac{\partial^2 B}{\partial Z^2} + P_m^B\cdot B\tanh\left(\frac{\alpha I}{P_m^B}\right)$$

$$- R_m\left\{1-\exp[-\rho(B-B_o)]\right\}. \qquad (14.9)$$

Here, $\underset{\to}{V}$ and W are the mean horizontal and vertical water velocities, and the last term on the right is a modified Ivlev expression (Parsons, et al., 1967) describing the grazing by zooplankton in terms of three parameters: a rate constant ρ, a maximum ration R_m and a threshold biomass B_o below which no grazing takes place. Introducing the dimensionless variables x, y, z, t, b, $\underset{\to}{v}$, w, i defined by

$$\begin{aligned} B &= \phi b \\ X &= L_H x \\ Y &= L_H y \\ Z &= L_V z \\ T &= (P_m^B)^{-1} t \\ W &= (\nu-\upsilon) w \\ \underset{\to}{V} &= \nu(L_H/L_V)\underset{\to}{v} \\ I &= P_m^B \alpha^{-1} i \end{aligned} \qquad (14.10)$$

where ϕ is a characteristic biomass; L_H and L_V are horizontal and vertical length scales; ν is a characteristic velocity and υ is a mean sinking rate for phytoplankton, we can write the scaled general equation as

$$P_m^B\frac{\partial b}{\partial t} + \frac{\nu}{L_V}(\underset{\to}{v}\cdot\nabla_H b + w\frac{\partial b}{\partial z}) - \frac{\upsilon}{L_V}w\frac{\partial b}{\partial z} = \frac{K_H}{L_H^2}\nabla_H^2 b + \frac{K_V}{L_V^2}\frac{\partial^2 b}{\partial z^2}$$

$$+ [P_m^B\tanh(i) - \rho R_m]b + R_m\rho b_o + R_m\sum_{n=2}^{\infty}(-1)^n\frac{\rho^n}{n!}\phi^{n-1}(b-b_o)^n. \qquad (14.11)$$

Note that this equation is not as general as possible in that it contains no reference to the influence of environmental parameters such as nutrients on α and P_m^B, and it ignores the inhibitory effect of very high light intensities on photosynthesis.

The relative magnitudes of the coefficients in this equation may now be calculated (Table 14.1) using data from the literature (Platt and Denman, 1975). The tabulated ranges are rather broad, extending over several orders of magnitude in each case. This reflects the wide variety of situations studied in oceanography, from estuarine to deep ocean conditions. The table shows that, given the appropriate conditions, it is possible for any one of the processes listed (phytoplankton growth, upwelling, sinking, diffusion and grazing) to dominate the equation for $b(x, y, z, t)$.

TABLE 14.1

Relative sizes of coefficients in the general equation for biomass. These coefficients each have the dimensions of a frequency, and they may be compared with the fundamental frequency which is the phytoplankton turnover rate. The reciprocals of the coefficients give the characteristic time-scales of the processes corresponding to the terms which they multiply.

QUANTITY	RANGE (s^{-1})	PROCESS
P^B_m	$(0.5-3.0) \times 10^{-5}$	Phytoplankton turnover
ν/L_V	$10^{-9} - 10^{-3}$	upwelling
υ/L_V	$10^{-8} - 10^{-4}$	sinking
$^*K_H/L^2_H$	$10^{-8} - 10^{-3}$	horizontal diffusion
K_V/L^2_V	$10^{-7} - 4 \times 10^{-3}$	vertical diffusion
$P^B_m \tanh(i)$	$0-3.0 \times 10^{-5}$	growth at suboptimal light intensity (depth)
ρR_m	$10^{-12} - 10^{-5}$	grazing
$^{**}\phi_o/\phi$	0-1 (dimensionless)	proximity to grazing threshold

*The possible range of K_H/L^2_H is limited since the diffusion coefficient is not independent of the length-scale (e.g., see Bowden, et al., 1974).

**Note that in case $\phi_o\phi^{-1} \geq 1$, grazing is zero, by definition.

In the construction of models for particular situations (such as an upwelling system), the parameters could be specified much more closely and the various terms could be ordered, leading to a possible simplification of the equation. As an example, for an open ocean mixed layer ecosystem, Wroblewski and O'Brien (1976) have formulated a simple spatial model for phytoplankton biomass of the Kierstead and Slobodkin genre but with nutrient cycling and grazing included. They found from a sensitivity analysis that the rate of grazing by zooplankters (ρR_m in our equations) was the most important parameter in the model.

14.7 Discontinuous systems

The models which we have described treat the physical environment as continuous. A large proportion of the most productive communities, however, are found in places where the physical regime is discontinuous in some way (the major upwelling areas; most continental shelf areas; Gulf stream rings; transitory oceanic fronts, and so on). In the mathematical description of such systems it will not be easy to parameterize spatial variability in terms of average quantities such as diffusion or dissipation coefficients. Explicit modelling will be required of events at boundary regions, first for the physical fields then for the dependent biological processes. This job has barely been started.

14.8 Concluding remarks

Further development of predictability models for phytoplankton will require the close cooperation of physical oceanographers. One obvious area in which progress is required is in the time-dependent description of the response of the upper layer to changes in the surface wind stress. Again, there is accumulating evidence that in the open ocean, modelling the biological system is synonymous with modelling the response to physical transients (intermittent upwelling, passage of storms, etc.). A theoretical study of Puget Sound (Winter, et al., 1975) showed that several consecutive days of bright sunshine were sufficient to promote massive development, but that horizontal advection due to sustained winds would remove the products of growth and that a succession of cloudy days would cause phytoplankton blooms to decline. The apparently close control of phytoplankton growth by meteorological conditions suggests that a stochastic or transfer-function approach to modelling it might be rewarding. Stochastic description might help to resolve the apparent contradiction between the observed persistence of anomalous features in the ocean and the short predictability times calculated for models that use forward integration of so-called exact partial differential equations. A stochastic approach might also be the best way to model the adaptive response of the biological parameters, such as α and P^B_m, to changing environmental conditions.

Part V.

EXPERIMENTAL CONSIDERATIONS

Chapter 15

INFORMATION THEORY RELATED TO EXPERIMENTS IN THE UPPER OCEAN

J. D. Woods

15.1 Introduction

A theoretical model remains hypothetical until it has been tested against field data and shown to be valid, to within a measured accuracy, and within a clearly identified range of conditions. Yet, because of the difficulty of making adequate observations at sea, many models of the upper ocean have not yet been properly tested and any faith we may have in them rests mainly on a supposed physical analogy with better observed situations in the atmosphere or laboratory. It is scarcely necessary to dwell on the dangers of false analogy. One example will suffice to illustrate the point. Perhaps the most important error that has crept into the upper ocean models by this route has been the assumption that vertical heat and momentum transports in the top hundred or so meters are carried by turbulence, which resembles in all essentials the isotropic, homogeneous turbulence described theoretically, for example, by Batchelor (1960), with the effect of buoyancy parameterized in terms of the Richardson number (Taylor, 1931) as in Munk and Anderson's (1948) model.

The quarter of a century that has passed since that first pioneering model has seen little advance in our understanding of the physics of transport through the upper ocean and the crude parameterizations adopted in contemporary models are still justified by analogy with laboratory turbulence, rather than physical models derived from field observations. And when authors (e.g., Denman and Miyake, 1973b; Gill and Turner, 1976) compare the predictions of their models with field observations they usually follow the approach of the engineer, who asks whether (given a suitable adjustment of parameters) his model adequately predicts such useful quantities as sea surface temperature or layer depth, rather than the approach of the physicist, who seeks a critical test of the underlying assumptions of his model.

These considerations lead us to identify the following three main purposes of field observations.

1. To explore the nature of transport processes in the upper ocean in order to stimulate improvements in the physical models of the upper ocean.

2. To obtain experimental data to test the physical assumptions in models of the upper ocean.

3. To obtain climatological data against which to test the usefulness of individual models for a particular purpose.

There is overlap between these three requirements, and data sets collected for one purpose may go some way to serving the others. But it is so expensive to collect data in the upper ocean that we cannot afford the luxury of haphazard experimentation. The objectives of each investigation must be specified carefully in advance, in the context of one or more of the three aims listed above. The theory of experiments in the upper ocean is concerned with translating these objectives into practical schemes of measurement and analysis.

15.2 The information content of a data set

As we shall see in the next section, there is strong evidence that the distributions of physical, chemical, and biological properties of the upper ocean are highly intermittent on all scales in space and time, so a data set that records the variations of a selection of these properties at a particular geographical location and time must be treated as a sample of a non-stationary phenomenon. It is necessary to emphasize this non-stationarity, which is typical of most geophysical variability, in order to sound a note of warning against the overly hasty application of signal theory as developed in the standard texts such as that by Bendat and Piersol (1971), who were concerned primarily with stationary, random signals in electrical engineering. There is some evidence from meteorology (e.g., Hide, 1976) that even such cherished assumptions as ergodicity may be invalid for motions in which global quantization occurs and this effect may extend to smaller, topographically controlled flows. In the absence of adequate field evidence from the upper ocean we had better avoid relying on the ergodic hypothesis when designing experiments.

These warnings apply not only to large scale motions affected by the dimensions of the globe or ocean basin and bottom topography. Even at the smallest scales present (less than, say, 10 cm), the motions and the associated distributions of physical, chemical, and biological properties are far from the ideal of statistical homogeneity adopted as a useful assumption in early theories of turbulence (e.g., Batchelor, 1960).

Finally, in developing methods for analyzing the information content of an upper ocean data set it is necessary to deal with sharp frontal discontinuities which occur commonly on all scales from the viscous subrange (Tennekes, 1973) to the ocean (e.g., the Gulf Stream). The presence of sharp discontinuities in a data set poses problems that are seldom discussed in text books on signal theory, but they can often dominate treatment of upper ocean data. In designing an experiment we must allow for the likelihood that the resulting data set will represent a sample from a non-stationary, non-ergodic field containing sharp discontinuities. Practical considerations often force the investigator to gamble that these effects will not seriously limit his ability to interpret the data, just as he gambles on the occurrence of the desired weather conditions, the alternative strategy of designing a more elaborate experiment that would cover all possibilities being impossible.

In practice, therefore, even if all the apparatus worked perfectly during a particular experiment, it is unlikely that Nature would offer a perfectly representative example of her variability under the prevailing conditions. The data set is no more than a sample which we seek to use as a touchstone in testing our theoretical models. The construction of such a test must take account of the limitations - the information content - of the data set. The following parts of this section are devoted to reviewing techniques used in the analysis of upper ocean data sets. One important aim will be to point out some of the assumptions made (but not often stated) when experimental data are compared with theoretical predictions, i.e., when theoretical models are tested against observations of nature.

a. Local values of mean and total variance and spectrum

Consider an ideal data set consisting of a synoptic sample of some variable V at each of a series of n measuring points distributed along a straight line at equal intervals δx. The total length of the data set is $(n-1)\delta x$. Let us select i adjacent samples from within this data set centered on position x. The length of the selected section is $(i-1)\delta x=\Delta x$. The mean value of the variable V in this section is given by

$$\overline{V}(x \pm \frac{\Delta x}{2}) = \sum_{x-\frac{\Delta x}{2}}^{x+\frac{\Delta x}{2}} \frac{V(x)}{i},$$

and the total variance of the variable E_V in this section is given by

$$E_V(x \pm \frac{\Delta x}{2}) = \sum_{x-\frac{\Delta x}{2}}^{x+\frac{\Delta x}{2}} \frac{(\overline{V}(x \pm \frac{\Delta x}{2}) - V(x))^2}{i}.$$

The distribution of this total variance in the wavenumber band

$$\frac{1}{\Delta x} \leqslant k \leqslant \frac{1}{2\delta x}$$

can be estimated by Fourier analysis, which gives $\frac{i}{2}$ estimates of the form

$$E_V(x \pm \frac{\Delta x}{2} \; ; \; k \pm \delta k) = \left\{\left[\sum_{x-\frac{\Delta x}{2}}^{x+\frac{\Delta x}{2}} V'(x) \sin k\delta x\right]^2 + \left[\left(\sum_{x-\frac{\Delta x}{2}}^{x+\frac{\Delta x}{2}} V'(x) \cos k\delta x\right)\right]^2\right\}^{1/2},$$

where $V'(x)=V(x)-bx-c$

$$b = \frac{V(x + \frac{\Delta x}{2}) - V(x - \frac{\Delta x}{2})}{\Delta x},$$

$$c = V(x - \frac{\Delta x}{2}).$$

These local estimates of the mean, total variance and variance concentration at $i/2$ locations in Fourier space represent the average conditions in the sample taken from the total data set. In the absence of other information, or hypotheses, they cannot be assumed to be at all representative of the corresponding statistics of the whole data set (length $(n-1)\delta x$) and do not reveal anything about the spatial distribution within the sample $x + \frac{\Delta x}{2}$.

b. The maximum resolution spectrum

In the preceding section it was stated that the Fourier transform of a sample $(x \pm \frac{\Delta x}{2})$ of the data set yields an estimate of the average distribution of variance within the band $(\frac{1}{\Delta x} < k < \frac{1}{2\delta x})$ for the sample as a whole, but does not tell us anything about the spatial distribution of these spectral components within the sample. In order to discover this, it is necessary to subdivide the sample into a series of sub-samples, each of which is analyzed in the same way as the original sample. This technique of subdivididing a data set into sections in order to observe any variation in the variance density in any narrow spectral band is often carried out in the analysis of oceanographic time series.

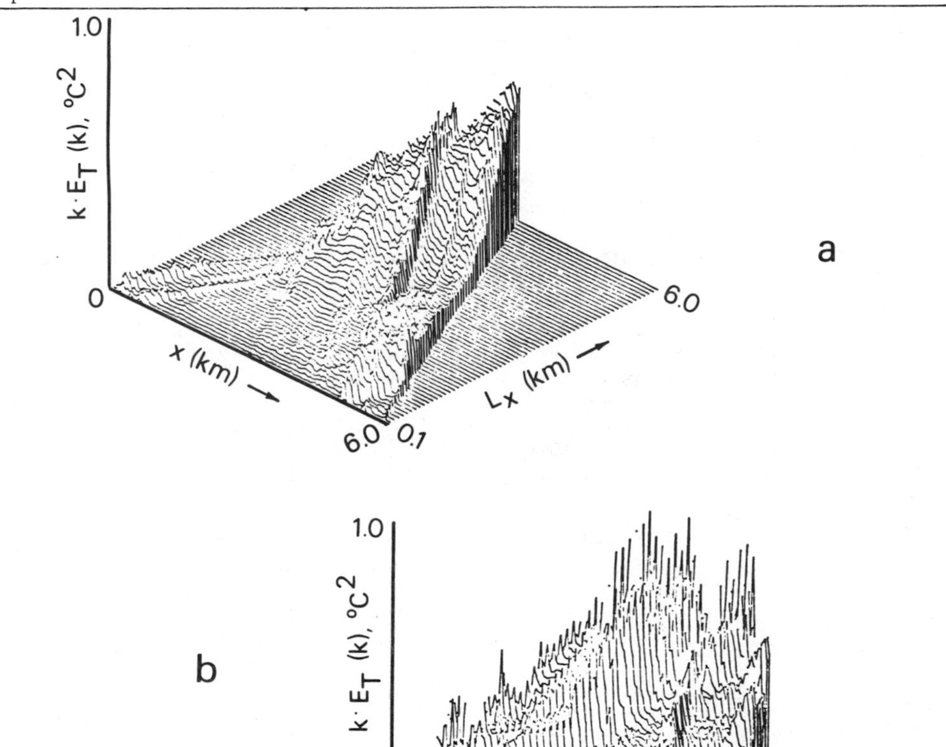

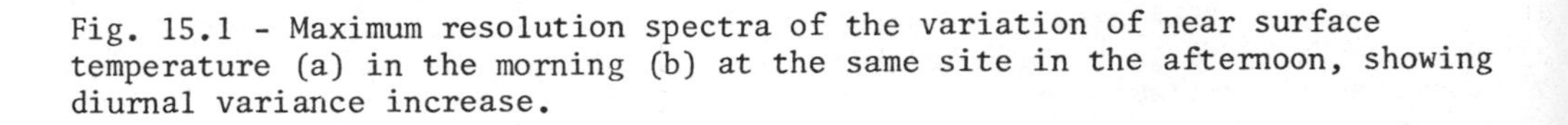

Fig. 15.1 - Maximum resolution spectra of the variation of near surface temperature (a) in the morning (b) at the same site in the afternoon, showing diurnal variance increase.

Usually the choice of sub-sample lengths is chosen to suit the needs of a particular test, but Woods (1975) has proposed a method of systematic scanning of the full data set with a sample whose size is increased incrementally before each scan. For any given sample size it is only necessary to keep the first few spectral estimates (i.e., those for low wavenumbers) since the higher wavenumber data will be given later as the first few spectral estimates during subsequent scans through the data set at smaller sampler sizes. The distribution $E(x; k)$ is thus built up for all positions, $0 < x < (n-1)\delta x$, and wave numbers, $\frac{1}{(n-1)\delta x} \leq k \leq \frac{1}{2\delta x}$. This <u>maximum resolution spectrum</u> contains all the information in the original data set which can readily be resynthesized from the distribution $E(x;k)$. But if the fast Fourier transform is used to derive the lowest wavenumber spectral estimates for every sample size Δx and position x, the estimates are distributed unevenly in the x-k plane. This can be overcome by smoothing and interpolation, using a method that takes account of the uncertainty relation to yield a surface of known ΔE (e.g., $\frac{\Delta E}{E}$ = constant is obtained by smoothing with Δx, Δk both constant). The use of the fast Fourier transform in calculating the maximum resolution spectrum involves much unnecessary computation, but it is not unacceptably long on a large computer. For example, the calculation, with samples stepped through the data at one-eighth of their size, took five seconds on an ICL 1906A computer (Fig. 15.1). If only the first spectral estimate is required for each sample size and position, the algorithm given in Appendix 15.1 can be used.

c. The uncertainty relationship

It was emphasized in the preceding section that the distribution of variance in wavenumber space calculated by Fourier analysis gave <u>i/2 estimates</u> of variance concentration. Signal theory (e.g., Bendat and Piersol, 1971, p. 199) tells us that the probable error $\frac{\Delta E}{E}$ of these estimates is given by the uncertainty relationship

$$\frac{\Delta E}{E} (\Delta x \cdot \Delta k)^{1/2} \geq 1^*$$

where Δk is the bandwidth over which the estimate was averaged. If we can put

$$\Delta k = j\delta k$$

then j is number of original spectral estimates that were averaged to achieve the bandwidth Δk. If $j=1$, there is no band smoothing and the uncertainty $\frac{\Delta E}{E} \geq 1$, since $\delta k = 1/\Delta x$. So the individual spectral estimates have an uncertainty of at least 100% of their value.

The error can be reduced by averaging a number of estimates, derived either from the same sample (i.e., estimates at different wavenumbers) or from a number of samples. It is common practice to average over adjacent estimates in wavenumber space, so that a continuous wave band of width $j\delta k$

---

* = 1 if one neglects bias error, which depends upon the local spectral slope.

is covered, or in physical space, so that a continuous section of the original data set, of length $k\Delta x$ is covered, but contiguity of samples is not necessary. Any ensemble of samples from physical or wavenumber space can be selected for averaging with a view to reducing ΔE. The success of this "ensemble averaging" depends upon the statistical properties of the data set and upon clear specification of what one is trying to achieve by averaging.

d. Conditional sampling

Selection of samples from physical or Fourier space-time for averaging to reduce the uncertainty ΔE must be based on clearly specified conditions. The selection process, called conditional sampling, is often used in studies of laboratory turbulence (see, e.g., Bradshaw, 1972). The condition is specified by the terms of the proposed test of a particular model.

To illustrate this point let us consider an "engineering" test of the Pollard, Rhines, and Thompson (1973) model, in which a comparison is made between the predicted transient deepening of the mixed layer associated with the passage overhead of a mid-latitude depression, with data from ocean weather stations. Assuming that the mixed layer depth h can be calculated from the experimental data to sufficient accuracy according to some rule that is consistent with the theory, then the data can be reduced to a time series $h(t)$. The distribution is then sampled according to the condition required by the test, namely that a storm was passing overhead. The local spectrum for a particular sample may have insufficient information content to provide an unambiguous test, so the individual spectra are averaged (taking care to keep them in phase with the changing wind field, as required by the theory) to produce an ensemble spectrum $\overline{E}(t, \omega)$ with far smaller uncertainty, and therefore a better chance of providing a convincing test of the theory.

The corresponding "physical" test of the Pollard, Rhines, and Thompson model in which attention is focused on the correlation between the mixed layer depth, on the one hand, and the phase relation between the mixed layer "slab-like" current and the wind, on the other hand, requires similar conditional sampling. And the same approach would be followed in more sophisticated tests in which a data set of the JASIN type (Pollard, 1975) is used to investigate the physical basis for the parameterization of turbulent transport adopted in a particular model.

e. Extension to four dimensions

The analysis described above for a sample from a one-dimensional data set is easily extended to a time series of synoptic observations made at equal time intervals δt throughout a rectangular* array of instruments in three dimensions, with equal spacings δx, δy, δz respectively.** The local distribution of variance $E(\underset{\to}{x},t \ ; \ \underset{\to}{\kappa},\omega)$ is calculated for the sample in physical space-time

$$x \pm \frac{\Delta x}{2} \ ; \ y \pm \frac{\Delta y}{2} \ ; \ z \pm \frac{\Delta z}{2} \ ; \ t \pm \frac{\Delta t}{2}$$

---

* or an equivalent spherical array of instruments with equal spacings in radius, δr, latitude, $\delta\phi$, and longitude, $\delta\theta$.

** often it is more convenient to work in potential density rather than depth increments, in which case σ_θ replaces z in this section.

The analysis provides no knowledge of internal variance distribution within this region.

The distribution of variance in Fourier space-time is limited to the following spectral window

$$\frac{1}{\Delta x} \leq k \leq \frac{1}{2\delta x}; \quad \frac{1}{\Delta y} \leq l \leq \frac{1}{2\delta y}; \quad \frac{1}{\Delta z} \leq m \leq \frac{1}{2\delta z};$$

and $\frac{1}{\Delta t} \leq \omega \leq \frac{1}{2\delta t}$

The uncertainty principle in four dimensions is*

$$\frac{\Delta E}{E}(\Delta x \cdot \Delta y \cdot \Delta z \cdot \Delta t \cdot \Delta k \cdot \Delta l \cdot \Delta m \cdot \Delta \omega)^{1/2} = 1.$$

The frequency ω can be either the encounter frequency in an Eulerian analysis, or the wave frequency or reciprocal of the wave packet lifetime or turbulent eddy lifetime in a suitable Lagrangian analysis phase locked to a particular variance concentration, as proposed by Woods (1975).

f. Analysis of a vector variable

The technique for resolving the variability into physical and Fourier space-time described above for a scalar variable may be extended to vector variables. The vector series can be analyzed as a complex variable using the fast Fourier transform method, or it can be resolved into a pair of orthogonal component (scalar) series.

An interesting alternative approach has been developed by Gonella (1972). It is supposed that a vector time series of, say, horizontal current can be synthesized by summing a series of vectors each of constant amplitude $|u(\omega)|$ and rotating about a vertical axis at a constant frequency ω, with the sign convention that positive ω represents clockwise and negative anti-clockwise rotation. Ther series is calculated by the Fourier transform

$$\underline{U}(\omega) = \frac{1}{\Delta t}\int_0^{\Delta t} \underline{u}(t)e^{-i\omega t}\,dt = |u(\omega)|e^{i\psi(\omega)},$$

where $\psi(\omega)$ is the phase of the ω component at the start of the sample. Gonella (1972) presents a variety of statistical relationships for the rotating vector series produced by this analysis. These are useful mainly in studying the generation of inertial oscillations, which appear in the spectrum at ω=+f (i.e., a clockwise rotation of the current with a frequency f) but not at ω= - f.

15.3 Inadequate data

Almost invariably the information content of the data set at one's

---

* neglecting bias error which depends on the shape of the spectrum.

disposal is inadequate for the test that one has in mind. By inadequate it is meant that the data cannot be used to measure the performance of the model because the uncertainty in the test parameter derived from the data is larger than the difference between it and the prediction of the model. It becomes necessary to incorporate into the test a series of assumptions about the statistical properties of experimental data and about the variability of Nature as a whole, from which the sample was obtained. Many of the assumptions depend upon analogy with the atmosphere or laboratory experiments rather than direct evidence from the upper ocean. And some of them are so deeply entrenched into the minds of oceanographers that conclusions are often drawn without a clear statement of the assumptions that have been adopted. When challenged about this, many boundary layer meteorologists and oceanographers object that it would be tedious to repeat continuously the assumptions made in interpreting the data. Yet the inclusion of a brief statement about such assumptions need to be no more tedious than the theoretical author's statement that he has made, for example, the Boussinesq assumption in his model.

This section is devoted to describing some of the more commonly encountered assumptions of experimental data analysis, together with the problems they seek to overcome.

a. The limited size of the spectral window of a data set

While natural variability in the upper ocean extends over the wide ranges of space and time given in Table 15.1, experimental data often cover only a restricted set of the four dimensions and only a small fraction of these ranges. The spectral windows for a selection of common used data sets are summarized in Table 15.2. The first four-dimensional data set with more than a nominal bandwidth in the horizontal directions was collected during the GATE C-scale experiment (Woods, 1976), but its spectral window is still only a tiny fraction of the whole range (Fig. 15.2). This situation will remain little changed while observations depend primarily upon _in situ_ sensors: the long term possibilities offered by penetrative remote sensing (e.g., Apel, Byrne, Proni, and Charnell, 1975; Munk and Woods, 1973), while encouraging, remain to be clarified. Meanwhile, experiments must be designed and data sets analyzed and interpreted in the light of their limited spectral windows.

TABLE 15.1

Ranges of Scales of Variability in the Upper Ocean

Dimension	Maximum Size	Minimum Size	Ratio
Meridional	~1000 km	1 mm	$\sim 10^9$
Zonal	~1000 km	1 mm	$\sim 10^9$
Vertical	100 m	1 mm	10^5
Temporal	10 years	1 second	3×10^8

Clearly an experimental data set contains no information about variability lying beyond the boundaries of its spectral window. Furthermore, variability

TABLE 15.2

Spectral Windows for Some Data Sets from the Upper Ocean

Data set	n_x	δx	n_y	δy	n_z	δz	n_t	δt	Variable
Tide gauges	-	-	-	-	-	-	10^5	1 h	ζ
WHOI Site D	-	-	-	-	-	-	10^6	15 m	$\underset{\to}{u}$, T
OWS	2	500 km	2	500 km	100	1 m	10^4	1/2 d	T
GATE Equatorial Experiment					400	1 m	5	2.5 days	$\underset{\to}{u}$,T,S
GATE C-scale Experiment:									
outer array	8	25 km	8	25 km	6	10 m	10^3	10 min	$\underset{\to}{u}$,T
inner array	9	10-175 m	-	-	10	≃10 m	10^3	10 min	$\underset{\to}{u}$,T
mobile survey*	40	1/2 km	6	4 km	10^3	10 m	7	10 h	T,S

*after application of Taylor assumption and interpolation.

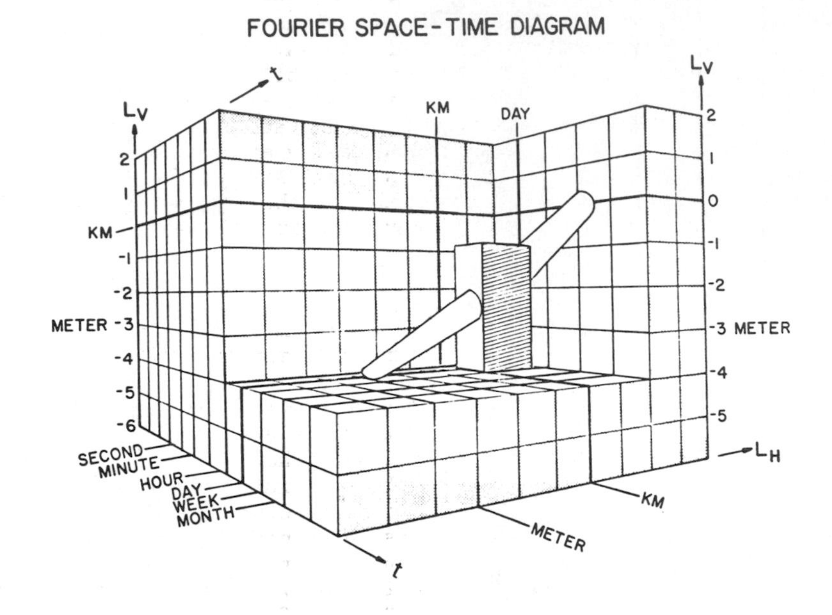

Fig. 15.2 - Comparison of the spectral window for the Lagrangian time series of batfish surveys during GATE (the inner box) with the zone occupied by turbulence (the cylinder) according to Woods (1974a).

encountered in the various dimensions is uncorrelated except in the case of those motions of nature that occupy positions inside the spectral window. This is important because the appropriate test of a physical model of variability in the upper ocean normally depends upon identifying the correct correlation between space and time variables.

For example, discrimination between variability due to internal waves à la Garrett and Munk (1975) and turbulence à la Woods (1974a) depends critically upon their respective dimensional correlations, i.e., the dispersion and characteristic relations of waves and their equivalents for eddies. The spectral window must cover the combinations (k, l, m, ω) predicted by the two theories if it is to test them.

b. The effect of variability outside the spectral window

Assuming, for the time being, that the data set does have a spectral window at the right location, it is necessary to consider the effect of natural variability outside it. Perhaps the most familiar effect is aliasing due to unresolved small-scale variations. In general it is necessary to assume that significant fluctuations of velocity, temperature and salinity occur on all scales down to a limit set by molecular viscosity, conduction and diffusion. Gregg and Cox (1972) and Grant, Moilliet, and Stewart (1959) have achieved remarkably successful results in extending the resolution of sensors down to these limiting microscales, which lie at around a millimeter, but such microstructure observations are extremely difficult and have so far not been combined into multi-dimensional data sets of the kind needed to test theoretical models of the upper ocean. Normally the small-scale boundaries of the spectral window of a data set fall a long way short of these molecular limits leaving a more or less broad range of variability unresolved. This small scale structure lies aliased in the data set, with the possibility that it significantly contaminates estimates of the variance distribution inside the spectral window. The magnitude of the aliasing error depends upon the magnitude and distribution of the unresolved small-scale variations. A variety of assumptions are involked in support of the hypothesis that the error is not serious in the analysis of a particular data set.

c. The red spectrum assumption

The most common assumption is that the spectrum of the small-scale variations do not significantly depart from a known climatological spectrum in which normally the variance density decreases rapidly with increasing wavenumber/frequency, i.e., the climatological spectrum is "red". The benefit to be gained from this assumption is that any aliasing error will decrease rapidly as wavenumber /frequency decreases inside the window, and the fractional error will decrease even faster if the spectrum inside the window is also red and continuous like the climatological spectrum. This assumption is more likely to be valid for analysis that is concerned with an ensemble estimate of the variance distribution in the window, but it is likely to be unreliable when the analysis is concerned with the local distribution, for which intermittent departures of the small-scale "noise" from the mean may significantly increase the aliasing error, and in particular the fractional error where the "signal" in the window is locally reduced at a transient spectral gap.

d. The spectral gap assumption

When the bandwidth of the data set is so narrow that serious aliasing errors would arise from even a rapidly decreasing spectrum beyond the spectral window, an alternative assumption is made, namely that there is no variance on scales that would, if present, be aliased. In other words, during the period of the experiment there was a spectral gap with a width at least equal to the width of the spectral window of the data set and adjacent to it on the small-scale side. This assumption echoes the spectral distribution of highly intermittent turbulence pictured by Woods (1975) and conjecture by many authors that internal waves may occur as intermittent packets in the upper ocean, but it remains virtually untested except in special circumstances (e.g., Moen, 1973). Nevertheless, the spectral gap assumption lies at the heart of many published analyses of upper ocean data sets. To quote a recent example, Duing, et al. (1975) conclude in a preliminary report that variations in the velocity-temperature-density distribution encountered during the GATE equatorial experiment indicate the presence of a wave on the equatorial undercurrent with λ=1800±200 km, τ=16±2 days and c=4.5±1.5 knots westward. The data are consistent with this conclusion, but it depends on the rejection of alternative conclusions on the basis of the spectral gap hypothesis. In order to justify this assumption it will be necessary to put forward evidence from other supporting data sets not included in the preliminary analysis, or to supply theoretical arguments for the absence of variability on scales that might have aliased the analysis. Of course, the assumption that the variability calculated on the straightforward assumption of no aliasing is more likely to be correct than the assumption that an apparently clear signal is all due to aliasing. Nevertheless, while the possibility of an alternative interpretation remains, the data set is less valuable for testing theoretical models.

e. Transformation from time to space

Synoptic data from arrays of instruments provide the ideal spatial series, but their bandwidths are inevitably narrow simply because the number of instruments is limited. In contrast, the frequency bandwidths of time series from rapidly sampling instruments are very much broader. The temporal changes recorded by such an instrument are due in part to the advection past it of spatial inhomogeneities and it is often useful to use the following simple transformation from time to spatial series: $\lambda = u/\omega$, where λ is the wavelength, u the advection velocity and ω the encounter frequency.

This transformation is based on the famous Taylor hypothesis that the time scale τ for evolution of structures of dimension λ is so long that there is no significant change during its passage through the instrument, i.e., $u >> \lambda/\tau$. The ratio λ/τ for turbulent eddies is of order 1 cm s^{-1} in the buoyancy/rotation ranges, so a relatively weak current meets the condition for moored instruments, but this is not the case for internal waves whose phase speeds of order 1 m s^{-1} usually exceed the current speed. When the instrument is towed through the water by a ship moving at up to 5 m s^{-1} there remains a significant Doppler distortion due to the wave propagation. Sometimes, when the waves occur as isolated packets (e.g., the tidally generated wave packets observed during GATE by Proni, see Woods, 1976), it is possible to measure the wave speed by running the ship back and forth through them along reciprocal courses, and then to reconstruct the synoptic picture by applying a Doppler correction $\lambda = 2u/(\omega_1 + \omega_2)$, there ω_1 and ω_2 are the encounter frequencies along the reciprocal runs. The component of wave celerity c along the ship's track is $c = u(\omega_2 - \omega_1)/(\omega_2 + \omega_1)$.

f. The Taylor hypothesis in two and three dimensions

So far we have considered extraction of spatial information in one dimension from a time series obtained from an instrument moving through the water. A natural extension is to tow the instrument back and forth over some kind of regular grid, with a view to mapping the spatial distribution in two dimensions, or in three if the instrument undulates (batfish) or comprises a line of sensors (e.g., a thermistor chain). The aim is again to make the Taylor hypothesis that the structure does not significantly change during the survey and hence to interpret the time series as a synoptic spatial array of measurements.

In practice it may take several hours to complete a survey of this kind, and during this interval the structure may change significantly because of internal wave propagation, steady current shear and temporal changes in the wave field current due to internal interactions and surface forcing. Furthermore, if the ship navigates relative to a fixed (Eulerian) framework, the structure of the water in the survey area will also change due to advection through the boundaries.

The last source of error can be minimized by advecting the survey area with the average speed of the water; Doppler distortion can be reduced by the method described for analysis in one dimension, and it should also be possible to correct a steady shear distortion, given appropriate current measurements.

This leaves the modified Taylor condition that there should be no temporal changes in the structure due to internal interactions or boundary exchange during the survey. Any scheme designed to recreate a synoptic picture from the survey data contaminated by these errors would involve assumptions about the physical processes causing the changes and this means that the data cannot be used to test models of these processes.

g. Interpolation

Analysis of field data is greatly simplified if they are arranged in regularly spaced arrays, but it is seldom possible to collect the data in this fashion. However carefully the experiment was designed, moorings invariably end up in the wrong place and ships fail to navigate precisely along the prescribed tracks. The technique of interpolation is used to estimate the most probable value of the variable at the intersections of the regular array needed for analysis from nearby measurements in the actual, rather irregular array. It is equally well suited to the time domain, where interpolation can be used to yield estimates of synoptic distributions from non-synoptic data collected either haphazardly or in a serial scan through a survey area (see above).

An ideal interpolation scheme for a particular investigation must steer the difficult course between the Scylla of losing information contained in the original data set and the Charybdis of introducing new information other than as the direct result of explicit assumptions concerning the physical nature of the system under study. The last phrase is all important. If we knew nothing at all about the nature of variability in the upper ocean, it would be impossible to find a safe passage between the dangers of losing detail by oversmoothing the data, on the one hand, and corrupting the data with artifacts, on the other. Fuglister (1955) has drawn attention to the great differences of interpretation that arise from making different assumptions about the structure of, in his case, the Gulf Stream. The art of the physicist is to make an inspired guess about the structure present in the system he is studying, in other words, to propose a model that is susceptible to experimental verification. Interpolation is an essential step in testing the model, but to be successful, the interpolation scheme must be based on physical assumptions which may share features with the model. The possibilities of self delusion are obvious; to extend the metaphor, we must program our computer to follow strict rules ignoring the Sirens' call, while we keep our minds open to the possibilities their song suggests. In this section we consider a number of different methods of interpolation and the assumptions on which they rest. The discussion is necessarily brief; for a more detailed account readers are referred to recent meteorological literature (e.g., Gandin, 1965; Bengtsson, 1975; Morel, 1973).

i. Bandwidth consideration

To illustrate the essentials of interpolation let us consider an irregularly spaced scalar data set in one dimension. In regions where the data are relatively more closely spaced, the information available extends locally to higher wavenumbers than elsewhere and it is possible to gain access to this high wavenumber information through such methods as maximum resolution spectral analysis (see Section 15.2b). As yet no objective interpolation technique has been developed to retain such local variations of spectral content in creating a regularly spaced data set.

At present the standard methods assume that the spectral properties of the data set are homogeneous; that is to say, local variations of data concentration produce corresponding variations of bandwidth in samples drawn from the same spectrum. A simple approach is then to design the interpolation scheme to produce a regularly spaced distribution with a constant bandwidth. The high wavenumber cutoff of the band is then chosen to fall in the range covered by the original data. The pessimist steers close to Scylla by choosing his cutoff to occur at the lowest value in the original data set, (i.e., where the data are sparsest) thereby losing information everywhere else. The optimist steers close to Charybdis by choosing his cutoff to occur at the highest value in the original data set (i.e., where the data are most densely packed), thereby introducing high wavenumber noise - albeit with the universal spectrum - everywhere else. A better method would be to design an interpolation scheme that adjusts the bandwidth to the correct local value everywhere. Even this improvement does not produce perfect interpolation while the assumption of spectral homogeneity is retained, but our discussion of the principles of interpolation will be helped by doing so for the time being.

ii. Errors due to aliasing

The regular distribution produced by a method following the general principles outlined above will have no information beyond some high wavenumber cutoff, which may be constant or locally adjusted. But the original data set consisted of a series of irregularly spaced samples of a natural distribution which, of course, contained fluctuations at higher wavenumbers. These are aliased in the data set and it should be remembered that the interpolated distribution will be equally in error unless some criterion is adopted for the reduction of aliasing in the original data set by eliminating "bad" data values.

iii. Elimination of "bad" data

Because the stage of interpolation is so crucial to correct interpretation of the data, it is important that the criteria adopted for rejecting "bad" data should be consistent with the assumptions incorporated into the interpolation scheme. The usual reason for rejecting a measurement is that it lies outside the range of values expected at that position on the basis of a comparison with one of the following:

1. neighboring values in the same data set;

2. a forecast value for that position;

3. the climatological mean value for that position.

The acceptable range is determined on the basis of what is physically possible given the known spectrum of the observed variable and other related variables at that position. The same spectrum is used in objective interpolation.

h. Adjustment

The interpolated data set is only an estimate of the distribution of measured parameters presented in a convenient form for analysis. It contains errors that were in the original data due to instrumental limitations and aliasing, and the interpolation routine may have added noise. These errors may lead to internal inconsistencies in the interpolated data set, according to some assumption about the physical properties of the phenomena it represents. For example, we may have reason to believe that the motion in the spectral window of the interpolated data set should be in geostrophic balance, whereas inspection may reveal that this is not so of the data set. A number of schemes may have been proposed by meteorologists to modify the interpolated data set automatically to establish the required balance. While existing data do not yet justify such adjustment in upper ocean data sets it seems likely that this situation will be changed in the near future.

15.4 Experimental design

a. Strategy for testing models

The starting point in any experimental design is to consider the predictions to be tested. In an engineering test, these take the form of correlations between two "engineering" variables, e.g., wind speed and layer depth. But for a physical test, it is also necessary to consider a third "physical" variable representative of the physical process assumed in the model to link cause (e.g., wind speed) and effect (deepening of the mixed layer). And when the test is designed to discriminate between the predictions of two or more models in which different physical assumptions are made, the number of physical variables must be correspondingly increased. Each of these physical variables may be derived from a whole set of field measurements.

The specification for the experimental data set is that it should contain sufficient information to permit calculation of these correlations with an uncertainty that is less than the differences between the correlations predicted by the various models. Methods of estimating the uncertainties of experimental correlations have been presented above in Sections 15.2 and 15.3. These form the basis for transforming experimental strategy into the tactics of experimental design.

b. Experimental tactics

Let us start this final section by briefly reviewing the various schemes that have been used to measure space-time variations in the upper ocean.

i. Moored arrays

Eulerian time series from moored arrays sampled synoptically (or interpolated in the time domain to synoptic times) provide unambiguous spatial distributions in physical and/or Fourier space. But they are heavy consumers of ship time and often involve losses of data and equipment. The bandwidth of the array is usually narrow, because of the limited number of instruments, but it can often be extended by transforming encounter frequency to wavelength using the Taylor hypothesis. But when this is true the array cannot measure the lifetimes of inhomogeneities advected through it. So, moored arrays can seldom provide simultaneous information in Fourier space and time, making discrimination between internal waves and turbulence difficult. Nevertheless, this may be possible with consistency tests of the type developed for IWEX.

Fourier line. In this extreme case, all instruments are arranged along a straight line with the interval between successive pairs increased according to some systematic progression (e.g., logarithmic). Since each spacing occurs only once there is no information about the spectral distribution in physical space, but the mean spectrum from the whole data set has the widest possible bandwidth for the number of instruments in the array. Uncertainty in the spectral estimate can only be reduced by smoothing over wavenumber . Example: Albers, 1965, p. 72.

Fourier cross in two or three dimensions. The local extension of the Fourier line to two or three dimensions, this array has the same advantages and disadvantages as its one-dimensional counterpart. Examples: GATE F1 mooring; Soviet Atlantic 1970 experiment.

Nested Fourier triangles. This requires one third more instruments than the Fourier cross for the same spectral information, and the redundancy provides some information of spectral distribution in physical space, which can, if desired, be averaged to reduce the uncertainty in spectral estimates. Example: GATE C-scale outer array.

Regularly spaced arrays. This array provides the optimum compromise between information in physical and Fourier space, when it is desired to avoid making any assumption about the structures that will be encountered. It permits interpolation from spatial to other (e.g., isentropic) coordinates. Example: Meteorological synoptic network.

Clumped arrays. These are the spatial equivalent of reqular bursts of rapid sampling in the time domain. There is a spectral gap between scales resolved by the whole array and within the clumps. Example: possible design for POLYMODE.

ii. Lagrangian tracers

The ideal of tracking an ensemble of Lagrangian tracers is particularly attractive, since it overcomes the major objection to a moored array; namely, that eddies resolved in space by the array drift through it too swiftly to be resolved in time. In principle the data return per unit ship time should also be higher than for moored arrays.

Neutrally buoyant floats. Swallow floats, developed for use in the deep ocean, might equally well be used to measure the flow in the upper ocean (Pollard, 1975) but this has not yet been tried. Assuming that the technical problems of float design and tracking can be solved, the major question is what is the optimum injection routine. Real time analysis of float positions is needed to identify any unacceptably large gaps that appear in the array as the floats are dispersed by the turbulent fluctuation in current. A particularly attractive possibility is to inject the initial array and to replenish it by aircraft. The array would be tracked acoustically from moored hydrophones. The distribution of the floats in the array will change as they disperse and will, of course, have to be transformed to a regular grid by interpolation. Otherwise, the experimental design and data interpretation are essentially identical to that for the fixed array. Example: in deep water, MODE (Freeland,Rhines,and Rossby, 1975).

Drogues. Parachutes, crosses, and square sails suspended at the required depth on a line hanging from a surface float which can be tracked visually, or by radar or LOCATE transonder, provide a convenient tracer for flow in the upper ocean, but a number of studies have shown that even the best of them are dragged through the water by the support wire and surface float at speeds of a few kilometers per day, the precise value depending upon the magnitude of current shear above the drogue, and the wind speed. However, Saunders (1976) has pointed out this error may be less than the errors in current meter measurements near the surface.

Dye tracers. A number of investigators have measured the dispersion of a patch of dyed water by surveying the dye concentration with fluorimeters lowered from roving ships. Although, in principle, this technique can be used to interpolate the velocity field between moored current meters or Swallow floats using the method of Saunders (1974) described below, it has so far only been used to estimate diffusion parameters from the bulk extension of the dye patch (see, e.g., Okubo, 1971) and to gain qualitative insight into the relative importance of lateral intrusion versus vertical mixing (e.g., Kullenberg, 1974). The design of ship tracks to achieve efficient surveys of the dye concentration follows essentially the same arguments as those presented below in paragraphs *iii*.

Sea surface temperature patterns. Saunders (1973b) has developed a method for calculating the flow field from displacements of isotherms mapped on successive aircraft surveys of sea surface temperature and limited number of spot measurements of current velocity at moorings. Morrice (1974) has used a similar approach to estimate the convergence at the surface outcrops of fronts. In both methods, corrections are applied for changes of sea surface temperature due to insolation and and heat exchange with the atmosphere.

Isotherms and isohalines below the surface. Woods (1974c) has proposed methods of extending the Saunders technique to motions along isentropic (constant σ_θ) surfaces in the batfish surveys described below.

iii. Surveys by roving ships and aircraft

Ships dropping XBTs or towing thermistor chains or undulating vehicles such as batfish instruments with CTD and, in the future, current meters, are often used to survey the three-dimensional distribution of the measured variables by criss-crossing the area in a series of (usually straight) sections. As was pointed out above (Section 15.3f), the main problem is that the pattern may be advected or distorted during the time taken to complete the survey.

However, roving ship surveys produce far more data per unit of ship time than any other method* and the risk of data and instrument losses are minimal, so there is a large incentive to develop effective tactics for measurement and analysis.

A series of sections fixed relative to the seabed. This method, the natural descendant of the ocean wide sections of classical oceanography, runs into difficulty when the spacing between sections becomes less than 100 km, as is required by testing upper ocean models. The problem is that the water mass being surveyed may be displaced by a significant fraction of the spacing between sections during the time taken to complete each one. At worst, as happened in operation Mayfrost, the fraction may exceed unity, and the ship's track relative to a frozen turbulence advection of the water bears no resemblance to the ground track. On the other hand, with a launch moving at 30

---

*During GATE the ratio between moorings and roving data returns per unit ship time was 1 : 10,000 for the RRS "Discovery".

knots or an aircraft moving at 100 knots (Woods and Watson, 1970) the fraction is small and may be easily corrected in analysis.

A fixed pattern of parallel sections advected with the water. During GATE, Woods (1974c) made three time series of water mass surveys, each of which consisted of a series of parallel legs two nautical miles apart and 20 km long, navigating the ship relative to the water (which was drifting at a mean speed of about 35 km/day). Each survey in the series took approximately half a day to complete, so the current displaced the water by an amount roughly equal to the size of the survey area; in the terms used above, the fraction was almost exactly unity. The (Nyquist) spatial resolution along the tracks, determined by the batfish cycle distance, was approximately 1 km.

Zig-zag pattern adjusted on the basis of real time analysis. When the aim is to survey a front or some other feature that has a signature that can be readily identified each time it is encountered, the alignment, position and length of each successive section can be adjusted in the light of real time analysis of the data. This is a technique that has frequently been adopted in surveys of the Gulf Stream (e.g., Fuglister, 1963). The obvious drawback is that the spectral window of the data set cannot easily be controlled, so unexpected small-scale displacements of the front may be aliased or at worst a false continuity may be assumed between similar but separate features encountered during successive sections.

Controlled pattern with real time phase locking. Given real time data analysis, it should be possible to improve the method of fixed parallel sections described above, so that navigation errors or distortion of the pattern by steady or time varying shear may be recognized and the survey area (or shape) adjusted to phase lock onto a particular feature of the distribution being measured.

15.5 The experimental test of a model

The ingredients of an experimental test of a theoretical model are the following:

i. An experimental data set;

ii. The model to be tested;

iii. A standard against which the model is to be tested.

The test consists of demonstrating that the model predicts certain critical correlations occurring in the data set more accurately than does the standard method, or another competing model. If there is no standard to compare the model with then the test is to demonstrate that the model prediction comes significantly closer to the experimental correlations than one could have achieved by taking the climatological mean or, in the case of a time series, better than persistence, where the word "significantly" is interpreted in terms of the climatological statistics for the test correlation.

It is necessary to restate this familiar definition of a test, if only because it has never been satisfactorily applied to tests of models of the upper ocean. It is not sufficient to present curves of, for example, sea surface temperature variation or mixed layer depth, (i) observed during an experiment and (ii) predicted by a particular model for the wind and other conditions encountered during that experiment. The experimental and model curves may appear similar, but until they have been compared statistically according to the criterion written above, they do not constitute a test of the model.

Ultimately the test reduces to a comparison of differences between distributions in space-time of:

i. The experimental value of the test variable after assumptions A, B, C . . . have been made in processing the data,

ii. the theoretical value of the test variable derived from the model under test,

iii. the theoretical value of the test variable derived from the standard method, and

iv. the climatological value of the test variable, with its climatological mean spectrum.

An experimental test is only valid in those regions of Fourier space-time within which the spectral windows of the *model* and *standard* distributions overlap that of the *experimental* one. This is often difficult to achieve; and, indeed, our knowledge of the climatological statistics of variability in the upper ocean is so scanty that the available spectral window of the *climatological* distribution may not overlap that of the *model* distribution.

The *model* and *standard* distributions can now be analyzed into variance distributions in physical and Fourier space-time using the maximum resolution spectrum method. Corresponding to each variance distribution will be an error distribution in the same physical and Fourier dimensions. The magnitude of this error depends solely upon the *uncertaintly relationship* as expressed in Section 15.2c above. By averaging and smoothing in physical or Fourier space-time, it is possible to reduce this uncertainty at the expense of resolution in those dimensions.

The *experimental* distribution can be similarly analyzed using the maximum resolution spectrum method, but in this case the corresponding error distribution depends not only upon the uncertainty introduced during this stage of analysis, but also the uncertainty in the *experimental* distribution itself deriving from previous stages of data processing. These two sources of error should be combined into an *experimental total error* distribution in physical and Fourier space.

The *climatological* distributions can be analyzed in the same way as the *experimental* distribution to give variance and total error distributions in physical and Fourier space-time.

The first experimental test consists of comparing the differences between the four variance distributions with the error distributions. Clearly the test cannot succeed if the errors are larger than the differences, because that implies that the uncertainties in model and standard predictions and experimental and climatological descriptions are too large for any definite conclusions to be drawn. If this is the case, further averaging/smoothing or the introduction of more restrictive assumptions in data processing may sufficiently reduce the errors. When the errors are smaller than the differences, it is possible to compare the differences between the experimental distribution, on the one hand, and the model, standard, and climatological distributions, on the other, to discover whether or not the model under test is superior to the standard or climatological method of prediction.

So far the test has explored whether or not the model produces better predictions of changes in the distribution of, for example, sea surface temperature or mixed layer depth, from the viewpoint of the length and time scales of the variations. Such a test might reveal that model X is better than model Y at predicting mixed layer depth changes that occur on the seasonal time scale, while Y is better than X in coping with changes close to the inertial frequency. Our ability to close the gap between these two widely separate frequencies (1 year versus 1 day) depends upon the errors in the test distribution and the degree of smoothing needed to reduce them to acceptable levels.

After this basic test one can explore the ability of the model to predict the correct phase relationship between, say, wind and mixed layer depth at frequencies that pass the variance test. Other detailed tests can be devised to suit particular models.

15.6 Conclusion

Models of the upper ocean are designed to predict correlations between two or more variables such as wind and mixed layer depth. Inevitably, any particular model can only produce useful predictions over a limited range of space and time scales. The aim of this paper has been to develop a theoretical framework for testing models, and for collecting and processing experimental data in such a way that they can be used in such tests. While no satisfactory tests of upper ocean models have been published at the time of writing (1975), it seems likely that improvements in experimental techniques may make it possible to do so within the next decade. Meanwhile, many of the ideas presented in this paper can be developed using the best available data, such as the GATE C-scale data set.

Appendix 15.1 Maximum resolution spectrum

This algorithm, due to Moen (unpublished, 1975) can be used to obtain the first spectral estimate for each sample size and position. It is computationally much faster than the complete maximum resolution spectrum method discussed in Section 15.3b.

Let a_n^j and b_n^j be the nth Fourier cosine and sine coefficients, respectively, for the jth iteration; the window size is N. a_n^{j+1} and b_n^{j+1} are then given by

$$a_n^{j+1} = \cos\left(\frac{2\pi n}{N}\right)\left[a_n^j + 2\left(\frac{y_{N+1} - y_j}{N}\right)\right] + \sin\left(\frac{2\pi n}{N}\right) b_n^j;$$

$$b_n^{j+1} = \cos\left(\frac{2\pi n}{N}\right) b_n^j - \sin\left(\frac{2\pi n}{N}\right)\left[a_n^j + 2\left(\frac{y_{N+1} - y_j}{N}\right)\right].$$

The spectral estimate for the jth iteration is given by

$$[k\,E_T(k)]^j = \frac{j+1}{2} \sum_{n=1}^{j} \left[\frac{(a_n^j + \delta a_n^j)^2 + (b_n^j + \delta b_n^j)^2}{2}\right],$$

where the slope terms δa_n^j and δb_n^j are given by

$$\delta a_n^j = S_j \left(\frac{N-1}{N}\right);$$

$$\delta b_n^j = S_j \left(\frac{N-1}{N}\right) \left[\frac{\sin\left(\frac{2\pi n}{N}\right)}{1 - \cos\left(\frac{2\pi n}{N}\right)}\right];$$

and the slope S_j is calculated by joining the end points of the window

$$S_j = \frac{y_{N+j-1} - y_j}{(N-1)\delta y}.$$

LIST OF SYMBOLS

Due to the wide range of topics covered in these papers, it is not possible to have a set of symbols which is unambiguously one-to-one with the physical quantities discussed. Therefore, this list is neither complete nor absolute; there are exceptions to virtually every character listed here and many symbols are not included. The aim was to use common symbols, which are listed below, whenever possible. Some of these have more than one meaning, and these are listed.

A	albedo; wave amplitude; horizontal area
A_c	latent heat of condensation of water vapor
a	absorption coefficient; constant of proportionality
B	buoyancy flux; Bowen ratio
b	buoyancy
C	drag coefficient, with various subscripts
c	specific heat of sea water ; wave velocity
c_p	specific heat of air at constant pressure
c_i	internal wave velocity
$\vec{c}_g$	wave group velocity
D	undisturbed ocean depth
d	friction coefficient
E	variance; wave energy density
F	flux
$\vec{F}$	frictional accelerations
f	Coriolis (inertial) frequency ($=2\Omega \sin \phi$)
g	gravitational acceleration
H	sensible heat flux; herbivore biomass
h	mixed layer depth
I	solar irradiance, with various subscripts
i	$\sqrt{-1}$; index
$\vec{i}$	(1,0,0): x-direction unit vector

J	radiance
j	index
$\vec{j}$	(0,1,0): y-direction unit vector
K	eddy viscosity or diffusivity, with various subscripts
k	x-direction wavenumber
$\vec{k}$	(0,0,1): z-direction unit vector
L	length scale, with various subscripts; in particular, Monin-Obukhov length
l	length scale, in particular mixing length; y-direction wavenumber
m	empirical proportionality constant, with various subscripts; z-direction wavenumber
N	Brunt-Vaisala (buoyancy) frequency
n	proportionality constant; cloud area fraction
P	precipitation rate; phytoplankton biomass
P_e	Péclet number
p	pressure
Q	evaporation rate
q	turbulence kinetic energy (various exact formulations)
R	reflectance; vertical radiative flux distribution
Re	Reynolds number
Ri	Richardson number
Ro	Rossby number
r	specific humidity
r_e	mean earth radius
S	salinity; stability functions
s	proportionality constant
T	temperature; time period or scale
T_V	virtual temperature
t	time

U	velocity scale
u	eastward velocity; general horizontal velocity
u_*	friction velocity
$\vec{u}$	(u,v,w) three-dimensional velocity
v	northward velocity
$\vec{v}$	(u,v,0) horizontal two-dimensional velocity
w	upward velocity
w_e	entrainment velocity
x	eastward coordinate; general horizontal coordinate
$\vec{x}$	(x,y,z): radius vector
y	northward coordinate
z	upward coordinate
z_o	roughness length
α	thermal expansion coefficient for sea water; index; specific volume; proportionality constant
β	df/dy; saline expansion coefficient for sea water; index; proportionality constant
γ	extinction coefficient for solar radiation; proportionality constant
Δ	finite-difference jump of a quantity
δ	small increment
δ_{ij}	Kronecker delta function
ε	dissipation
ε_{ijk}	alternating tensor
ζ	surface displacement; interfacial displacement; solar zenith angle
η	angle with respect to vertical
$\vec{\eta}$	unit vector parallel to earth rotation axis
θ	potential temperature; angle with respect to vertical; scattering angle

κ	von Kármán constant
$\underset{\rightarrow}{\kappa}$	(k,l,m) wavenumber vector
λ	wavelength; longitude
μ	diffusivity
ν	kinematic viscosity
ξ	solar elevation angle
ρ	density, with various subscripts
σ	standard deviation; density anomaly
τ	stress; time scale; transmissivity
$\underset{\rightarrow}{\tau}$	stress vector
υ	apparent (Doppler shifted) frequency
ϕ	latitude; geopotential
ψ	streamlines
Ω	earth rotation rate
ω	frequency; solid angle

SUBSCRIPTS

$(\ )_a$	relating to atmosphere, e.g., u_a is wind speed
$(\ )_b$	bottom quantity
$(\ )_H$	horizontal quantity
$(\ )_M$	relating to momentum, e.g., K_{MH} is horizontal momentum mixing coefficient (eddy viscosity)
$(\ )_r$	reference quantity; relating to humidity
$(\ )_V$	vertical quantity
$(\ )_\theta$	relating to sensible heat, e.g., C_θ is bulk coefficient for sensible heat flux
$(\ )_o$	surface/mixed layer quantity; reference quantity

OPERATORS

∇	$(\frac{\partial}{\partial x}, \frac{\partial}{\partial y}, \frac{\partial}{\partial z})$

∇_H $\left(\frac{\partial}{\partial x}, \frac{\partial}{\partial y}, 0\right)$

$\nabla_{\vec{\kappa}}$ $\left(\frac{\partial}{\partial k}, \frac{\partial}{\partial l}, \frac{\partial}{\partial m}\right)$

Any quantity can be represented as the sum of an average and the corresponding deviation:

$(\ ) = \overline{(\ )} + (\ )'; \quad \overline{(\ )'} \equiv 0$

ACRONYMS

AMTEX	Air-Mass Transformation Experiment
APEX	Atlantische Passatwind Experiment
ATEX	Atlantic Tradewind Experiment
AXBT	Air Drop Expendable Bathythermograph
BOMEX	Barbados Oceanographic and Meteorological Experiment
CUE	Coastal Upwelling Experiment
FNWC	Fleet Numerical Weather Central
GARP	Global Atmospheric Research Program
GATE	GARP Atlantic Tropical Experiment
GFDL	Geophysical Fluid Dynamics Laboratory
GISS	Goddard Institute for Space Studies
IAPSO	International Association for the Physical Sciences of the Ocean
IWEX	Internal Wave Experiment
JASIN	Joint Air/Sea Interaction Project
JONSWAP	Joint North Sea Wave Analysis Project
MODE	Mid-Ocean Dynamics Experiment
NATO	North Atlantic Treaty Organization
NCAR	National Center for Atmospheric Research
OWS	Ocean Weather Station
STD	Salinity, Temperature, Depth Recorder
WHOI	Woods Hole Oceanographic Institution
WMO	World Meteorological Organization
XBT	Expendable Bathythermograph

REFERENCES

Cross-references within this volume are not included here. Numbers in square brackets [ ] indicate page citations in this volume.

Albers, V. M. (ed.), 1967: Underwater Acoustics, 2, Plenum Press, New York. [278]

Allen, J. S., 1973: Upwelling and coastal jets in a continuously stratified ocean, J. Phys. Oceanogr., 3, 245-257. [179]

Allen, J. S., 1975: Coastal trapped waves in a stratified ocean, J. Phys. Oceanogr., 5, 300-325. [181]

Apel, J. R., H. M. Byrne, J. R. Proni, R. L. Charnell, 1975: Observations of oceanic internal and surface waves from the earth resources technology satellite, J. Geophys. Res., 80, 865-881. [119], [270]

Arthur, R. S., 1965: On the calculation of vertical motion in eastern boundary currents from determinations of horizontal motion, J. Geophys. Res., 70, 2799-2803. [200]

Assaf, G., R. Gerard, and A. L. Gordon, 1971: Some mechanisms of oceanic mixing revealed in aerial photographs, J. Geophys. Res., 76, 6550-6572. [103], [104], [111], [114]

Atkins, W. R. G. and H. H. Poole, 1952: An experimental study of the scattering of light by natural waters, Proc. Roy. Soc. London, Ser. B, 140, 321-338. [62]

Badgley, F. I., C. A. Paulson, and M. Miyake, 1972: Profiles of Wind, Temperature, and Humidity over the Arabian Sea, The Univ. Press of Hawaii. [84]

Baines, P. G., 1974: On the drag coefficient over shallow water, Boundary-Layer Meteorol., 6, 299-303. [88]

Ball, F. K., 1960: Control of inversion height by surface heating, Quart. J. Roy. Meteorol. Soc., 86, 483-494. [149], [160], [163], [164]

Banner, M. L. and O. M. Phillips, 1974: On the incipient breaking of small scale waves, J. Fluid Mech., 65, 647-656. [235]

Barnett, T. P. and J. C. Wilkerson, 1967: On the generation of wind waves as inferred from airborne measurements of fetch-limited spectra, J. Mar. Res., 25, 292-328. [233]

Bartlett, M. S., 1957: Competitive and predatory biological systems, Biometrika, 44, 29-31. [244]

Batchelor, G. K., 1960: The Theory of Homogenous Turbulence, The Univ. Press, Cambridge. [263]

Begis, D. and M. Crepon, 1975: On the generation of currents by winds - an identification method to determine oceanic parameters, (unpublished manuscript) Mus. Nat. d'Hist. Naturelle. [107]

Belyaev, V. S., M. M. Lubimtzev, and R. V. Ozmidov, 1975: The rate of dissipation of turbulent energy in the upper layer of the ocean, J. Phys. Oceanogr., 5, 499-505. [256]

Bendat, J. S., and A. G. Piersol, 1971: Random data: analysis and measurement procedures, John Wiley & Sons, New York. [264], [267]

Bengtsson, L., 1975: Four-dimensional assimilation of meteorological observations, GARP Pub. No. 15, WMO, Geneva. [275]

Berliand, T., 1960: Metodika climatologicheskih raschetov radiatsii, Meteorologia i Hydrologia, n°6, 9-16. [50], [51]

Bethoux, J. P., 1968: Adaptation d'une thermopile à la mesure de 'éclairement sousmarin, Thèse 3ème cycle, Faculté des Sciences de Paris. [70], [71]

Bjerknes, J., 1966: A possible response of the atmosphere Hadley circulation to equatorial anomalies of ocean temperature, Tellus, 18, 820-829. [23]

Bjerknes, J., 1969: Atmospheric teleconnections from the equatorial Pacific, Monthly Weather Rev., 97, 163-172. [34]

Black, P. G. and W. D. Mallinger, 1972: The mutual interaction of hurricane Ginger and the upper mixed layer of the ocean, 1971 Project Stormfury Annual Report, Dept. of Commerce, 63-87. [222]

Blackburn, M., 1973: Regressions between biological oceanographic measurements in the eastern tropical Pacific and their significance to ecological efficiency, Limnol. Oceanogr., 18, 552-563. [245]

Boston, N. E. J., 1970: An investigation of high wave-number temperature and velocity spectra in air, Ph.D. thesis, Univ. of British Columbia. [80]

Boston, N. E. J. and R. W. Burling, 1972: An investigation of high-wavenumber temperature and velocity spectra in air, J. Fluid Mech., 55, 473-492. [79]

Bowden, K. F., M. R. Howe, and R. I. Tait, 1970: A study of the heat budget over a seven-day period at an oceanic station, Deep-Sea Res., 17, 401-411. [103], [105]

Bowden, K. F., D. P. Krauel, and R. E. Lewis, 1974: Some features of turbulent diffusion from a continuous source at sea, Adv. Geophys., 18, 315-329. [259]

Bradshaw, P., 1972: An introduction to conditional sampling of turbulent flow, IC Aero Report 72-18, 1-11. [268]

Bradshaw, P., D. H. Ferris, and N. P. Atwell, 1967: Calculation of boundary-layer development using the turbulent energy equation, J. Fluid Mech., 28, 593-616. [233]

Brand, S., 1971: The effects on a tropical cyclone of cooler surface waters due to upwelling and mixing produced by a prior tropical cyclone, J. Appl. Meteorol., 10, 865-874. [217]

Brennecke, W., 1921: Die ozeanographischen Arbeiten der Deutschen Antarktischen Expedition, 1911-1912, Arch. dt. Seewarte, 39, 206ff. [103], [104]

Bretherton, F. P., 1971: The general linearized theory of wave propagation. Mathematical Problems in the Geophysical Sciences. Am. Math. Soc., 61-102. [230]

Bretherton, F. P., 1975: Recent developments in dynamical oceanography (Symons Memorial Lecture), Quart. J. Roy. Meteorol. Soc., 101, 705-722. [178]

Bretherton, F. P. and C. J. R. Garrett, 1968: Wave-trains in inhomogeneous moving media, Proc. Roy. Soc. London, Ser. A, 302, 529-554. [117], [230]

Briscoe, M., 1975: Preliminary results from the tri-moored internal wave experiment (IWEX), J. Geophys. Res., 80, 3872-3884. [119]

Brocks, K., and L. Krügermeyer, 1970: Die hydrodynamische Rauhigkeit der Meeresoberfläche, Berichte Nr. 14 des Inst. Für Radiometeorol. und Maritime Meteorol., Univ. Hamburg. [87]

Bryan, K., 1969: A numerical method for the study of the circulation of the world ocean, J. Comp. Phys., 3, 347-376. [9]

Bryan, K. and M. D. Cox, 1967: A numerical investigation of the oceanic general circulation, Tellus, 19, 54-80. [13]

Bryan, K. and M. D. Cox, 1968a: A nonlinear model of an ocean driven by wind and differential heating: Part I. Description of the three dimensional velocity and density fields, J. Atmos. Sci., 25, 945-967. [11], [12], [19], [21]

Bryan, K. and M. D. Cox, 1968b: A nonlinear model of an ocean driven by wind and differential heating: Part II. An analysis of the heat, vorticity and energy balance, J. Atmos. Sci., 25, 968-978. [11], [19], [21]

Bryan, K. and M. D. Cox, 1970: Vertical mixing and ocean circulation, Presented at IAPSO Symposium, Tokyo. [135]

Bryan, K., S. Manabe, and R. C. Pacanowski, 1975: A global ocean-atmosphere climate model. Part II. The oceanic circulation, J. Phys. Oceanogr., 5, 30-46. [16], [19], [32]

Bryant, P. J., 1966: Ph. D. Dissertation, Univ. of Cambridge. [233]

Budyko, M. I. 1956: The heat balance of the Earth's surface, Hydrometeorologicheskoe izdatelstvo, Leningrad. [50]

Budyko, M. I., 1963: Atlas of the heat balance of the world, Mezhd. Geofiz. Komitet Prezidiume Akad. Nauk SSSR. [57], [58]

Burt, W. V., 1953: A note on the reflection of diffuse radiation by the sea surface, Trans. Am. Geophys. Un., 34, 199-200. [55], [56]

Busch, N. E., 1973a: The surface boundary layer, Boundary-Layer Meteorol., 4, 213-240. [72], [79], [81]

Busch, N. E., 1973b: On the mechanics of atmospheric turbulence, Workshop on Micrometeorology (D. A. Haugen, ed.), Am. Meteorol. Soc., Boston, 1-65. [73], [76], [78]

Busch, N. E. (ed.), 1973c: Turbulence spectra at scales smaller than 1 meter, Boundary-Layer Meteorol., 5, 211-217. [82]

Businger, J. A., 1973: Turbulent transfer in the atmospheric surface layer, Workshop on Micrometeorology (D. A. Haugen, ed.), Am. Meteorol. Soc., Boston, 67-100. [81]

Businger, J. A., 1975: Interactions of sea and atmosphere. Rev. Geophys. and Space Phys., 13, 720-726, 817-822. [72], [79], [86]

Businger, J. A., J. C. Wyngaard, Y. Izumi, and E. F. Bradley, 1971: Flux-profile relationships in the atmospheric surface layer, J. Atmos. Sci., 28, 181-189. [81]

Calathas, J., 1970: Contribution à l'étude de la réflexion de la lumière du jour à la surface de la mer, Thèse 3ème cycle, Université de Paris, VI. [56], [58]

Carl, D. M., T. C. Tarbell, and H. A. Panofsky, 1973: Profiles of wind and temperature from towers over homogenous terrain, J. Atmos. Sci., 30, 788-794. [81]

Charney, J. G., 1955: The generation of oceanic currents by wind, J. Marine Res., 14, 477-498. [182]

Charnock, H., 1958: Recent advances in meteorology, Science Progress, 46, 470-487. [87]

Chervin, R. M., W. M. Washington, and S. H. Schneider, 1976: Testing the statistical significance of the response of the NCAR general circulation model to north Pacific Ocean surface temperature anomalies, J. Atmos. Sci., 33, 413-423. [34]

Clancy, R. M. 1975: The mesoscale interaction between coastal upwelling and the sea breeze circulation. M.S. thesis, Univ. of Miami. [190]

Clarke, A. J., 1975: Coastal upwelling and coastally trapped long waves, Ph.D. thesis, Cambridge Univ. [199]

Clarke, R. H., and G. D. Hess, 1975: On the relation between surface wind and pressure gradient, especially in lower latitdues, Boundary-Layer Meteorol., 9, 325-339. [89]

Coantic, M., and B. Seguin, 1971: On the interaction of turbulent and radiative transfers in the surface layer, Boundary-Layer Meteorol., 1, 245-263. [74]

Collins, J. R., 1925: Change in the infra-red absorption spectrum of water with temperature, Phys. Rev., 26, 771-779. [60]

Corrsin, S., 1943: Investigation of flow in an axially symmetrical heated jet of air, NACA, WP. W-94. [92], [93]

Cox, C., and Munk, W. 1956: Slopes of the sea surface deduced from photographs of sun glitter, Bull. Scripps Inst. Oceanogr., Univ. Calif., 6, 401-487. [54], [55]

Craik, A. D. D., 1970: A wave-interaction model for the generation of windrows, J. Fluid Mech., 41, 801-821. [117]

Crepon, M., 1969: Hydrodynamique marine en regime impulsionnel, Pt 3, Chap. 1, Formation d'ondes internes de Longues periodes dans un océan à deux couches, Cah. Oceanogr., 221, [108]

Crepon, M., 1972: Generation of internal waves of inertial period in a two-layer ocean, in Physical variability in the North Atlantic CIEM-PV162, Oct. 1972, 85-88. [182]

Crepon, M., 1974: Genèse d'ondes internes dans un milieu a deux couches, La Houille Blanche, 29, no. 7-8, 631-636. [182]

Csanady, G. T., 1965: Windrow studies, Rep. No. PR26, Great Lakes Inst., Univ. Toronto, Ontario. [113]

Curcio, J. A. and C. C. Petty, 1951: The near infrared absorption spectrum of liquid water, J. Opt. Soc. Amer., 41, 302-304. [60]

Davidson, K. L., 1974: Observational results on the influence of stability and wind-wave coupling on momentum transfer and turbulent fluctuations over ocean waves, Boundary-Layer Meteorol., 6, 305-331. [88]

Davies, J. M., J. C. Gamble, and J. H. Steele, 1975: Preliminary studies with a large plastic enclosure, in Estuarine research, ed. L. Eugene Cronin, Academic Press, New York, 251-264. [250]

Davis, R. E., 1972: On prediction of the turbulent flow over a wavy boundary, J. Fluid Mech., 52, 287-306. [233]

Davis, R. E., 1974: Perturbed turbulent flow, eddy viscosity and the generation of turbulent stresses, J. Fluid Mech., 63, 673-93. [235]

Davis, R. E., 1976: Predictability of sea surface temperature and sea level pressure anomalies over the North Pacific Ocean. J. Phys. Oceanogr., 6, 249-266. [35]

Day, C. G. and F. Webster, 1965: Some current measurements in the Sargasso Sea, Deep-Sea Res., 12, 805-814. [106]

Deardorff, J. W., 1968: Dependence of air-sea transfer coefficients on bulk stability, J. Geophys. Res., 73, 2549-2557. [86]

Deardorff, J. W., 1970: A three-dimensional numerical investigation of the idealized planetary boundary layer, Geophys. Fluid Dyn., 1, 377-410. [144]

Deardorff, J. W., G. E. Willis, and D. K. Lilly, 1969: Laboratory investigation of non-steady penetrative convection, J. Fluid Mech., 35, 7-31. [165], [177]

de Brichambault, Ch., and G. Lamboley, 1968: Le rayonnement solaire au sol et ses mesures, Cahiers de L'A.F.E.D.E.S.,n°1, 11-109. [48], [49]

Denman, K. L., 1973: A time-dependent model of the upper ocean, J. Phys. Oceanogr., 3, 173-184. [164], [185], [218]

Denman, K. L., 1976: Covariability of chlorophyll and temperature in the sea, Deep-Sea Res.,23, 539-550. [254], [255]

Denman, K. L., and M. Miyake, 1973a: The behaviour of the mean wind, the drag coefficient, and the wave field in the open ocean, J. Geophys. Res., 78, 1917-1931. [82]

Denman, K. L., and M. Miyake, 1973b: Upper layer modification at Ocean Station Papa: observations and simulation, J. Phys. Oceanogr., 3, 185-196 [253],[263]

Denman, K. L., and T. Platt, 1975: Coherences in the horizontal distributions of phytoplankton and temperature in the upper ocean, Memoires Société Royale des Sciences de Liège, ed. J. Nihoul, 6e série, tome VII: 19-30. [254]

de Szoeke, R. A. and P. B. Rhines, 1976: Asymptotic regimes in mixed-layer deepening, J. Mar. Res., 34, 111-116. [166]

Dobson, F. W., 1971: Measurements of atmospheric pressure on wind-generated sea waves, J. Fluid Mech., 48, 91-127. [76], [108], [235]

Donelan, M. A. and M. Miyake, 1973: Spectra and fluxes in the boundary layer of the trade wind zone, J. Atmos. Sci., 30, 444-464. [76], [83]

Donelan, M. A., F. C. Elder, and P. F. Hamblin, 1974: Determination of the aerodynamic drag coefficient from wind set-up, The 17th Conf. on Great Lakes Res., Internat. Assoc. Great Lakes Res. Hamilton, 12-14, Aug. 1974. [73]

Dreyer, G. F., 1974: Comparison of momentum, sensible heat and latent heat fluxes over the open ocean determined by the direct covariance, inertial and direct dissipation techniques, Ph.D. thesis, Univ. of Calif., San Diego. [76]

Druyan, L. M., R. C. J. Somerville, and W. J. Quirk, 1975: Extended-range forecasts with the GISS model of the global atmosphere, Monthly Weather Rev., 103, 779-795. [33]

Düing, W., P. Hisard, E. Katz, J. Meincke, L. Miller, K. V. Moroshkin, G. Philander, A. A. Ribnikov, K. Voigt, and R. Weisberg, 1975: Meanders and long waves in the equatorial Atlantic, Nature, 257, 280-284. [273]

Dunckel, M., L. Hasse, L. Krügermeyer, D. Schriever, and J. Wucknitz, 1974: Turbulent fluxes of momentum, heat and water vapor in the atmospheric surface layer at sea during ATEX, Boundary-Layer Meteorol., 6, 81-106. [76], [84], [86], [89]

Duntley, S. Q., 1963: Light in the sea, J. Opt. Soc. Am. 53, 214-233. [62]

Durst, C. S., 1924: The relationship between current and wind, Quart. J. Roy. Meteorol. Soc., 50, 113-119. [105]

Ekman, V. W., 1905: On the influence of the earth's rotation on ocean currents, Ark. Mat. Astr. Fys., 2, 1-52. [102], [108]

Elliott, J. A., 1972a: Microscale pressure fluctuations measured within the lower atmospheric boundary layer, J. Fluid Mech., 53, 351-383. [76], [77]

Elliott, J. A., 1972b: Microscale pressure fluctuations near waves being generated by the wind, J. Fluid Mech., 54, 427-448. [76]

Ellison, T. H., 1957: Turbulent transport of heat and momentum from an infinite rough plane, J. Fluid Mech., 2, 456-466. [133]

Elsberry, R. L., N. A. S. Pearson, and L. B. Corgnati, Jr., 1974: A quasiempirical model of the hurricane boundary layer, J. Geophys. Res., 79, 3033-3040. [220]

Elsberry, R. L., T. S. Fraim, and R. N. Trapnell, Jr., 1976: A mixed-layer model of the oceanic thermal response to hurricanes, J. Geophys. Res., 81, 1153-1162. [218]

Elsberry, R. L., and S. L. Grigsby, 1976: Response of a two-layer hydro-thermodynamic ocean model to a moving hurricane. Manuscript in preparation. [220], [221]

Faller, A. J., 1963: An experimental study of the instability of the laminar Ekman boundary layer, J. Fluid Mech., 15, 560-576. [115]

Faller, A. J., 1964: The angle of windrows in the ocean, Tellus, 16, 363-370. [111], [115], [116]

Faller, A. J., 1969: The generation of Langmuir circulations by the eddy pressure of surface waves, Limnol. Oceanogr., 14, 504-513. [113], [116], [117]

Faller, A. J., 1971: Oceanic turbulence and the Langmuir circulations. Ann. Rev. of Ecology and Systematics, ed. R. F. Johnston, et al., 2, 201-236.

Faller, A. J, and A. H. Woodcock, 1964: The spacing of windrows of Sargassum in the ocean, J. Mar. Res., 22, 22-29. [111], [112], [113], [114]

Faller, A. J. and R. E. Kaylor, 1966: A numerical study of the instability of the laminar Ekman boundary layer, J. Atmos. Sci., 23, 466-480. [115]

Farmer, D. M., 1975: Penetrative convection in the absense of mean shear, Quart. J. Roy. Meteorol. Soc., 101, 869-891. [165], [177]

Fasham, M. J. R. and P. R. Pugh, 1976: Observations on the horizontal coherence of chlorophyll α and temperature. Deep Sea Res., 23, 527-538. [254]

Fedorov, K. M., 1973: The effect of hurricanes and typhoons on the upper active ocean layers, Oceanology, 12, 329-332. [218], [220]

Forristall, George Z., 1974: Three-dimensional structure of storm-generated currents, J. Geophys. Res., 72, 2721-2729. [220]

Fox, M. J. H., 1976: On the nonlinear transfer of energy in the peak of a gravity wave spectrum - II, Proc. Roy. Soc. London, Ser. A 348, (1655), 467-483. [232]

Frankignoul, C. J., and E. J. Strait, 1971: Coherence and vertical propagation of high-frequency energy in the deep sea, (unpublished manuscript), Woods Hole Oceanographic Institution. [109]

Freeland, H. J., P. B. Rhines, and T. Rossby, 1975: Statistical observations of the trajectories of neutrally buoyant floats in the North Atlantic, J. Mar. Res., 33, 383-404. [278]

✓ Friehe, C. A. and K. F. Schmitt, 1976: Parameterization of air-sea interface fluxes of sensible heat and moisture by the bulk aerodynamic formulas, J. Phys. Oceanogr., (in press). [76], [83], [86], [88], [89]

Fritz, S., 1951: Solar radiant energy and its modification by the earth and its atmosphere, in Compendium of Meteorology, Am. Meteorol. Soc., Boston, 13-33. [48]

Fuglister, F. C., 1955: Alternative analyses of current surveys, Deep-Sea Res., 2, 213-229. [275]

Fuglister, F., 1963: Gulf Stream '60. Progress in Oceanography, 1, 263-373. [280]

Gammelsrød, T., 1975: Instability of Couette flow in a rotating fluid and origin of Langmuir circulations, J. Geophys. Res., 80, 5069-5075. [115], [116]

Gammelsrød, T., M. Mork, and L. P. Røed, 1975: Upwelling possibilities at an ice edge: Homogeneous model, Mar. Sci. Communications, 1, 115-145. [216]

Gandin, L. S., 1965: Objective Analysis of Meteorological Fields, Israel Programme for Scientific Translations, Jerusalem. [275]

Garratt, J. R., 1972: Studies of turbulence in the surface layer over water (Lough Neagh). Part II: Production and dissipation of velocity and temperature fluctuations, Quart. J. Roy. Meteorol. Soc., 98, 642-657. [80]

Garrett, C. J. R., 1976: Generation of Langmuir circulations by surface waves - a feedback mechanism, J. Mar. Res., 34, 117-130. [115ff]

Garrett, C. J. R. and W. Munk, 1971: Internal wave spectra in the presence of fine structure, J. Phys. Oceanogr., 1, 196-202. [131], [132]

Garrett, C. J. R. and W. H. Munk, 1972: Oceanic mixing by breaking internal waves, Deep-Sea Res., 19, 823-832. [109]

Garrett, C. J. R. and W. Munk, 1975: Space-time scales of internal waves: a progress report, J. Geophys. Res., 80, 291-297. [272]

Garvine, R. W., 1971: A simple model of coastal upwelling dynamics, J. Phys. Oceanogr., 1, 169-179. [183]

Garvine, R. W., 1974: Ocean interiors and coastal upwelling models, J. Phys. Oceanogr., 4, 121-125. [183], [206]

Garwood, R. W., Jr., 1976: Ph.D. thesis, Univ. of Washington, Seattle. [146], [156]

Gavrilov, A. S. and D. L. Laykhtman, 1973: Influence of radiative heat transfer on the conditions in the atmospheric surface layers, Izv. Atmos. Oceanic Phys., 9, 12-14. [74]

Geisler, J. E., 1970: Linear theory of the response of a two layer ocean to a moving hurricane, Geophys. Fluid Dyn., 1, 249-272. [216], [218], [220], [222]

Gent, P. and P. A. Taylor, 1976: A numerical model of air flow over waves, J. Fluid Mech., (in press). [233], [235]

Gibson, C. H., G. R. Stegen, and R. B. Williams, 1970: Statistics of the fine structure of turbulent velocity and temperature fields measured at high Reynolds number, J. Fluid Mech., 41, 153-167. [79], [80]

Gilbert, K. D., 1974: The non-linear transient response of a continuously stratified, baroclinic ocean to stationary and transitory axially-symmetric atmospheric cyclones, Report P-5236, Rand Corporation, Santa Monica. [218]

Gill, A. E., 1975: Models of equatorial currents, in Numerical Models of Ocean Circulation, ed. R. O. Reid, A. R. Robinson, K. Bryan, Nat. Acad. of Sci., 181-201. [179]

Gill, A. E. and A. J. Clarke, 1974: Wind-induced upwelling, coastal currents, and sea level changes, Deep-Sea Res., 21, 325-345. [181], [200]

Gill, A. E., and J. S. Turner, 1976: A comparison of seasonal thermocline models with observation, Deep-Sea Res., 23, 391-401. [149], [163], [165], [171], [263]

Global Atmospheric Research Program, 1972: Parameterization of sub-grid scale processes, GARP Pub. No. 8, WMO, Geneva. [135]

Global Atmospheric Research Program, 1974: Modelling for the First GARP Global Experiment, GARP Pub. No. 14, WMO, Geneva. [33]

Global Atmospheric Research Program, 1975: The physical basis of climate and climate modelling, GARP Pub. No. 16, WMO, Geneva. [33]

Goldberg, E. D., I. N. McCave J. J. O'Brien, and J. H. Steele (ed.), 1976: The Sea Vol. VI, Wiley-Interscience, New York (in press). [243]

Gonella, J., 1970: Le courant de derive d'après les observations effectuées à la Bouée Laboratoire, (unpublished manuscript), Mus. Nat. d'Hist. Naturelle. [102], [103], [110], [111]

Gonella, J., 1971: A local study of inertial oscillations in the upper layers of the ocean, Deep-Sea Res., 18, 775-788. [106], [108]

Gonella, J., 1972: A rotary-component method for analysing meteorological and oceanographic vector time series, Deep-Sea Res., 19, 833-846. [269]

Gordon, A. L., 1970: Vertical momentum flux accomplished by Langmuir circulations, J. Geophys. Res., 75, 4177-4179. [104], [114], [117]

Grant, H. L., A. Moilliet, and R. W. Stewart, 1959: A spectrum of turbulence at very high Reynolds number, Nature, 184, 808-810. [272]

Grant, H. L., A. Moilliet, and W. M. Vogel, 1965: Some observations of the occurence of turbulence in and above the thermocline, J. Fluid Mech., 34, 443-448. [156]

Gregg, M. C. and C. S. Cox, 1972: The vertical microstructure of temperature and salinity, Deep-Sea Res., 19, 355-376. [272]

Grichtchenko, D. L., 1959: Zvistmost' albedo morja ot vysoty solnea i volnenija morskoj poverchnosti, Trudy, Glavnoja Geofis Observa A. I. Boejhova, 80, 32-38. [56]

Grossman, R. L. and B. R. Bean, 1973: An aircraft investigation of turbulence in the lower layers of a marine boundary layer, NOAA Tech. Rep, ERL 291-WMPO 4, Boulder, Colorado, (COM-74-50337). [76]

Hale, G. M. and M. R. Querry, 1973: Optical constants of water in the 200 nm to 200 μm wavelength region, Appl. Optics, 13, 555-563. [60]

Halpern, D., 1974a: CUE II wind stress at Station B, CUEA Newsletter, 3, 18-22. [207]

Halpern, D., 1974b: Observations of deepening of the wind-mixed layer in the northeast Pacific Ocean, J. Phys. Oceanogr., 4, 454-466. [164]

Halpern, D., 1976: Structure of a coastal upwelling event observed off Oregon during July 1973, Deep Sea Res., 23, 495-508

Hamilton, P. and M. Rattray, Jr., 1975: A numerical model of the depth dependent wind driven upwelling circulation on a continental shelf. Submitted to J. Phys. Oceanogr. [179]

Haney, R. L., 1971: Surface thermal boundary condition for ocean circulation models, J. Phys. Oceanogr., 1, 241-248. [19], [20]

Haney, R. L., 1974: A numerical study of the response of an idealized ocean to large-scale surface heat and momentum flux, J. Phys. Oceanogr., 4, 145-167. [19]

Harris, F. P. and J. N. A. Lott, 1973: Observations of Langmuir circulations in Lake Ontario, Limnol. Oceanogr., 18, 584-589. [111], [112], [114]

Hasse, L., 1970: On the determination of the vertical transports of momentum and heat in the atmospheric boundary layer at sea, Tech. Rep. No. 188, Dept. of Oceanography, Oregon State Univ., Corvallis, (AD-711 376). [72], [76], [89]

Hasse, L., 1971: The sea surface temperature deviation and the heat flow at the sea-air interface, Boundary-Layer Meteorol., 1, 368-379. [89]

Hasse, L. and M. Dunckel, 1974: Direct determination of geostrophic drag coefficients at sea, Boundary-Layer Meteorol., 7, 323-329. [89]

Hasse, L. and V. Wagner, 1971: On the relationship between geostrophic and surface wind at sea, Monthly Weather Rev., 99, 255-260. [89]

Hasselmann, K., 1962: On the non-linear energy transfer in a gravity wave spectrum - Part 1, J. Fluid Mech., 12, 481-500. [232]

Hasselmann, K., 1963: On the non-linear energy transfer in a gravity wave spectrum - Part 2, J. Fluid Mech., 15, 273-281; Part 3, Ibid., 15,385-398. [232]

Hasselmann, K., 1970: Wave-driven inertial currents, Geophys. Fluid Dyn., 1, 463-502. [108]

Hasselmann, K., D. B. Ross, P. Muller, and W. Sell, 1976: A parametrical wave prediction model, J. Phys. Oceanogr., 6, 200-228. [236]

Haurwitz, B., 1948: Insolation in relation to cloud type, J. Meteorol., 5, 110-113. [52]

Hicks, B. B., R. L. Drinkrow, and G. Grauze, 1974: Drag and bulk transfer coefficients associated with a shallow water surface, Boundary-Layer Meteorol., 6, 287-297. [88]

Hide, R., 1976: Motions in planetary atmospheres, Quart. J. Roy. Meteorol. Soc., 192, 1-24. [264]

Hoeber, H., 1972: Eddy thermal conductivity in the upper 12 m of the tropical Atlantic, J. Phys. Oceanogr., 2, 303-304. [103], [105]

Högström, U., 1974: A field study of the turbulent fluxes of heat, water vapour and momentum at a "typical" agricultural site, Quart. J. Roy. Meteorol. Soc., 100, 624-639. [81]

Holladay, C. G. and J. J. O'Brien, 1975: Mesoscale variability of sea surface temperatures, J. Phys. Oceanogr., 5, 761-772. [187], [207]

Hollan, E., 1969: Die Veränderlichkeitder Strömungsverteilung im Gotland - Becken am Beispiel von Strömungsmessungen im Gotland, Tief. Kieler Meeresforsch., 25, 19-70. [108]

Holland, J. Z. and E. M. Rasmussen, 1973: Measurements of the atmospheric mass, energy and momentum over a 500 km square of tropical ocean, Monthly Weather Rev., 101, 44-55. [73]

✓ Holland, W. R. and L. B. Lin, 1975a: On the generation of mesoscale eddies and their contribution to the oceanic general circulation. I. A preliminary numerical experiment, J. Phys. Oceanogr., 5, 642-657. [26ff]

√ Holland, W. R. and L. B. Lin, 1975b: On the generation of mesoscale eddies and their contribution to the oceanic general circulation. II. A parameter study, J. Phys. Oceanogr., 5, 658-669. [26ff]

Holling, C. S., 1965: The functional response of predators to prey density and its role in mimicry and population regulation, Mem. Ent. Soc. Can., 45, 5-60. [243]

Hoskins, B. J. and F. P. Bretherton, 1972: Atmospheric frontogenesis models: mathematical formulation and solution, J. Atmos. Sci., 29, 11-37. [223]

Houghton, D. D., J. E. Kutzbach, M. McClintock, and D. Suchman, 1974: Response of a general circulation model to a sea temperature perturbation, J. Atmos. Sci., 31, 857-868. [34], [35]

Hsu, S. A., 1974a: A dynamic roughness equation and its application to wind stress determination at the air-sea interface, J. Phys. Oceanogr., 4, 116-120. [88]

Hsu, S. A., 1974b: On the log-linear wind profile and the relationship between shear stress and stability characteristics over the sea, Boundary-Layer Meteorol., 6, 509-514. [84]

Huffaker, C. B., 1958: Experimental studies on predation: dispersion factors and predator-prey oscillations, Hilgardia, 27, 343-383. [243]

Hunkins, K., 1966: Ekman drift currents in the Arctic Ocean, Deep-Sea Res., 13, 607-620. [104]

Hunkins, K., 1967: Inertial oscillations of Fletcher's ice island (T-3), J. Geophys. Res., 72, 1165-1174. [103], [105]

Hurlburt, H. E., 1974: The influence of coastline geometry and bottom topography on the eastern ocean circulation, Ph.D. Thesis, The Florida State University. [181], [199ff], [205], [207], [209], [211ff]

Hurlburt, H. E., and J. D. Thompson, 1973: Coastal upwelling on a β-plane, J. Phys. Oceanogr., 3, 16-32. [181], [183], [191], [201], [205], [209], [212]

Hurlburt, H. E., J. C. Kindle and J. J. O'Brien, 1976: A numerical simulation of the onset of El Niño, J. Phys. Oceanogr., 6 (in press). [179]

Hussain, A. K. M. F. and W. C. Reynolds, 1970: The mechanics of an organized wave in turbulent shear flow, J. Fluid Mech., 41, 241-58. [233]

Huyer, A., 1974: Observations of the coastal upwelling region off Oregon during 1972, Ph. D. Thesis, Oregon State University. [201], [202], [204], [209], [211ff]

Huyer, A., R. D. Pillsbury, and R. L. Smith, 1975: Seasonal variation of the alongshore velocity field over the continental shelf off Oregon, Limnol. Oceanogr., 20, 90-95. [201], [209]

Ichiye, T., 1967: Upper ocean boundary-layer flow determined by dye diffusion, Phys. Fluids Suppl., 5270-5277. [104], [111ff]

Iverson, R. L., 1976: Mesoscale oceanic phytoplankton patchiness caused by hurricane effects on nutrient distribution in the Gulf of Mexico, in Oceanic Sound Scattering Prediction, ed. N. R. Anderson and B. V. Zahuranec, Plenum Press, New York (in press). [217]

Iverson, R. L., H. C. Curl, Jr., and J. L. Saugen, 1974: Simulation model for wind-driven summer phytoplankton dynamics in Auke Bay, Alaska, Mar. Biol. 28, 169-177. [257]

Jacobsen, J. P., 1913: Beitrag zur Hydrographie der dänischen Gewasser, Medd. Komm. Havunders Ser Hyd., 2, No. 2, 1-94, Copenhagen. [133]

Jassby, A. D. and T. Platt, 1976: Mathematical formulation of the relationship between photosynthesis and light for phytoplankton, Limnol. Oceanogr., 21, 540-547. [252]

Jensen, C. E., 1975: A review of federal meteorological programs for fiscal years 1965-1975, Bull. Am. Meteorol. Soc., 56, 208-224. [32]

Jerlov, N. G., 1961: Optical measurements in the eastern North Atlantic, Medd. Oceanogr. Inst. Gotenborg, Ser. B., 8. [62]

Jerlov, N. G., 1968: Optical Oceanography, Elsevier, Amsterdam. [65]

Judd, D. B., 1942: Fresnel reflection of diffusely incident light, J. Res. Nat'l. Bureau of Standards, 29, 329-332. [55]

Kajiura, K., 1956: A forced wave caused by atmospheric disturbances in deep water, Technical Report 133-1, A. & M. College of Texas. [220]

Kalle, K., 1938: Zum Problem der Meereswasserfarbe, Ann. Hydrol. u. Marit, Meteor., 66, 1-13. [59]

Kantha, L. H., 1975: Turbulent entrainment at the density interface of a two-layer stably stratified fluid system, Geophysical Fluid Dynamics Lab. Tech. Rept. 75-1, Johns Hopkins Univ. [95], [98], [163], [177]

Kantha, L. and O. M. Phillips, 1976: On turbulent entrainment at a stable density interface, J. Fluid Mech., (in press). [95]

Kasahara, A. and W. M. Washington, 1967: NCAR global general circulation model of the atmosphere, Monthly Weather Rev., 95, 389-402. [21]

Kasahara, A. and W. M. Washington, 1971: General circulation experiments with a six-layer NCAR model, including orography, cloudiness and surface temperature calculations, J. Atmos. Sci., 28, 657-701. [34]

Kato, H. and O. M. Phillips, 1969: On the penetration of a turbulent layer into stratified fluid, J. Fluid Mech., 37, 643-55. [96], [163], [177], [184]

Katsaros, K. and J. A. Businger, 1973: Comments on the determination of the total heat flux from the sea with a two-wavelength radiometer system as developed by McAlister, J. Geophys. Res., 78, 1964-1970. [73]

Katz, B., R. Gerard, and M. Costin, 1965: Responses of dye tracers to sea surface conditions, J. Geophys. Res., 70, 5505-5513. [104], [111], [114]

Kendall, J. M., 1970: The turbulent boundary layer over a wall with progressive surface waves, J. Fluid Mech., 41, 259-281. [234]

Kierstead, H. and L. B. Slobodkin, 1953: The size of water masses containing plankton blooms, J. Mar. Res., 12, 141-147. [255], [256], [259]

Kim, Jeong-Woo, 1975: Oceanic mixed layer parameterization: II Theory, Rand Corp. Rept. WN-8901-ARPA, Santa Monica, Calif. [156]

Kimball, H. H., 1928: Amount of solar radiation that reaches the surface of the earth on the land and on the sea, and methods by which it is measured, Monthly Weather Rev., 56, 393-399. [55]

Kitaigorodskii, S. A., 1960: On the computation of the thickness of the wind-mixing layer in the ocean, Izv. Akad. Nauk, S.S.S.R. Geophys. Ser. 3, 425-431. (English ed. pp. 284-287). [160]

Kitaigorodskii, S. A., 1970: Pizika vzaimodestviya atmosferi i okeana (Physics of air-sea interaction), Gidromet Izdatel'stvo, Leningrad; Eng. Trans. (1973), Israel Prog. Sci. Trans., Jerusalem. [72], [86], [87], [236], [237]

Kitaigorodskii, S. A., O. A. Kutznetsov, and G. N. Panin, 1973: Coefficients of drag, sensible heat, and evaporation in the atmosphere over the surface of a sea, Izv. Atmos. Oceanic Phys., 9, 644-647. [88]

Kolmogorov, A. N., 1941a: The local structure of turbulence in incompressible viscous fluid for very large Reynolds numbers, C. R. Acad. Sci. URSS, 30, 301ff. [129]

Kolmogorov , A. N., 1941b: On degeneration of isotropic turbulence in an incompressible viscous liquid, C. R. Acad. Sci. URSS, 31, 538ff. [129]

Kondo, J., 1975: Air-sea bulk transfer coefficients in diabatic conditions, Boundary-Layer Meteorol., 9, 91-112. [87ff]

Kozlyanimov, M. V., 1957: New instrument for measuring the optical properties of sea water, Tr. Inst. Biol. Vodochran., Akad., Nauk SSR. [62]

Kraus, E. B., 1967: Organized convection in the ocean surface layer resulting from slicks and wave radiation stress, Phys. Fluids, Suppl., 5294-5297. [113], [115], [116]

Kraus, E. B., 1972: Atmosphere-Ocean Interaction, Clarendon Press, Oxford. [2], [72], [86], [146], [147], [171]

Kraus, E. B. and C. Rooth, 1961: Temperature and steady state vertical heat flux in the ocean surface layers, Tellus, 13, 231-239. [159]

Kraus, E. B. and J. S. Turner, 1967: A one-dimensional model of the seasonal thermocline: II. The general theory and its consequences, Tellus, 19, 98-106. [19], [149], [162], [218]

Krauss, W., 1972: Wind-generated internal waves and inertial-period motions, Deutschen Hydrog. Zeit, 25, 241-250. [168]

Krügermeyer, L., 1975: Vertikale Transporte von Impuls, sensibler und latenter Wärme aus Profilmessungen über dem tropischen Atlantik während APEX, Berichte Nr. 29 des Inst. für Radiometeorol. und Maritime Meteorol., Univ. Hamburg. [84], [87], [89]

Kuettner, J. P., 1971: Cloud bands in the earth's atmosphere, Tellus, 23, 404-425. [113], [116]

Kullenberg, G., 1966: (Internal report) University of Copenhagen. [62]

Kullenberg, G., 1972: Apparent horizontal diffusion in stratified vertical shear flow, Tellus, 24, 17-28. [249]

Kullenberg, G., 1974: Diffusion in stratified vertical shear flow, Mém. Soc. Roy. des Sciences de Liège, 6e ser. 4, 41-45. [279]

Laevastu, T., 1960: Factors affecting the temperature of the surface layer of the sea, Societas Scientiarum Fennica, Commentatisnes Physico-Mathematicae, Helsinki, vo. 25, n°1. [56]

Landsberg, H., 1961: Solar radiation at the Earth's surface, Solar Energy, 5, 95ff. [51]

Langmuir, I., 1938: Surface motion of water induced by wind, Science, 87, 119-123. [111ff]

Leavitt, E., 1975: Spectral characteristics of surface layer turbulence over the tropical ocean. J. Phys. Oceanogr., 5, 157-163. [83]

Leavitt, E. and C. A. Paulson, 1975: Statistics of surface layer turbulence over the tropical ocean, J. Phys. Oceanogr., 5, 143-156. [77]

LeBlond, P. H. and L. A. Mysak, 1976: Trapped coastal waves and their role in shelf dynamics, in The Sea, Vol. VI, ed. E. D. Goldberg, I. N. McCave, J. J. O'Brien and J. H. Steele, Wiley-Interscience, New York, (in press). [193]

Lee, J. D., 1973: Numerical simulation of the planetary boundary layer over Barbados, W. I., Ph.D. Thesis, Florida State Univ. [189]

Leibovich, S. and D. Ulrich, 1972: A note on the growth of small-scale Langmuir circulations, J. Geophys. Res., 77, 1683-1688. [117]

Leipper, D. F., 1967: Observed ocean conditions and hurricane Hilda, 1964, J. Atmos. Sci., 24, 182-196. [217], [218], [222]

Leith, C. E., 1975: Numerical weather prediction, Rev. Geophys. Space Phys., 13, 681-684. [33]

Lenschow, D. H., 1973: Two examples of planetary boundary layer modification over the Great Lakes, J. Atmos. Sci., 30, 568-581. [76]

Leslie, D. C., 1974: Developments in the theory of turbulence, Oxford University Press. [128]

Levanon, N., 1971: Determination of the sea surface slope distribution and wind velocity using sun glitter viewed from a synchronous satellite, J. Phys. Oceanogr., 1, 214-220. [73]

Levine, E. R. and W. B. White, 1972: Thermal front zones in the eastern Mediterranean Sea, J. Geophys. Res., 77, 1081-1086. [227]

Lilly, D. K., 1968: Models of cloud topped mixed layers under a strong inversion, Quart. J. Roy. Meteorol. Soc., 94, 292-307. [160]

Linden, P. F., 1973: The interaction of a vortex ring with a sharp density interface: a model for turbulent entrainment, J. Fluid Mech., 60, 467-480. [174]

Linden, P. F., 1975: The deepening of a mixed layer in a stratified fluid, J. Fluid Mech., 71, 385-405.[109], [176]

Long, R. R., 1975: The influence of shear on mixing across density interfaces, J. Fluid Mech., 70, 305-20. [97]

Longuet-Higgins, M. S., 1976: On the nonlinear transfer of energy in the peak of a gravity-wave spectrum, Proc. Roy. Soc. London, Ser. A, (in press). [232]

Longuet-Higgins, M. S. and R. H. Stewart, 1962: Radiation stress and mass transport in gravity waves, with applications to 'surf beats,' J. Fluid Mech., 13, 481-504. [236]

Longuet-Higgins, M. S., D. E. Cartwright, and N. D. Smith, 1963: Observations of the directional spectrum of sea waves using the motions of a floating buoy, in Ocean Wave Spectra, Prentice-Hall, Inc., 111-136 [234], [235]

Lorenz, E. N., 1969: The predictability of a flow which possesses many scales of motion, Tellus, 21, 289-307. [32], [256]

Lumb, F. E., 1964: The influence of clouds on hourly amounts of total solar radiation at the sea surface, Quart. J. Roy. Meteorol. Soc., 90, 43-56. [52], [53]

Lumley, J. L., 1970: Towards a turbulent constitutive relation, J. Fluid Mech. 41, 413-34. [233]

Lumley, J. L. and B. Khajeh-Nouri, 1974: Modeling of turbulent fluxes of momentum and heat in a stratified flow, Izv. Atmos. Oceanic Phys., 10, 636-645 (Eng. Ed. 388-393). [146]

Manabe, S., J. Smagorinsky, and R. F. Strickler, 1965: Simulated climatology of a general circulation model with a hydrologic cycle, Monthly Weather Rev., 93, 769-798. [21], [33], [34]

Manabe, S., K. Bryan, and M. J. Spelman, 1975: A global ocean-atmosphere climate model. Part I. The atmospheric circulation, J. Phys. Oceanogr., 5, 3-29. [16], [32]

Math. Modelsea, 1975: Mathematical Models of Continental Seas, I.C.E.S. Hydrography Committee, C. M. 1975-C:21, unpublished manuscript. [240]

Matsuike, K., T. Morinaga, and S. Sasaki, 1970: The optical characteristics of the water in the three oceans, Part IV, J. Oceanogr. Soc. Japan, 26, 52-60. [50], [51]

May, R. M., 1973: Stability and complexity in Model Ecosystems, Monographs in Population Biology, 6. Princeton Univ. Press. [243]

Maynard-Smith, J., 1974: Models in ecology, Cambridge Univ. Press. [243]

McAlister, E. D., W. Mcleish, and E. A. Corduan, 1971: Airborne measurements of the total heat flux from the sea during BOMEX, J. Geophys. Res., 76, 4172-4180. [73]

McBean, G. A., R. W. Stewart, and M. Miyake, 1971: The turbulent energy budget near the surface, J. Geophys. Res., 76, 6540-6549. [82]

McGrath, J. R., and M. F. M. Osborne, 1973: Some problems with wind drag and infra-red images of the sea surface, J. Phys. Oceanogr., 3, 318-327. [73]

McLeish, W., 1968: Small scale structures in ocean fronts, Trans. AGU, 49, 199ff. [227]

McNider, R. and J. J. O'Brien, 1973: A multi-layer transient model of a coastal upwelling, J. Phys. Oceanogr., 3, 258-273. [181], [182]

McPhee, M. A., 1975: The effect of ice motion on the mixed layer under Arctic pack ice, AIDJEX Bulletin, 30, 1-27. [165]

Mellor, G. L., 1973: Analytic prediction of the properties of stratifed planetary surface layers, J. Atmos. Sci., 30, 1061-1069. [145], [146]

Mellor, G. L. and T. Yamada, 1974: A hierarchy of turbulence closure models for planetary boundary layers, J. Atmos. Sci., 31, 1791-1806. [147]

Mellor, G. L. and P. A. Durbin, 1975: The structure and dynamics of the ocean surface mixed layer, J. Phys. Oceanogr., 5, 718-728. [134], [148]

Merry, M. and H. A. Panofsky, 1976: Statistics of vertical motion over land and water, Quart. J. Roy. Meteorol. Soc., 102, 255-260. [85]

Miles, J. W., 1957: On the generation of surface waves by shear flows, J. Fluid Mech., 3, 185-204. [233]

Miles, J. W., 1959: On the generation of surface waves by shear flows. Part 2, J. Fluid Mech., 6, 568-582. [234]

Mintz, Y., 1964: Very long-term global integration of the primitive equations of atmospheric motion, WMO/IUGG Symp. on Research and Development Aspects of Long-Range Forecasting, Boulder, Colo.; WMO Tech. Note No. 66, 141-167. [21]

Mitsuta, Y. and T. Fujitani, 1974: Direct measurement of turbulent fluxes on a cruising ship, Boundary-Layer Meteorol., 6, 203-217. [76]

Miyakoda, K., G. D. Hembree, R. F. Strickler, and I. Shulman, 1972: Cumulative results of extended forecast experiments. I. Model performance for winter cases, Monthly Weather Rev., 100, 836-855. [33]

Moen, J., 1973: The spectrum of horizontal variability of sea surface temperature, M. Sc. dissertation, Southampton Univ. [121], [273]

Monin, A. S. and A. M. Yaglom, 1971: Statistical Fluid Mechanics: Mechanics of Turbulence Vol. 1, The MIT Press. [81]

Monin, A. S. and A. M. Yaglom, 1975: Statistical Fluid Mechanics: Mechanics of Turbulence Vol. 2, The MIT Press. [81]

Mooers, C. N. K. and J. S. Allen, 1973: Final Report of the Coastal Upwelling Ecosystems Analysis Summer 1973 Theoretical Workshop, School of Oceanography, Oregon State University. [199], [208]

Mooers, C. N. K., C. A. Collins, and R. L. Smith, 1976: The dynamic structure of the frontal zone in the coastal upwelling region off Oregon, J. Phys. Oceanogr., 6, 3-21. [187], [209], [211]

Moore, D. W. and S. G. H. Philander, 1976: Modelling of the tropical oceanic circulation, in The Sea, Vol VI, ed. Goldberg, McCave, O'Brien, and Steele, Wiley-Interscience, New York (in press). [179]

Moravek, D., H. A. Panofsky, and A. Weber, 1976: Determination of surface stress from vertical-velocity spectra, Quart. J. Roy. Meteorol. Soc., 102, 260-263. [85]

Morel, A., 1967: Etude pour diverses longueurs d'ondes de l'indicatrice de diffusion de la lumière des eaux de mer, XVI General Assembly I.A.P.S.O. (Berne), Proc. Verb. n°10, 203-204. [59]

Morel, A. and L. Prieur, 1975: Analyse spectrale des coefficients d'atténuation diffuse, de réflexion diffuse, d'absorption et de rétrodiffusion pour diverses régions marines, Centre de recherches océanogr. de Villefranche sur mer, rapp. n°17. [60], [61]

Morel, P., 1973: Dynamic Meteorology, Reidel, Dordrecht [275]

Morrice, A. M., 1974: Analysis and interpretation of temperature distribution in the vicinity of upper ocean fronts, M.Sc. dissertation, Southampton Univ. [122], [279]

Mosby, H., 1936: Verdunstung und Strahlung auf dem Meere, Ann. d. Hydrogr. u. Mar. Meteor., 54, 281-286. [50]

Muller, P., 1974: On the interaction between short internal waves and larger scale motions in the ocean, Hamburger Geophysikalische Einzelschriften, Heft 23, Hamburg. [119]

Müller-Glewe, J. and H. Hinzpeter, 1974: Measurements of the turbulent heat flux over the sea. Boundary-Layer Meteorol., 6, 47-52. [76], [89]

Munk, W. H., 1950: On the wind-driven ocean circulation, J. Meteorol., 7, 79-83. [8]

Munk, W. H. and E. R. Anderson, 1948: Notes on a theory of the thermocline, J. Mar. Res., 7, 276-295. [133], [147], [148], [263]

Munk, W. H. and J. D. Woods, 1973: Remote sensing of the ocean, Boundary-Layer Meteorol., 5, 201-209. [270]

Murray, J. D., 1975: Non-existence of wave solutions for the class of reaction-diffusion equations given by the Volterra interacting-population equations with diffusion, J. Theor. Biol., 52, 459-469. [248]

Myer, G. E., 1971: Structure and mechanics of Langmuir circulations on a small inland lake, Ph.D. Dissertation, State Univ. N.Y., Albany. [117]

Namias, J., 1962: Influences of abnormal surface heat sources and sinks on atmospheric behavior, Proc. Intl. Symp. Numerical Weather Prediction, Tokyo, Japan, 7-13 November 1960, Meteorol. Soc. Japan, 615-627. [34]

Namias, J., 1963: Large-scale air sea interactions over the north Pacific from summer 1962 through subsequent winter, J. Geophys. Res. 68, 6171-6186. [34]

Namias, J., 1965: On the nature and cause of climatic fluctuations lasting from a month to a few years, WMO Tech. Note No. 66, 46-62. [34]

Namias, J., 1968: Long-range weather forecasting - history, current status and outlook, Bull. Am. Meteorol. Soc., 49, 438-470. [33], [34]

Namias, J., 1969: Seasonal interactions between the north Pacific Ocean and the atmosphere during the 1960's, Monthly Weather Rev., 97, 173-192. [34]

Namias, J., 1970: Macroscale variations in sea surface temperature in the north Pacific, J. Geophys. Res., 75, 565-582. [34]

Namias, J., 1971: The 1968-69 winter as an outgrowth of sea and air coupling during antecedent seasons, J. Phys. Oceanogr., 1, 65-81. [34]

Nasmyth, P. W., 1970: Oceanic turbulence. Thesis, Univ. British Columbia. [130]

National Academy of Sciences, 1975: Understanding Climatic Change, Washington, D. C. [33]

Neiburger, M., 1948: The reflection of diffuse radiation by the sea surface, Trans. Am. Geophys. Union, 29, 647-652. [56]

Neumann, G. and W. J. Pierson, Jr., 1966: Principles of Physical Oceanography Prentice-Hall, New Jersey. [17]

Nihoul, J. C. K. and F. C. Ronday, 1975: The influence of the tidal stress on the residual circulation, Tellus, 27, 484-490. [239], [240]

Niiler, P. P., 1975: Deepening of the wind-mixed layer, J. Mar. Res., 33, 405-422. [149], [156], [166]

O'Brien, J. J., 1967: The non-linear response of a two-layer, baroclinic ocean to a stationary, axially-symmetric hurricane: Part II, J. Atmos. Sci., 24, 208-215. [218]

√ O'Brien, J. J., 1968: The response of the ocean to a slowly moving cyclone, Annalen der Meteorologie, 4, 60-66. [218]

O'Brien, J. J. and R. O. Ried, 1967: The non-linear response of a two-layer baroclinic ocean to a stationary, axially-symmetric hurricane: Part I, J. Atmos. Sci., 24, 197-207. [217], [218]

O'Brien, J. J. and H. E. Hurlburt, 1972: A numerical model of coastal upwelling, J. Phys. Oceanogr., 2, 14-26. [220]

O'Brien, J. J. and H. E. Hurlburt, 1974: Equatorial jet in the Indian Ocean: Theory. Science, 184, 1075-1077. [179ff], [209]

Okubo, A., 1971: Horizontal and vertical mixing in the sea, in Impingement of Man on the Oceans, ed. D. W. Hood, Wiley-Interscience, New York. [134], [279]

Okubo, A., 1974: Diffusion-induced instability in model ecosystems: another possible explanation of patchiness, Technical Report 86, Chesapeake Bay Inst., The Johns Hopkins Univ. [248], [255]

Olbers, D. J., 1975: On the energy balance of small-scale internal waves in the deep sea, Hamburger Geophysikalische Einzelschriften. [119], [136]

Osborn, T. R., 1974: Vertical profiling of velocity microstructure, Journal of Physical Oceanography, 4, 109-115. [135]

Orlanski, I. and K. Bryan, 1969: Formation of the thermocline step structure by large-amplitude internal gravity waves, J. Geophys. Res., 74, 6975-6983.

Ostapoff, F., Y. Tarbeyev, and S. Worthem, 1973: Heat flux and precipitation estimates from oceanographic observations, Science, 180, 960-962. [166]

Ostapoff, F. and S. Worthem, 1974: The intradiurnal temperature variation in the upper ocean layer, J. Phys. Oceanogr., 4, 601-612. [103], [105]

Otchakovsky, Yu.E., 1966: On the dependence of the total attenuation coefficient upon suspensions in the sea, U. S. Dept. Comm, Joint Publ. Res. Serv., Rept. 36, 16-24. [62]

Owen, R. W., Jr., 1966: Small-scale horizontal vortices in the surface layer of the sea, J. Mar. Res., 24, 56-65. [113], [114]

Ozmidov, R. V., 1965: On the turbulent exchange in a stably stratified ocean, Izv. Atmos. Oceanic. Phys., 1, 493-497. [129]

Panchev, S., 1971: Random functions and turbulence, Pergamon, London. [129]

Pandolfo, J. P. and C. A. Jacobs, 1972: Numerical simulations of the tropical air-sea planetary boundary layer, Boundary-Layer Meteorol., 3, 15-46. [148]

Panofsky, H. A., 1969: The spectrum of temperature, Radio Sci., 4, 1143-1146. [77]

Panofsky, H. A., 1973: Tower micrometeorology, Workshop on Micrometeorology (D. A. Haugen, ed.), Am. Meteorol. Soc., Boston, 151-176. [81]

Paquin, J. E. and S. Pond, 1971: The determination of the Kolmogoroff constants for velocity, temperature and humidity fluctuations from second- and third-order structure functions, J. Fluid Mech., 50, 257-269. [79]

Parsons, T. R., R. J. LeBrasseur, and J. D. Fulton, 1967: Some observations on the dependence of zooplankton grazing on the cell size and concentration of phytoplankton blooms, J. Oceanogr. Soc. Japan, 23, 10-17. [258]

Patzert, W. C., 1969: Eddies in Hawaiian waters, Hawaii Institute of Geophysics Report No. HIG-69-8 ONR/NONR-3748(06). [126], [130]

Paulson, C. A., 1970: The mathematical representation of wind and temperature profiles in the unstable surface layer, J. Appl. Meteorol., 9, 857-861. [84]

Paulson, C. A., E. Leavitt, and R. G. Fleagle, 1972: Air-sea transfer of momentum, heat and water determined from profile measurements during BOMEX, J. Phys. Oceanogr., 2, 487-497. [85]

Paulson, C. A., 1969: Comments on a paper by J. W. Deardorff "Dependence of air-sea transfer coefficients on bulk stability", J. Geophys. Res., 74, 2141-2142. [86]

Paulson, C. A. and T. W. Parker, 1972: Cooling of a water surface by evaporation, radiation and heat transfer, J. Geophys. Res., 77, 491-495. [89]

Payne, R. E., 1972: Albedo of the sea surface, J. Atmos. Sci., 29, 959-970. [55ff]

Pedlosky, J., 1974a: On coastal jets and upwelling in bounded basins, J. Phys. Oceanogr., 4, 3-18. [179], [200]

Pedlosky, J., 1974b: Longshore currents, upwelling and bottom topography, J. Phys. Oceanogr., 4, 214-226. [179], [200], [212]

Pedlosky, J., 1974c: Longshore currents and the onset of upwelling over bottom slope, J. Phys. Oceanogr., 4, 310-320. [179], [200], [212]

Peffley, M. B. and J. J. O'Brien, 1976: A three-dimensional simulation of coastal upwelling off Oregon, J. Phys. Oceanogr., 6, 164-180. [181], [199ff]

Perse, St. John, 1958: Seamarks (trans. by Wallace Fowlie) Harper & Bros., New York. [1]

Phelps, G. T. and S. Pond, 1971: Spectra of the temperature and humidity fluctuations and of the fluxes of moisture and sensible heat in the maritime boundary layer, J. Atmos. Sci., 28, [76], [83]

Philander, S. G. H., 1973: Equatorial undercurrent: Measurements and theories, Rev. Geophys. Space Phys., 2, 513-570. [179]

✓ Phillips, O. M., 1969: The Dynamics of the Upper Ocean, The Univ. Press, Cambridge. [230ff]

Phillips, O. M., 1972: The entrainment interface, J. Fluid Mech., 51, 97-118. [93]

Phillips, O. M. and Banner, M. L., 1974: Wave breaking in the presence of wind drift and swell, J. Fluid Mech., 66, 625-40. [235]

Pielou, E. C., 1969: An introduction to mathematical ecology, Wiley-Interscience, London. [244]

Pierson, W. J., F. C. Jackson, R. A. Stacy, and E. Mehr, 1971: Research on the problem of the radar return from a wind-roughened sea. Appl. Flight Experimental Program, NASA Langley Res. Center, Virginia, 83-114. [73]

Pietrafesa, L., 1973: Steady baroclinic circulation on a continental shelf, Ph.D. Dissertation. University of Washington, Seattle. [179]

Plate, E. J., 1971: Aerodynamic Characteristics of Atmospheric Boundary Layers, U. S. Atomic Energy Commission TID-25465. [81]

Platt, T., 1972: Local phytoplankton abundance and turbulence, Deep-Sea Res., 19, 183-188. [254]

Platt, T., L. M. Dickie, and R. W. Trites, 1970: Spatial heterogeneity of phytoplankton in a near-shore environment, J. Fish. Res. Bd. Canada, 27, 1453-1473. [254]

Platt, T. and C. Filion, 1973: Spatial variability of the productivity: biomass ratio for phytoplankton in a small marine basin, Limnol. Oceanogr., 18, 743-749. [253]

Platt, T. and K. L. Denman, 1975: A general equation for the mesoscale distribution of phytoplankton in the sea, Mémoires Société Royale des Sciences de Liège, J. Nihoul (ed.), 6e serie, tome VII: 31-42. [255], [258]

Platt, T., K. L. Denman, and A. D. Jassby, 1975: The mathematical representation and prediction of phytoplankton productivity, Fish. Mar. Serv. Tech. Rpt. 523. [252]

Platt, T. and A. D. Jassby, 1976: The dependence of photosynthesis on light for natural populations of coastal marine phytoplankton, J. Phycology, (in press). [253]

Pollard, R. T., 1969: Ph.D. Thesis, University of Cambridge. [108]

Pollard, R. T., 1970a: On the generation by winds of inertial waves in the ocean, Deep-Sea Res., 17, 795-812. [106ff], [216], [220]

Pollard, R. T., 1970b: Surface waves with rotation: an exact solution, J. Geophys., 75, 5895-5898. [108]

Pollard, R. T., 1971: Properties of near-surface inertial oscillations, (Unpublished manuscript). [107], [108], [110]

Pollard, R. T., 1973: The Joint Air-Sea Interaction Trial - JASIN 1972, Mém. Soc. Roy. des Sciences de Liège, 6, 17-34. [111]

Pollard, R. T. (ed.), 1975: The Joint-Air-Sea Interaction Experiment, Royal Society, London. [268], [278]

Pollard, R. T. and R. C. Millard, Jr., 1970: Comparison between observed and simulated wind-generated inertial oscillations, Deep-Sea Res., 17, 813-821. [106ff], [152, [164]

Pollard, R. T., P. B. Rhines, and R. O. R. Y. Thompson, 1973: The deepening of the wind mixed layer, Geophys. Fluid Dyn., 3, 381-404. [99], [100], [110], [161], [162], [169], [268]

Pollard, R. T. and L. R. Wyatt, 1975: Deepening of the wind-mixed layer, (unpublished manuscript). [164]

Pollard, R. T. and S. Tarbell, 1975: A compilation of moored current meter and wind observations: Vol. VIII (1970 array experiment), Woods Hole Oceanographic Institution, Ref. 75-7. [107], [108]

Pond, S., 1971: Air-sea interaction, Trans. Amer. Geophys. Union (IUGG), 52, 389-394. [72], [76]

Pond, S., 1972: The exchange of momentum, heat and moisture at the ocean-atmosphere interface, in Proceedings of a symposium on numerical models of ocean circulations, Washington, D. C., October 1972, ed. N. H. Durham, U. S. Nat'l Acad. Sciences, Washington, D. C., 26-36. [72], [86]

Pond, S., G. T. Phelps, J. E. Paquin, G. A. McBean, and R. W. Stewart, 1971: Measurements of the turbulent fluxes of momentum, moisture and sensible heat over the ocean, J. Atmos. Sci., 28, 901-917. [76], [82], [83]

Pond, S., D. B. Fissel, and C. A. Paulson, 1974: A note on bulk aerodynamic coefficients for sensible heat and moisture fluxes, Boundary-Layer Meteorol., 6, 333-339. [84], [88], [89]

Powell, T. M., P. J. Richerson, T. M. Dillon, B. A. Agee, B. J. Dozier, D. A. Godden, and L. O. Myrup, 1976: Spatial scales of current speed and phytoplankton biomass fluctuations in Lake Tahoe, Science, (in press). [254]

Powell, W. M. and G. L. Clarke, 1936: Reflection and absorption of daylight at the surface of the ocean, J. Opt. Soc. Am., 26, 111-120. [56]

Preisendorfer, R. W., 1957: Exact reflectance under a cardiodal luminance distribution, Quart. J. Roy. Meteorol. Soc., 83, 540ff. [55]

Price, J. F., C. N. K. Mooers, and J. C. Van Leer, 1976: Observations and simulation of storm induced mixed layer deepening, submitted to J. Phys. Oceanogr. [164], [169]

Proudman, J., 1953: Dynamical oceanography, Methuen, London. [133]

Prümm, D., 1974: Height dependence of diurnal variations of wind velocity and water temperature near the air-sea interface of the tropical Atlantic, Boundary-Layer Meteorol., 6, 341-347. [103], [105]

Pruvost, P., 1972: Contribution à l'étude des échanges radiatifs Atmosphère-Océan. Calcul des flux dans la mer. Thèse 3ème cycle, Universite de Lillé, n°341. [67], [70], [71]

Radach, G. and E. Maier-Reimer, 1975: The vertical structure of phytoplankton growth dynamics, a mathematical model, Mémoires Société Royale des Sciences de Liège, J. Nihoul (ed.) 6e serie, tome VII: 113-146. [246], [257]

Ramage, C. S., 1974: The typhoons of October 1970 in the South China Sea: Intensification, decay and ocean interaction, J. Applied Meteorol., 13, 739-751. [217]

Robinson, A., 1975: Preliminary results of Polymode, (unpublished lecture at IUGG), Grenoble. [130]

Robinson, G. D., 1971: The predictability of a dissipative flow, Quart J. Roy. Meteorol. Soc., 97, 300-312. [256], [257]

Roll, H. U., 1965: Physics of the Marine Atmosphere, Academic Press, New York. [72]

Ronday, F. C., 1975: Modeles de circulation hydrodynamique en Mer du Nord, Ph.D. dissertation Liège University. [239], [240]

Rotta, J. C., 1951: Statistische Theorie Nichthomogener Turbulenz, Z. fuer Physik, 129, 547-572. [146]

Rouse, H. and J. Dodu, 1955: Turbulent diffusion across a density interface, La Houille Blanche, 10, 522-532. [94], [184]

Rowlands, P. B., 1975: Ph.D. thesis, Cambridge Univ. [199]

Rowntree, P. R., 1972: The influence of tropical east Pacific Ocean temperatures on the atmosphere, Quart. J. Roy. Meteorol. Soc., 98, 290-321. [34], [35]

Salmon, R. and M. C. Hendershott, 1976: Large scale air-sea interactions with a simple general circulation model, Tellus, 28, 228-242. [35]

Sarkisyan, A. S. and V. F. Ivanov, 1971: Joint effect of baroclinicity and bottom relief as an important factor in the dynamics of sea currents, Izv. Atmos. Oceanic Phys., 7, 173-188 (Eng. Ed. 116-124). [205]

Saunders, P. M., 1967: Shadowing on the ocean and the existence of the horizon, J. Geophys. Res., 72, 4643-4649. [54]

Saunders, P. M., 1973a: The skin temperature of the ocean: A review, Mém. Soc. Roy. des Sci. de Liège, 6e Serie, 6, 93-98. [89]

Saunders, P. M., 1973b: Tracing surface flow with surface isotherms. Mém. Soc. Roy. des Sci. de Liège, 6e Serie, 6, 99-108. [225], [279]

Saunders, P. M., 1976: Near-surface current measurements, Deep-Sea Res., 23, 249-257. [102], [279]

Schedvin, J., G. R. Stegen, and C. H. Gibson, 1974: Universal similarity at high grid Reynolds numbers, J. Fluid Mech., 65, 561-579. [79]

Schneider, S. H. and R. E. Dickinson, 1974: Climate modeling, Rev. Geophys. Space Phys., 12, 447-493. [33]

Schott, F., 1971: Spatial structure of inertial period motions in a two-layered sea, based on observations, J. Mar. Res., 29, 85-102. [109], [110]

Schule, J. J., L. S. Simpson, and P. S. DeLeonibus, 1971: A study of fetch limited wave spectra with an airborne laser, J. Geophys. Res., 76, 4160ff. [73]

Scott, J. T., G. E. Myer, R. Stewart, and E. G. Walther, 1969: On the mechanism of Langmuir circulations and their role in epilimnion mixing, Limnol. and Oceanogr., 14, 493-503. [111ff],

Shaffer, G., 1974: On the North West African coastal upwelling system, Ph.D. Thesis, Institut für Meereskunde, Universität Kiel. [200], [207], [210], [211]

Sheets, R. C., 1974: Unique data set obtained in hurricane Ellen, Bull. Amer. Meteorol. Soc., 55, 144-146. [222]

Shei, C. M., H. Tennekes, and J. L. Lumley, 1971: Airborne measurements of the small-scale structure of atmospheric turbulence, Phys. Fluids, 14, 201-215. [79]

Sheldon, R. W., A. Prakash, and W. H. Sutcliffe, Jr., 1972: The size distribution of particles in the ocean, Limnol. Oceanogr., 17, 327-340. [251]

Sheldon, R. W., W. H. Sutcliffe, Jr., and A. Prakash, 1973: The production of particles in the surface waters of the ocean with particular reference to the Sargasso Sea, Limnol. Oceanogr., 18, 719-733. [251]

Shemdin, O. H. and E. Y. Hsu, 1967: The dynamics of wind in the vicinity of progressive water waves, J. Fluid Mech., 30, 403-416. [234], [235]

Shukla, J., 1975: Effect of Arabian sea-surface temperature anomaly on Indian summer monsoon: A numerical experiment with the GFDL model, J. Atmos. Sci., 32, 503-511. [34]

Simpson, J., 1972: A free fall probe for the measurement of velocity microstructure. Deep-Sea Res., 19, 331-336. [135]

Simpson, R. W. and W. K. Downey, 1975: The effect of a warm mid-latitude sea surface temperature anomaly on a numerical simulation of the general circulation of the southern hemisphere, Quart. J. Roy. Meteorol. Soc., 101, 847-867. [34]

Smagorinsky, J., 1974: Global atmospheric modeling and the numerical simulation of climate, in Weather and Climate Modification, W. N. Hess, ed., Wiley, 633-686. [33]

Smedman-Högström, A. S., 1973: Temperature and humidity spectra in the atmospheric surface layer, Boundary-Layer Meteorol., 3, 329-347. [80]

Smith, R. L., 1974: A description of current, wind, and sea level variations during coastal upwelling off the Oregon coast, July-August 1972, J. Geophys. Res., 79, 435-443. [208]

Smith, R. L., C. N. K. Mooers, and D. B. Enfield, 1971: Mesoscale studies of the physical oceanography in two coastal upwelling regions: Oregon and Peru, in Fertility of the Sea, 2, ed. J. D. Costlow, Jr., Gordon and Breach, New York. [207]

Smith, S. D., 1974: Eddy flux measurements over Lake Ontario, Boundary-Layer Meteorol., 6, 235-255. [76]

Smith, S. D. and E. G. Banke, 1975: Variation of the sea surface drag coefficient with wind speed, Quart. J. Roy. Meteorol. Soc., 101, 665-673. [76], [87]

Snyder, R. L., 1974: A field study of wave-induced pressure fluctuations above surface gravity waves, J. Mar. Res., 32, 485-496. [76]

Somerville, R. C. J., 1975: Recent results from the GISS model of the global atmosphere, Lect. Notes Phys., 35, 373-378. [31]

Somerville, R. C. J., P. H. Stone, M. Halem, J. E. Hansen, J. S. Hogan, L. M. Druyan, G. Russell, A. A. Lacis, W. J. Quirk, and J. Tenebaum, 1974: The GISS model of the global atmosphere. J. Atmos. Sci., 31, 84-117. [33], [36]

Spar, J., 1973a: Some effects of surface anomalies in a global general circulation model, Monthly Weather Rev., 101, 91-100. [35]

Spar, J., 1973b: Transequatorial effects of sea surface temperature anomalies in a global general circulation model, Monthly Weather Rev., 101, 554-563. [35]

Spar, J., 1973c: Supplementary notes on sea surface temperature anomalies and model generated meteorological histories, Monthly Weather Rev., 101, 767-773. [35]

Spar, J., and R. Atlas, 1975: Atmospheric response to variations in sea-surface temperature, J. Appl. Meteorol., 14, 1235-1245. [35], [36]

Steele, J. H., 1974: The structure of marine ecosystems, Harvard Univ. Press. [245], [247]

Steele, J. H., 1976: Patchiness, in Ecology of the sea, ed. Cushing and Walsh, Blackwell, (in press). [243]

Steele, J. H. and E. W. Henderson, 1976: Simulation of vertical structure in a planktonic ecosystem, Scot. Fish. Res. Rep. No. 5. [243]

Steele, J. H. and M. Mullin, 1976: Zooplankton dynamics in The Sea, Vol. VI ed. Goldberg, McCave, O'Brien, and Steele, Wiley-Interscience, New York (in press). [246], [247], [249]

Stegen, G. R., C. H. Gibson, and C. A. Friehe, 1973: Measurements of momentum and sensible heat fluxes over the open ocean, J. Phys. Oceanogr., 3, 86-92. [79], [82]

Stevenson, M. R., R. W. Garvine, and B. Wyatt, 1974: Lagrangian measurements in a coastal upwelling zone off Oregon, J. Phys. Oceanogr., 4, 321-336. [209]

Stewart, R., 1970: On surface stress and Langmuir circulations, J. Geophys. Res., 75, 7635. [117]

Stewart, R. and R. K. Schmitt, 1968: Wave interaction and Langmuir circulations, Proc. 11th Conf. Great Lakes Res. [166], [117]

Stewart, R. W., 1974: The air-sea momentum exchange, Boundary-Layer Meteorol., 6, 151-167. [88]

Stewart, R. W. and H. L. Grant, 1962: Determination of the rate of dissipation of turbulent energy near the sea surface in the presence of waves, J. Geophys. Res., 67, 3177-3180. [111], [256]

Stommel, H., 1951: Streaks on natural water surfaces, Weather, 6, 72-74. [113], [114], [116]

Stommel, H., K. Saunders, W. Simmons, and J. Cooper, 1969: Observations of the diurnal thermocline, Deep-Sea Res., 16, Suppl., 269-284. [164]

Suginohara, H., 1973: Response of a two-layer ocean to typhoon passage in the western boundary region. J. Oceanogr. Soc. Japan, 29, 236-250. [220]

Suginohara, N., 1974: Onset of coastal upwelling in a two-layer ocean by wind stress with longshore variation, J. Oceanogr. Soc. Japan, 30, 23-33. [181], [199], [200]

Sutcliffe, W. H., E. R. Baylor, and D. W. Menzel, 1963: Sea surface chemistry and Langmuir circulations, Deep-Sea Res., 10, 233-243. [111], [114]

Swallow, J. C. and J. G. Bruce, 1966: Current measurements off the Somali coast during the southwest monsoon of 1964, Deep-Sea Res., 13, 861-888. [103], [104]

Swallow, J. C. and L. V. Worthington, 1961: An observation of a deep counter current in the North Atlantic, Deep-Sea Res., 8, 1-19. [12]

Taylor, G. I., 1931: Effect of variation in density on the stability of superposed streams of fluid, Proc. Roy. Soc. London, Ser. A., 132, 499-523. [133], [263]

Tennekes, H., 1973: Intermittency of the small-scale structure of atmospheric turbulence, Boundary-Layer Meteorol., 4, 241-250. [264]

Tennekes, H., 1973: The logarithmic wind profile, J. Atmos. Sci., 30, 234-238. [81]

Tennekes, H. and J. L. Lumley, 1972: A First Course in Turbulence, The MIT Press. [78], [79]

Thompson, J. D., 1974: The coastal upwelling cycle on a β-plane: Hydrodynamics and thermocynamics, Ph.D. Thesis, The Florida State University. [179], [181], [187], [189], [201], [212]

Thompson, J. D. and J. J. O'Brien, 1973: Time-dependent coastal upwelling, J. Phys. Oceanogr., 3, 33-46. [180], [181]

Thompson, S. M., 1969: Turbulent interfaces generated by an oscillating grid in a stably stratified fluid, Ph.D. Thesis, University of Cambridge. [94]

Thorpe, S. A., 1971: Experiments on instability of stratified shear flows: Miscible fluids, J. Fluid Mech., 46, 299-320. [96]

Thorpe, S. A., 1973: Turbulence in stably stratified fluids: a review of laboratory experiments, Boundary-Layer Meteorol., 5, 95-119. [174]

Townsend, A. A. 1966: Internal waves produced by a convective layer, J. Fluid Mech., 24, 307-319. [175]

Townsend, A. A., 1976: The structure of turbulent shear flow, 2nd Edit., The Univ. Press, Cambridge. [121], [128]

Turing, A. M., 1952: The chemical basis of morphogenesis, Phil. Trans. Roy. Soc. London, Ser. B, 237, 37-72. [248]

Turner, J. S., 1968: The influence of molecular diffusivity on turbulent entrainment across a density interface, J. Fluid Mech., 33, 639-656. [97]

Turner, J. S., 1969: A note on wind-mixing at the seasonal thermocline, Deep-Sea Res., 16, Suppl, 297-300. [94], [164]

Turner, J. S., 1973: Buoyancy effects in fluids, The Univ. Press, Cambridge. [94], [95], [133]

Turner, J. S. and E. B. Kraus, 1967: A one-dimensional model of the seasonal thermocline. I. A laboratory experiment and its interpretation, Tellus, 19, 88-97. [170]

Tyler, J. E., 1961: Scattering properties of distilled and natural waters, Limnol. Oceanogr., 6, 451-456. [62]

Tyler, J. E., and R. C. Smith, 1970: Measurements of Spectral Irradiance Underwater, Gordon and Breach Science Publishers. [63], [65]

Ursell, F., 1950: On the theoretical form of ocean swell on a rotating earth, Mon. Nat. Roy. Astron. Soc., Geophys. Suppl., 6, 1-8. [108]

Veronis, G., 1965: On parametric values and types of representation in wind-driven ocean circulation studies, Tellus, 17, 77-84. [216], [220]

Veronis, G. and H. Stommel, 1956: The action of variable wind stresses on a stratified ocean, J. Mar. Res., 15, 43-75. [183]

Voorhis, A. D. and J. B. Hersey, 1964: Oceanic thermal fronts in the Sargasso Sea, J. Geophys. Res., 69, 3809-3814. [1227]

Walin, G. S., 1972: Some observations of temperature fluctuations in the coastal region of the Baltic, Tellus, 24, 187-198. [193], [194], [199]

Weatherly, G. L., 1974: A numerical study of time-dependent turbulent Ekman layers over horizontal and sloping bottoms, J. Phys. Oceanogr., 5, 288-299. [150]

Webster, F., 1968: Observations of inertial-period motions in the deep sea, Rev. Geophys., 6, 473-490. [106], [119]

Welander, P., 1963: On the generation of wind streaks on the sea surface by action of surface film, Tellus, 15, 67-71. [11], [113], [114], [116]

Wetherald, R. T. and S. Manabe, 1972: Response of the joint ocean-atmosphere model to the seasonal variation of the solar radiation, Monthly Weather Rev., 100, 42-59. [15], [18]

Wieringa, J., 1974: Comparison of three methods for determining strong wind stress over Lake Flevo, Boundary-Layer Meteorol., 7, 3-19. [73], [76]

Williams, J., 1955, Ref. no. 55-4, Chesapeake Bay Inst., Johns Hopkins Univ. [55]

Winter, D. F., K. Banse, and G. C. Anderson, 1975: The dynamics of phytoplankton blooms in Puget Sound, a fjord in the northwestern United States, Mar. Biol., 29, 139-176. [257], [260]

Woodcock, A. H., 1950: Subsurface pelagic Sargassum, J. Mar. Res., 9, 77-92. [111], [114]

Woods, J. D., 1968: Wave-induced shear instability in the summer thermocline, J. Fluid Mech., 32, 791-800. [109], [125]

Woods, J. D., 1974a: Space-time characteristics of turbulence in the seasonal thermocline, Mém. Soc. Royale des Sciences de Liège, 6e ser. 6, 109-130. [132], [272]

Woods, J. D., 1974b: Diffusion due to fronts in the rotation sub-range of turbulence in the seasonal thermocline, La Houille Blanche, No. 7/8, 589-597. [223]

Woods, J. D., 1974c: Batfish experiments on RRS Discovery during Phase III of GATE, (unpublished report), Oceanography Dept., Univ. Southampton. [279], [280]

Woods, J. D., 1975: The local distribution in Fourier space-time of variability associated with turbulence in the seasonal thermocline. Mém. Soc. Royale des Sciences de Liège, 6e ser., 7, 171-189. [121], [137], [138], [267], [269], [273]

Woods, J. D. (Editor), 1976: The GATE Oceanic boundary layer experiment - a progress report, June 1975, WMO, Geneva. [119], [270], [274]

Woods, J. D. and N. R. Watson, 1970: Measurements of thermocline fronts from the air, Underwater J., 2, 90-99. [280]

Woods, J. D. and R. L. Wiley, 1972: Billow turbulence and ocean microstructure, Deep-Sea Res., 19, 87-121. [135], [139], [140]

Wroblewski, J. S. and J. J. O'Brien, 1976: A spatial model of phytoplankton patchiness, Mar. Biol., 35, 161-175. [259]

Wucknitz, J., 1974: Bestimmung der turbulenten Flüsse von Impuls und sensibler Wärme aus Fluktuationsmessungen und Struktur des Windfeldes über den Wellen über dem tropischen Atlantik während APEX, Berichte Nr. 25 des Inst. für Radiometeorol. und Maritime Meteorol., Univ. Hamburg. [76], [79], [82], [89]

Wunsch, C., 1972: The spectrum from two years to two minutes of temperature fluctuations in the main thermocline off Bermuda, Deep-Sea Res., 19, 577-594. [119], [122]

Wyngaard, J. C., 1973: On surface-layer turbulence, Workshop on Micrometeorology (D. A. Haugen, ed.), Am. Meteorol. Soc., Boston, 101-149. [74], [76]

Wyngaard, J. C., and O. R. Coté, 1971: The budgets of kinetic energy and temperature variance in the atmospheric surface layer, J. Atmos. Sci., 29, 190-201. [77]

Wyrtki, K., 1973: An equatorial jet in the Indian Ocean, Science, 181, 262-264. [179]

Yoshida, K., 1955: Coastal upwelling off the California coast, Rec. Ocn. Works in Japan, 2, 8-20. [181]

Yoshida, K., 1959: A theory of the Cromwell current and of the equatorial upwelling - an interpretation in a similarity to a coastal circulation, J. Oceanogr. Soc. Japan, 15, 159-170. [179], [181]

Yoshida, K., 1967: Circulation in the eastern tropical oceans with special reference to upwelling and undercurrents, Japan J. Geophys., 4, 1-75. [181], [183], [200]

INDEX